BASIC ELECTRONICS
FOR SCIENTISTS

BASIC ELECTRONICS FOR SCIENTISTS

Fifth Edition

James J. Brophy

University of Utah

McGraw-Hill Publishing Company

New York   St. Louis   San Francisco   Auckland   Bogotá   Caracas   Hamburg
Lisbon   London   Madrid   Mexico   Milan   Montreal   New Delhi
Oklahoma City   Paris   San Juan   São Paulo   Singapore   Sydney   Tokyo   Toronto

This book was set in Times Roman by Syntax International.
The editors were Denise T. Schanck and Scott Amerman;
the production supervisor was Salvador Gonzales.
The cover was designed by John Jeheber.
R. R. Donnelley & Sons Company was printer and binder.

BASIC ELECTRONICS FOR SCIENTISTS

1 2 3 4 5 6 7 8 9 0   DOCDOC   8 9 4 3 2 1 0 9

ISBN 0-07-008147-6

Library of Congress Cataloging-in-Publication Data

Brophy, James John, (date).
   Basic electronics for scientists/James J. Brophy.—5th ed.
      p.   cm.
   Includes bibliographies and index.
   ISBN 0-07-008147-6
   1. Electronics.   I. Title.
TK7815.B74   1990
621.381—dc20

89-12115

ABOUT THE AUTHOR

James J. Brophy was born in Chicago, Illinois, on June 6, 1926. After graduating from Steinmetz High School and serving in the Navy, he received his B.S. in Electrical Engineering and M.S. and Ph.D. in Physics from Illinois Institute of Technology. He was Vice President at the IIT Research Institute, Academic Vice President at Illinois Institute of Technology and Senior Vice President at the Institute of Gas Technology. He is presently Vice President for Research at the University of Utah where he is Professor of Physics, Professor of Electrical Engineering, and Adjunct Professor of Chemistry. His research interests include fluctuation phenomena annd condensed matter physics.

Dedicated to Muriel

CONTENTS

PREFACE

Electronic measurement and control pervade all corners of science and engineering. The great power and versatility of electronic devices, and consequently their widespread application, make it imperative that science and engineering students obtain a working familiarity with electronics. Yet this familiarity need not be as intensive as that achieved in the training of electrical engineers.

This text is written to provide undergraduate science and engineering students with a basic understanding of electronic devices and circuits. Their understanding should be sufficient for them to appreciate the operation and characteristics of the many electronic instruments they will use in their professional careers. Accordingly, the analysis, rather than design, of circuits is emphasized. It is assumed that students possess a general acquaintance with electricity and electrical properties of materials as covered in beginning physics courses. Circuit theory is treated by starting with direct currents, however, so that parts of the early chapters may be used as refresher material.

This fifth edition is an extensive modification of the earlier texts, but retains their thrust and focus on measurements. The treatment of digital circuits has been expanded into an entirely new chapter on microprocessor circuits. Concomitantly, integrated circuits are emphasized such that most discrete-component circuits are introduced primarily to explain the performance of their integrated-circuit counterparts. An appendix on vacuum tube circuits is retained both for historic pedagogical purposes and because vacuum tubes still remain in many laboratory instruments.

The text intentionally includes a greater scope of material than can be covered in most courses. This allows instructors to adjust their presentations to specific needs of their students. Introductory chapters on circuit theory and devices, Chaps. 1 through 5, are included to make the book self-contained, and some parts may be assigned as review material for well-prepared students. The order of presentation is basically historical and, in the author's view, pedagogically useful, but chapters may be taken up in a different order to meet specific goals. For example, digital electronics and microprocessors in Chaps. 9 and 11 can be introduced immediately following semiconductor devices in Chap. 4. In this edition, AC circuit analysis is placed in Chap. 5 in order to introduce nonlinear circuits and semiconductor devices at an earlier stage.

The inital portions of each chapter are written such that they might be included in courses surveying broad aspects of electronic measurements, with subsequent portions reserved for study in greater depth. The text is also designed to serve as a useful reference for both undergraduate and graduate students for whom this may be the only formal study of electronics.

Since this text is written from the point of view of an experimentalist, concurrent laboratory experience is highly recommended. Almost all the circuits analyzed give actual component values and can easily form the basis for suitable experiments. No attempt has been made to suggest such experiments in detail because of the diversity of equipment likely to be found in different university laboratories. On the other hand, the exercises have been chosen to demand quantitative answers and therefore may substitute in some measure for laboratory work where a separate laboratory course is not possible. Exercises particularly suitable for laboratory experiments are marked with an asterisk.

For many years I have felt that a working familiarity with electronics contributes immeasurably to a professional scientific or engineering career. If this text makes it possible for others to attain such familiarity, I will be well satisfied. I am deeply indebted to many colleagues who, through their publications, have provided much background material for this book. I must also express my sincere thanks to those who kindly read and criticized the manuscript, Dennis Barnaal, Luther College; John R. Brandenberger, Lawrence University; Thomas Cahill, University of California—Davis; Jai N. Dahiya, Southeast Missouri State University; and Thomas Wong, Illinois Institute of Technology, as well as to those instructors and students who have offered suggestions resulting from their use of previous editions. Lastly, only through the collaboration and encouragement of my wife, Muriel, has this project been completed.

James J. Brophy

DIRECT-CURRENT CIRCUITS

The operation of any electronic device, be it as complicated as a television receiver or as simple as a flashlight, can be understood by determining the magnitude and direction of electric currents in all parts of its functional unit, the circuit. In fact, it is not possible to appreciate how any given circuit functions without an understanding of the currents in its individual components.

Even the most complicated circuits can be examined in easy stages by first considering each part separately and subsequently noting how the various subcircuits fit together. Therefore, circuit analysis starts by treating elementary configurations under the simplest possible conditions. Circuits in which the currents are steady and do not vary with time are called direct-current circuits. Such dc circuits, which are considered in this chapter, are important and relatively simple to understand.

INTRODUCTORY CONCEPTS

Current, Voltage, and Resistance

The motion of electric charges (for example, electrons in a conducting material) constitutes an *electric current*. Specifically, the current I is the time rate at which charge Q passes a given point so that

$$I = \frac{dQ}{dt} \tag{1-1}$$

Electric current is measured in coulombs per second, which is termed an *ampere* (A) in honor of the French scientist Andre Marie Ampère.

A number of materials, most notably metals such as copper and silver, contain many free electrons which move in response to electric fields originating from other, external, electric charges and, accordingly, such materials are capable of carrying an electric current. Each free electron in a metal wire carrying current is accelerated by the electric field until it loses its velocity as a result of a collision within the metal. It then accelerates again until it suffers a subsequent collision. The energy required to effect the successive accelerations and move an electron from one point to another is called the electric *potential difference* between the two points. The potential difference V is measured in terms of work per unit charge, or

$$V = \frac{W}{Q} \tag{1-2}$$

Because potential difference is used so frequently in electric-circuit analysis, work per unit charge is termed a *volt* (V) in recognition of the early Italian worker in electricity, Alessandro Volta.

The resistance that each free electron encounters as a result of multiple collisions when moving in a conductor depends upon a material property called *resistivity* ρ. The range of resistivities typical of several metals and alloys at room temperatures is shown in Table 1-1. In addition to resistivity, the shape of the conductor is also important, so that the *resistance R* of a wire L meters long and A square meters in cross-sectional area is given by

$$R = \rho \frac{L}{A} \tag{1-3}$$

According to this expression, the resistance of long, thin wire is greater than that of a short, thick wire of the same material.

The unit of resistance is labeled the *ohm*, after Georg Simon Ohm, a German scientist who first discovered the relationship between current, voltage, and resistance discussed in the next section. The commonly accepted symbol for the resistance of a conductor in ohms is the Greek letter omega, Ω. It often proves convenient to describe the ability to conduct current in terms of the reciprocal of resistance, or *conductance*, which is measured in terms of reciprocal ohms, or *mhos*, which are also called *siemens*.

Table 1-1 Resistivities of metals and alloys

Material	Resistivity, $10^{-8} \, \Omega \cdot m$
Aluminum	2.6
Brass	6
Carbon	350
Constantan (Cu 60, Ni 40)	50
Copper	1.7
Manganin (Cu 84, Mn 12, Ni 4)	44
Nichrome	100
Silver	1.5
Tungsten	5.6

Ohm's Law

In order to maintain a large current in a conductor, more energy, hence a greater potential difference, is required than in the case of a small current in the same conductor. The constant of proportionality between current and potential difference is just the resistance of the conductor, or

$$V = RI \tag{1-4}$$

This equation is known as *Ohm's law*. According to Ohm's law, whenever a conductor of resistance R carries a current I, a potential difference, or *voltage V*, must be present across the ends of the conductor. This relation is fundamental to electric-circuit analysis and is used repeatedly in subsequent sections.

Joule's Law

The kinetic energy of the electrons in a conductor, which results from acceleration by the electric field, is dissipated in inelastic collisions within the conductor and converted to heat energy. Consequently the temperature of a conductor carrying a current must increase slightly, and it is apparent that electric power is expended in forcing a current through the resistance of the conductor.

The power P that must be supplied to the conductor to sustain the current is given by

$$P = \frac{dW}{dt} = V \frac{dQ}{dt} = VI \tag{1-5}$$

where the definitions of potential difference, Eq. (1-2), and current, Eq. (1-1), have been used. This expression may be written in terms of the resistance of the conductor using Ohm's law. The result,

$$P = I^2 R \tag{1-6}$$

is known as *Joule's law*, after Sir James Prescott Joule, the Englishman who discovered experimentally that the rate of development of heat in a resistance is proportional to the square of the current.

According to Joule's law, electric power is dissipated in a conductor whenever it carries an electric current. This effect is put to use in incandescent lamps,
where a thin metal filament is heated to white heat by the current, and also in
electric fuses, in which the conductor melts when the current exceeds a predeter-
mined value. On the other hand, the size of connecting wires, and therefore their
resistance, is selected so that the power loss is small and the temperature rise negligible when the wire is carrying less than the maximum design current. The joule
heat in a conductor is commonly spoken of as the "I-squared-R" loss. As usual,
the unit of power, according to Eq. (1-5), is a joule per second, which is called a
watt (W), in honor of another Englishman, James Watt, developer of the steam
engine.

CIRCUIT ELEMENTS

Resistors

An electrical component very frequently used in electronic circuits is the *resistor*,
which is a circuit element having a specified value of resistance. Resistance values
commonly encountered range from a few ohms to thousands of ohms, or *kilohms*
(abbreviated as kΩ), and even millions of ohms, or *megohms* (abbreviated as MΩ).
Lumped resistances that resistors introduce into a circuit are large compared
with those of wires and contacts between wires. According to Ohm's law, a
potential difference develops across the resistor as a result of current in it at the
place in the circuit where the resistor is inserted. The conventional symbol for a
resistor in a circuit diagram is a zigzag line, as illustrated in Fig. 1-1.

Some resistors are constructed from a long, very fine wire wound on an
insulating support. Resistance values can be increased by decreasing the cross-
sectional area of the wire and by increasing its length, as Eq. (1-3) shows, and by
selecting wire materials having a large resistivity (see Table 1-1). Such *wire-wound*
resistors commonly employ metal-alloy wires which have resistivities relatively
independent of temperature. Typical materials are manganin and constantan.

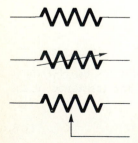

Figure 1-1 Conventional circuit symbols for fixed (top) and variable
resistors.

Wire-wound resistors are used where it may be necessary to dissipate sufficient joule heat for the temperature of the resistor to rise significantly. The resistance of wire-wound resistors can be determined quite precisely by choosing the proper wire length, so wire-wound resistors are also useful in applications where accurate resistance values are desired.

Thin-film resistors are made by depositing a thin film of a metal on an insulating support. High-resistance values are a consequence of the thinness of the film. Because of the difficulty in producing uniform films, it is not possible to control resistance values as precisely as in the case of wire-wound resistors. However, thin-film resistors are free of troublesome inductance effects common in wire-wound units (see Chap. 2), and this is important in high-frequency circuits. Thin-film resistors fabricated from nonmetallic materials, particularly finely divided carbon granules, are also common. Carbon itself has high resistivity, as do the points of contact between the granules. In fact, it is possible to achieve such high resistance values with carbon granules that in many situations it is unnecessary to employ thin films at all and the resistance element is a simple rod of pressed carbon granules. Such units are known as *composition resistors*.

Both thin-film and composition resistors are provided with insulation and wire leads to facilitate inserting them into circuits. It is common practice to provide colored markings which denote the resistance value of each unit according to a universal *resistor color code*. In addition, the physical size of the resistor is a rough indication of the maximum permissible power the unit is capable of dissipating without appreciable increase in temperature caused by joule heating. Thus, for example, common resistor power ratings are 1 W, $\frac{1}{2}$ W, and $\frac{1}{4}$ W, although other values are used as well. Examples of typical thin-film and composition resistors are shown in Fig. 1-2.

It is often necessary to alter the resistance of a resistor while it is permanently connected in a circuit. Such *variable resistors* employ a mechanical slider or arm which rides over the resistance element, thus selecting the length of the element included in the circuit. Both wire-wound and composition resistance elements are commonly made circular so that the position of the slider may be adjusted by rotation of a shaft. The circuit symbols for variable resistors are of two types, as in Fig. 1-1, depending upon whether two or three terminals are provided for

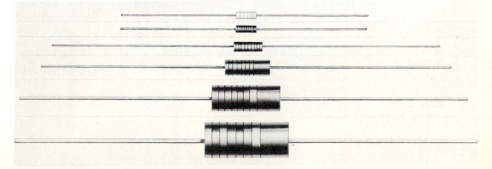

Figure 1-2 Typical composition resistors. (*Allen-Bradley Co.*)

external connections. A variable resistor having two terminals is called a *rheostat*, while one with three is known as a *potentiometer*. Obviously a potentiometer, with its terminals at each end of the resistance element and a third terminal attached to the slider, can be used as a rheostat if one of the resistance-element terminals is ignored.

Batteries

According to Joule's law, electric energy is dissipated in any conductor when it carries a current. In simple dc circuits the source of this energy, which must be supplied in order to maintain the current, is often a chemical *battery*. Other sources of dc electric power are considered in a later chapter. In a battery, chemical energy is converted into electric energy, and the chemical reactions maintain a potential difference between the battery terminals whether or not a current is present. This potential difference is commonly referred to as an *electro-motive force*, abbreviated *emf*, in order to distinguish it from the potential difference which appears across a resistance in accordance with Ohm's law. As a battery continues to supply the energy necessary to maintain current in a circuit, the chemical constituents eventually become depleted and the battery is said to be *discharged*. Depending upon the particular chemical nature of the battery, it may be possible to *charge* it, that is, return it to its original chemical composition, by passing a current between its terminals in a direction opposed to the internal emf. The symbol for a battery in circuit diagrams, Fig. 1-3, consists of a short heavy line parallel to a longer thin line. It is always assumed, if not explicitly indicated, that the longer line represents the higher potential, or positive, terminal of the internal emf. Since the internal emf is a potential difference, its unit is the volt.

The carbon-zinc battery is by far the most common, and least expensive, source of electric energy. Although it is conventionally referred to as a *dry cell*, it actually consists of a moist paste (called the *electrolyte*) of zinc chloride, ammonium chloride, and manganese dioxide contained between a zinc electrode and a carbon electrode. The zinc and carbon electrodes serve as the terminals of the battery. The operation of such a cell is briefly as follows. At the zinc electrode, zinc atoms are dissolved into solution as doubly charged zinc ions. The zinc electrode becomes negatively charged because each zinc atom leaves behind two electrons. At the carbon electrode, ammonium ions reacting with manganese dioxide withdraw electrons from the carbon, and thus it becomes charged positively. If the negative zinc electrode is connected externally through a circuit to

Figure 1-3 Conventional circuit symbol for a battery.

the positive carbon electrode, electrons can flow between them to complete the chemical reaction.

Notice that in order for the chemical reaction to continue, zinc ions must move away from the negative electrode and the reaction products near the positive terminal must likewise move away from the carbon electrode. Thus, current is carried internally to the battery by means of ions moving in the electrolyte, and this is a source of internal resistance. The voltage drop across the internal resistance has the effect of reducing the terminal voltage of the battery. The terminal voltage of the dry cell slowly decreases with use as the internal resistance increases because of depletion of the manganese dioxide. The internal resistance eventually becomes so large that the battery is useless.

If the dry cell is left idle for some time before it is completely discharged, the internal resistance gradually reduces because of internal diffusion of the ions. On the other hand, if a dry cell is allowed to age for extended periods (more than one year) internal ionic diffusion increases the internal resistance so much that the cell becomes inoperative, even though it may never have been used. The emf of a freshly prepared dry cell is 1.5 V. Higher voltages are conventionally obtained by connecting a number of individual units (Fig. 1-4); in fact the term *battery* originated from just such assemblies. Dry-cell batteries of 1.5, 9, 22.5, 45, 67.5, and 90 V are most commonly available.

The familiar lead-acid *storage battery* used in automobiles is an example of a battery that can be repeatedly recharged. The positive electrode of a fully charged storage battery is a porous coat of lead dioxide on a grid of metallic lead. The negative electrode is metallic lead, and both electrodes are immersed in a liquid sulfuric acid electrolyte at a specific gravity of about 1.3. During discharge the lead dioxide is converted to lead sulfate, which is poorly soluble and clings to the positive plate. This reaction withdraws electrons from the electrode, thus charging it positively. At the negative electrode, sulfate ions from solution produce lead sulfate and release electrons. Again the lead sulfate adheres to the electrode and at discharge both electrodes are nearly entirely converted to lead sulfate. The loss of sulfate ions from solution during discharge reduces the specific gravity to about 1.16, so that the condition of the battery may be determined by measuring the specific gravity of the electrolyte.

These chemical reactions are easily reversible, and current directed into the positive terminal acts to return the electrodes to their original chemical composition. Charging requires an external source to furnish electric energy, after which the battery again can supply energy during discharge. Thus, the storage battery may be said to store electric energy in chemical form. In addition, the internal resistance of the lead-acid battery is very low and the battery is capable of

Figure 1-4 Four batteries connected in series.

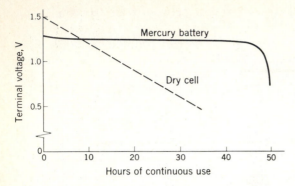

Figure 1-5 Discharge curve of carbon-zinc dry cell compared with that of mercury battery.

delivering currents of several hundred amperes for short times. The fully charged cell has an emf of about 2.1 V, and commercial units are available as 6-, 12-, and 24-V batteries. It is important to maintain an idle storage battery fully charged, for otherwise the electrodes slowly become converted to a sulfate which cannot be returned to the original chemical composition by a charging current. In this condition, the electric energy capacity of the battery is reduced.

The internal resistance of the more recently developed *mercury battery* does not change appreciably during discharge. This means that the terminal voltage remains essentially constant throughout the useful life. It then falls precipitously when the battery is exhausted, as illustrated in Fig. 1-5. The constant-voltage characteristic of mercury batteries is important in those electronic applications where the proper operation of a circuit depends critically upon the battery voltage. Such situations are not uncommon in transistor circuits. In addition, the constant-voltage feature means that the mercury battery is useful as a voltage standard in electrical measurement circuits. The mercury battery has a zinc amalgam for one electrode and mercuric oxide and carbon for the other. The chemical reactions at the electrodes are somewhat similar to those of the dry cell, and the potential developed is 1.35 V.

Other battery types include the *alkaline* battery and the *nickel-cadmium* battery. The alkaline battery is chemically quite similar to the dry cell, but has a strongly basic electrolyte between the electrodes. This, together with a modified electrode structure, lowers internal resistance, increases energy capacity, and improves shelf life. The nickel-cadmium battery can be repeatedly recharged like the lead storage battery, but is completely sealed, since gas evolution during charging acts as a self-regulating mechanism to prevent the buildup of a large gas pressure. This feature, and the fact that liquid electrolyte is not required, compensates for the high cost of this battery.

The *lithium* battery produces a large open-circuit voltage, 3.7 V, because of the high oxidation potential of this element. Various electrolyte compositions are possible, but thionyl chloride is common. Of particular interest is the extremely long shelf life, over 10 years, so that lithium cells are often incorporated permanently into electronic circuits to provide a backup power source in the event of power failure. Typical modern batteries are illustrated in Fig. 1-6.

Figure 1-6 Typical modern batteries. (*Eveready Battery Company, Inc*)

CIRCUIT ANALYSIS

Series and Parallel Circuits

If several electric components, such as resistors, are connected so that the current is the same in every one, the components are said to be in a *series* circuit. Consider the simple series circuit comprising the battery and three resistors illustrated in Fig. 1-7a. The current I results in a potential difference between the terminals of each resistor which is given by Ohm's law. That is

$$V_1 = R_1 I \qquad V_2 = R_2 I \qquad \text{and} \qquad V_3 = R_3 I \qquad (1\text{-}7)$$

Clearly, the sum of these voltages is equal to the battery emf, or

$$V = V_1 + V_2 + V_3 \qquad (1\text{-}8)$$

Equation (1-8) is a simple example of a principle of electronic circuits that is considered in greater detail in the next section. The equation states that the algebraic sum of the potential differences around any complete circuit is equal to

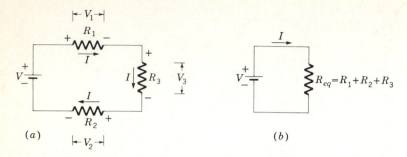

Figure 1-7 (a) Simple series circuit and (b) equivalent.

zero. Note the polarity distinction between the potential difference at the terminals of a resistor compared with that of a source of emf such as a battery: the current direction is *into* the positive terminal of a resistance while it is *out of* the positive terminal of an emf source. Since, the potential decreases in the direction of the current through a resistance, the potential difference is commonly referred to as the *IR drop* across the resistor.

If the *IR* drops of Eq. (1-7) are inserted into Eq. (1-8), the result is

$$V = IR_1 + IR_2 + IR_3 = I(R_1 + R_2 + R_3)$$

Thus, the current in the series circuit is

$$I = \frac{V}{R_1 + R_2 + R_3} = \frac{V}{R_{eq}} \tag{1-9}$$

where the equivalent resistance R_{eq} is defined as

$$R_{eq} = R_1 + R_2 + R_3 \tag{1-10}$$

Evidently the equivalent resistance of any number of resistors connected in series equals the sum of their individual resistances. Insofar as the current is concerned, the circuit of Fig. 1-7b containing the one single resistor R_{eq} is equivalent to that of Fig. 1-7a, which has three resistors.

A useful circuit based on the series connection of resistors is the *voltage divider*, Fig. 1-8, in which the junction between each pair of resistors is connected to a terminal of a *multiple-tap* selector switch. By positioning the switch on one of its various taps it is possible to present a given fraction of the battery voltage V at the output terminals. The division of the potential V among the various taps depends upon the magnitudes of the resistances in the potential divider. Obviously, if the series resistors are replaced by a potentiometer, the output voltage may be set at any desired fraction of V. This is the principle of the volume control in radio and television receivers.

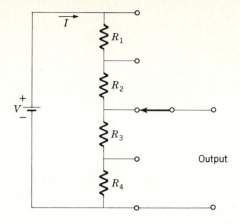

Figure 1-8 Voltage-divider circuit.

Another way of connecting electric components, such as resistors, is shown in Fig. 1-9. Here the potential difference across each resistor in the circuit is the same; this form of connection is called a *parallel circuit*. The current in each resistor is given by Ohm's law as

$$I_1 = \frac{V}{R_1} \qquad I_2 = \frac{V}{R_2} \qquad \text{and} \qquad I_3 = \frac{V}{R_3} \qquad (1\text{-}11)$$

In this case, the sum of the currents equals the battery current,

$$I = I_1 + I_2 + I_3 \qquad (1\text{-}12)$$

Substituting for the currents from Eq. (1-11), this becomes

$$I = \frac{V}{R_1} + \frac{V}{R_2} + \frac{V}{R_3} = V\left(\frac{1}{R_1} + \frac{1}{R_2} + \frac{1}{R_3}\right) \qquad (1\text{-}13)$$

Now, in order to determine the equivalent resistance for parallel resistors, we define R_{eq}, using Ohm's law, as

$$V = IR_{eq} \qquad (1\text{-}14)$$

Inserting Eq. (1-14) into Eq. (1-13)

$$I = \frac{V}{R_{eq}} = V\left(\frac{1}{R_1} + \frac{1}{R_2} + \frac{1}{R_3}\right) \qquad (1\text{-}15)$$

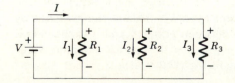

Figure 1-9 Parallel-connected resistors.

So that

$$\frac{1}{R_{eq}} = \frac{1}{R_1} + \frac{1}{R_2} + \frac{1}{R_3} \tag{1-16}$$

which states that for any number of resistors in parallel the reciprocal of the equivalent resistance equals the sum of the reciprocals of the individual resistances.

Networks

Network connections of series and parallel resistances can be analyzed by successive applications of Eqs. (1-10) and (1-16). Consider, for example, the network of Fig. 1-10a with the resistance values as marked on the circuit diagram. The parallel combination of R_5 and R_6, each of which is 10 Ω, may be replaced by a 5-Ω resistor since according to Eq. (1-16)

$$\frac{1}{R_{eq}} = \frac{1}{R_5} + \frac{1}{R_6} = \frac{1}{10} + \frac{1}{10} = \frac{2}{10}$$
$$R_{eq} = 5 \ \Omega \tag{1-17}$$

Therefore the network is reduced to that shown in Fig. 1-10b. Next, the combination of R_{eq} with R_4 ($= 10 \ \Omega$) is, by Eq. (1-10)

$$R'_{eq} = R_{eq} + R_4 = 5 + 10 = 15 \ \Omega \tag{1-18}$$

and the network is now Fig. 1-10c. R'_{eq} and R_3 are in parallel, so that their equivalent is

$$R''_{eq} = \frac{R'_{eq} R_3}{R'_{eq} + R_3} = \frac{15 \times 15}{15 + 15} = \frac{225}{30} = 7.5 \ \Omega \tag{1-19}$$

Lastly, the series combination of R''_{eq}, R_1, and R_2 is simply

$$R_T = R''_{eq} + R_1 + R_2 = 7.5 + 5 + 5 = 17.5 \ \Omega \tag{1-20}$$

and the entire network of Fig. 1-10a may now be replaced by its simple equivalent, Fig. 1-10e, where R_T represents the resistance of the entire network. The current in the battery is therefore

$$I = \frac{V}{R_T} = \frac{35}{17.5} = 2 \ \text{A} \tag{1-21}$$

Suppose it is desired to determine the current I_3 in R_3. This is accomplished by first calculating the potential difference V_3 between points b and c in the circuit diagram. The IR drop across R_1 is $IR_1 = 2 \times 5 = 10$ V, and a similar value applies to the IR drop across R_2. According to Eq. (1-8)

$$V = V_1 + V_3 + V_2 \tag{1-22}$$

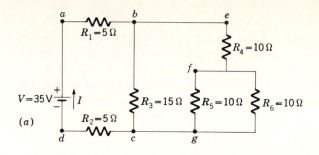

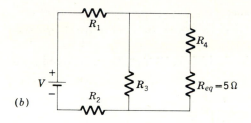

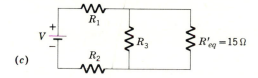

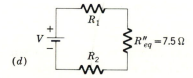

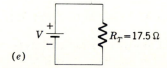

(e)

Figure 1-10 Network reduction by series and parallel equivalents.

Thus

$$V_3 = V - V_1 - V_2 = 35 - 10 - 10 = 15 \text{ V} \qquad (1\text{-}23)$$

The current in R_3 is therefore

$$I_3 = \frac{V_3}{R_3} = \frac{15}{15} = 1 \text{ A} \qquad (1\text{-}24)$$

By similar reasoning it is possible to determine the current in each resistor.

Kirchhoff's Rules

It is not possible to reduce many of the networks important in electronics to simple series-parallel combinations, so that more powerful analytical methods must be used. Two simple extensions of Eqs. (1-8) and (1-12), known as *Kirchhoff's rules*, named after Gustav Robert Kirchhoff, are most helpful in this connection. Consider first the simple parallel circuit, Fig. 1-9, redrawn as in Fig. 1-11 to illustrate the idea of a *branch point*, or *node*, of a circuit. A node is the point at which three (or more) conductors are joined. Kirchhoff's first rule is that the algebraic sum of the currents at any node is zero. Symbolically

$$\sum I = 0 \qquad (1\text{-}25)$$

Note that Eq. (1-25) is essentially a statement of continuity of current; it may also be looked upon as a result of the conservation of electric charge.

Kirchhoff's second rule has already been applied in analyzing the simple series circuit, Fig. 1-7a. It states that the algebraic sum of the potential difference around any complete loop of a network is zero. Symbolically

$$\sum V = 0 \qquad (1\text{-}26)$$

A loop of a network is understood to be any closed path such as *abcda* in Fig. 1-10a which returns to the same point. Other examples of complete loops in the same network are *befgcb* and *daefgd*. Equation (1-26) is a consequence of the conservation of energy.

In applying Kirchhoff's rules to any network the first step is to assign a current of arbitrary direction to each of the resistances in the network. The polarity of the potential difference across each resistor is then marked on the circuit diagram using the convention already noted that the current enters the positive terminal of a resistance. The polarity of emf sources are, of course, specified in advance from the circuit diagram itself. Kirchhoff's rules are then applied to the various nodes and circuit loops to obtain a sufficient number of simultaneous equations to solve for the total number of unknown currents.

It is true that if a network contains m nodes and n unknown currents there are $m - 1$ independent equations which result from Eq. (1-25). Similarly, there are

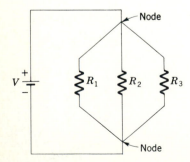

Figure 1-11 Nodes in simple parallel circuit.

$n - (m - 1) = n - m + 1$ independent equations derived from Eq. (1-26). The total number of independent equations obtained from Kirchhoff's rules applied to any network is therefore $(m - 1) + n - (m - 1) = n$. This is just the number of unknown currents and the network solution is therefore completely determined. It is generally possible to write down more node and loop equations than are needed but only n of them are truly independent.

The solution of these independent equations often results in certain of the currents being negative. This means that the original arbitrarily assigned current direction is, in fact, incorrect and the current is actually in the opposite direction. Thus, it is not necessary to know the true current direction in advance. Once the various currents have been calculated, the IR drops in any portion of the circuit are determined using Ohm's law.

The technique of applying Kirchhoff's rules to a network can best be illustrated with a few examples. Consider first the simple parallel-resistor circuit of Fig. 1-12. The current direction in each resistor has been arbitrarily selected and the polarity of the IR drops marked in accordance with these assigned directions. Note that this network has only two nodes, one at b and the other at e. Therefore, there is only $2 - 1 = 1$ independent node equation. Considering the branch point at b, Eq. (1-25) yields

$$I - I_1 + I_2 = 0 \qquad (1\text{-}27)$$

Notice at branch point e the current equation is

$$-I + I_1 - I_2 = 0 \qquad (1\text{-}28)$$

Clearly Eq. (1-28) is simply the negative of (1-27) and the two relations are therefore not independent. Either equation may be used in the solution of the network.

Consider now the loop $abef$. According to Eq. (1-26)

$$V - I_1 R_1 = 0 \qquad (1\text{-}29)$$

Similarly, around the loop $abcdef$

$$V + I_2 R_2 = 0 \qquad (1\text{-}30)$$

Since there are three unknown currents, there must be $3 - 2 + 1 = 2$ independent loop equations and these are (1-29) and (1-30). Note, however, that around loop $bcde$

$$I_1 R_1 + I_2 R_2 = 0 \qquad (1\text{-}31)$$

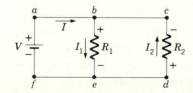

Figure 1-12

This is not an independent relation, as may be shown by subtracting Eq. (1-29) from (1-30). The result is Eq. (1-31). Thus these three loop equations are not independent, and any two may be used in solving the network.

Choose Eqs. (1-27), (1-29), and (1-31) as the three independent equations to solve for the three unknown currents. The solution is accomplished by first solving Eq. (1-29) for I_1,

$$I_1 = \frac{V}{R_1} \tag{1-32}$$

Next, I_2 is determined from (1-31),

$$I_2 = -\frac{I_1 R_1}{R_2} \tag{1-33}$$

Substituting (1-33) into (1-27),

$$I - I_1 - \frac{I_1 R_1}{R_2} = 0 \tag{1-34}$$

Substituting from (1-32) for I_1,

$$I - \frac{V}{R_1} - \frac{V R_1}{R_1 R_2} = I - V\left(\frac{1}{R_1} + \frac{1}{R_2}\right) = 0 \tag{1-35}$$

The current I is therefore

$$I = V\left(\frac{1}{R_1} + \frac{1}{R_2}\right) \tag{1-36}$$

which is quite equivalent to the solution corresponding to Eq. (1-13) arrived at by considering parallel resistors.

Finally, I_2 is determined by substituting for I_1 in Eq. (1-33),

$$I_2 = -\frac{V R_1}{R_1 R_2} \tag{1-37}$$

or

$$I_2 = -\frac{V}{R_2} \tag{1-38}$$

According to the minus sign in Eq. (1-38), this current is actually in the opposite direction to that assumed in Fig. 1-12. Correspondingly, the IR drop across R_2 has the opposite polarity from that shown on the circuit diagram.

More complicated networks require more than three equations, and it is sometimes desirable to employ the standard method of determinants to solve the set of simultaneous equations. This technique, illustrated in the following section, has the considerable advantage that it is possible to solve directly for only those currents that are of interest. Often, only one or two of the currents in a network are of direct concern, and in this case a complete solution for all the unknowns is superfluous.

Wheatstone Bridge

In this section Kirchhoff's rules are used to analyze the *Wheatstone bridge* circuit illustrated in Fig. 1-13. This extremely useful circuit was developed in 1843 by Charles Wheatstone and is widely used in electrical measurements to determine values of unknown resistances. The manner in which it is used may be understood from an analysis of the circuit. According to Kirchhoff's rule for branch points a, b, and d,

$$I - I_1 - I_2 = 0$$
$$I_1 - I_3 + I_5 = 0 \tag{1-39}$$
$$I_3 + I_4 - I = 0$$

Because there are four branch points in the Wheatstone bridge circuit, these three current equations are independent, so the fourth one, which could be written for branch point c, is not used.

Applying Kirchhoff's rule to loops *abdefa*, *acba*, and *bcdb*, the equations are

$$-I_1 R_1 - I_3 R_3 + V = 0$$
$$-I_2 R_2 - I_5 R_5 + I_1 R_1 = 0 \tag{1-40}$$
$$I_5 R_5 - I_4 R_4 + I_3 R_3 = 0$$

Note carefully the indicated polarities of the various IR drops as they are encountered in traversing each loop. Since there are six unknown currents, $6 - 4 + 1 = 3$ loop equations are needed here and any others are redundant.

Equations (1-39) and (1-40) are six equations in six unknowns. Therefore, in applying the method of determinants to these simultaneous equations it is necessary to evaluate two sixth-order determinants in calculating each current. The total solution involves seven different such determinants. While the evaluation of a sixth-order determinant is straightforward and a number of standard approaches exist to reduce the order before final evaluation, the complete solution

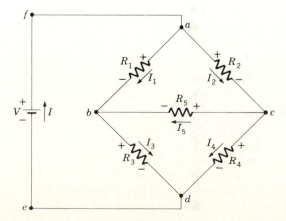

Figure 1-13 Wheatstone bridge.

of seven sixth-order determinants is quite laborious. Therefore, although the solution to the set of Eqs. (1-39) and (1-40) is in principle accomplished, it is useful to seek alternative methods.

The analysis of complex networks can usually be simplified by the use of *loop currents*. This technique, known as *Maxwell's method*, after James Clerk Maxwell, in effect applies both Kirchhoff's rules simultaneously and thereby reduces the number of simultaneous equations needed. The loop currents are drawn around any complete loop as the three illustrated for the case of the Wheatstone bridge in Fig. 1-14. After the polarity of the *IR* drops are indicated in accordance with the current directions, the usual voltage equations around each loop are written. Thus, with reference to Fig. 1-14

$$V - R_1(I_a - I_b) - R_3(I_a - I_c) = 0$$

$$-R_2 I_b - R_5(I_b - I_c) + R_1(I_a - I_b) = 0 \qquad (1\text{-}41)$$

$$R_3(I_c - I_a) + R_5(I_c - I_b) + R_4 I_c = 0$$

Here again, note the polarity of the *IR* drops and the current directions. Upon rearranging

$$-(R_1 + R_3)I_a + R_1 I_b + R_3 I_c = -V$$

$$R_1 I_a - (R_1 + R_2 + R_5)I_b + R_5 I_c = 0 \qquad (1\text{-}42)$$

$$-R_3 I_a - R_5 I_b + (R_3 + R_4 + R_5)I_c = 0$$

The solution of Eqs. (1-42) for any current, say I_b, using determinants is found by forming a ratio in which the denominator is the determinant of the coefficients of the currents and the numerator is a similar determinant with the coefficients of the unknown current replaced by the right side of the equation.

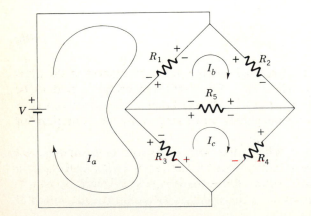

Figure 1-14 Loop-current analysis of Wheatstone bridge.

That is, the solution for I_b is

$$I_b = \frac{\begin{vmatrix} -(R_1 + R_3) & -V & R_3 \\ R_1 & 0 & R_5 \\ -R_3 & 0 & R_3 + R_4 + R_5 \end{vmatrix}}{\begin{vmatrix} -(R_1 + R_3) & R_1 & R_3 \\ R_1 & -(R_1 + R_2 + R_5) & R_5 \\ -R_3 & -R_5 & R_3 + R_4 + R_5 \end{vmatrix}}$$

$$= \frac{VR_5 R_3 + VR_1(R_3 + R_4 + R_5)}{\Delta} \tag{1-43}$$

where Δ signifies the denominator. Similarly, I_c is

$$I_c = \frac{\begin{vmatrix} -(R_1 + R_3) & R_1 & -V \\ R_1 & -(R_1 + R_2 + R_5) & 0 \\ -R_3 & -R_5 & 0 \end{vmatrix}}{\Delta}$$

$$= \frac{VR_1 R_5 + VR_3(R_1 + R_2 + R_5)}{\Delta} \tag{1-44}$$

Now, the current through R_5 is

$$I_5 = I_b - I_c$$

$$= \frac{V}{\Delta}(R_5 R_3 + R_1 R_3 + R_1 R_4 + R_1 R_5 - R_1 R_5 - R_1 R_3 - R_2 R_3 - R_5 R_3)$$

$$= \frac{V}{\Delta}(R_1 R_4 - R_2 R_3) \tag{1-45}$$

Equation (1-45) is a most important relation for the Wheatstone bridge. Note that if

$$R_1 R_4 = R_2 R_3$$

or

$$\frac{R_1}{R_2} = \frac{R_3}{R_4} \tag{1-46}$$

then I_5 is zero, independent of the applied voltage. If the resistances in the arms of the bridge obey the ratios indicated in Eq. (1-46), the bridge is said to be *balanced*. Thus, for example, if R_1, R_2, and R_3 are known resistances and I_5 is zero, the value of R_4 may be immediately calculated from the condition for balance, Eq. (1-46).

In the common version of a Wheatstone bridge, resistances R_1 and R_2 are connected to a switch to give decade values of the ratio R_2/R_1, and R_3 is a continuously variable calibrated resistor. Once the bridge is balanced by adjusting R_3, the value of the unknown R_4 is simply $(R_2/R_1)R_3$. The decade values of the ratio (R_2/R_1) may range from 10^{-3}, 10^{-2}, and 10^{-1} to 1, 10, 10^2, and 10^3, so that a very wide range of resistance values can be measured. In practice, a current-indicating instrument is connected in the position of R_5 to indicate balance. Note that this meter need not be calibrated, since it is used only to indicate the balance condition, that is, a zero current.

Equation (1-45) is an example of a case in which useful information concerning the circuit is derived without carrying through the complete solution for all currents. It is often possible to draw the loop currents in such a way that only one current need be determined. The facility of choosing loop currents that minimizes the effort required to solve any given network is attained with experience.

EQUIVALENT CIRCUITS

Thévenin's Theorem

Many times the analysis of electronic circuits is facilitated by replacing all or part of a network by an *equivalent circuit* which, for certain purposes, has the same characteristics as the original. An example of this possibility has already been discussed in connection with series and parallel combination of resistors. There, an entire network of resistances is replaced by a single equivalent resistance in order to calculate the current. In other situations, particularly in connection with transistor circuits, equivalent circuits may be used to represent the behavior of electronic devices.

One of the most useful equivalent circuits is the one that results from *Thévenin's theorem*, which states that any network of resistors and batteries having two output terminals may be replaced by the series combination of a resistor and a battery, as illustrated in Fig. 1-15. The form of the Thévenin equivalent circuit shows immediately how the proper values of V_{eq} and R_{eq} can be determined without knowing the actual configuration of the network itself. The equivalent emf is just the potential at the output terminals when the output current is zero, that is, the *open-circuit* voltage. The equivalent resistance is then the ratio of V_{eq} to the output current when $R_L = 0$, or to the *short-circuit* current.

Note also that R_{eq} is equal to the load resistance for which the voltage across the load is one-half of V_{eq}. This is useful in situations where the short-circuit current cannot be easily determined. The form of the Thévenin equivalent circuit, Fig. 1-15b, also shows that R_{eq} is the resistance of the network between the output terminals when V_{eq} is considered to be replaced by a short circuit. This way of determining R_{eq} analytically is useful when the configuration of the network is known since it usually involves only simple network reductions. Which of these alternative methods is employed to determine the equivalent emf and the equivalent internal resistance depends only upon which is easiest in any given situation.

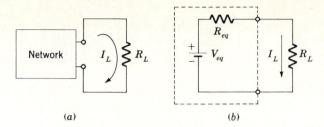

Figure 1-15 (*a*) Two-terminal network and (*b*) Thévenin equivalent circuit.

Consider, for example, the Thévenin equivalent of the simple circuit in Fig. 1-16. The equivalent emf is just

$$V_{eq} = \frac{VR_2}{R_1 + R_2} \qquad (1\text{-}47)$$

Replacing the battery with a short circuit, the resistance between the terminals is just the parallel combination of R_1 and R_2. Accordingly,

$$R_{eq} = \frac{R_1 R_2}{R_1 + R_2} \qquad (1\text{-}48)$$

why parallel?

The load current is then

$$I_L = \frac{V_{eq}}{R_L + R_{eq}} = \frac{VR_2}{R_1 R_2 + R_L(R_1 + R_2)} \qquad (1\text{-}49)$$

The right side of Eq. (1-49) may be derived by direct analysis of Fig. 1-16.

To illustrate the power of the equivalent-circuit method, consider the Wheatstone bridge circuit of Fig. 1-17*a*. The current through R_5 is analyzed by replacing the remainder of the circuit with its Thévenin equivalent. Replacing the battery with a short circuit puts R_3 in parallel with R_1 and this combination in series with the parallel combination of R_2 and R_4 across the output terminals, as illustrated in Fig. 1-17*b*. Therefore, R_{eq} in the equivalent circuit, Fig. 1-17*c*, is

$$R_{eq} = \frac{R_1 R_3}{R_1 + R_3} + \frac{R_2 R_4}{R_2 + R_4} \qquad (1\text{-}50)$$

The open-circuit voltage at the output terminals is simply the potential difference between the junction of R_1 and R_3 and the junction of R_2 and R_4. This potential difference is found by subtracting the *IR* drop across R_2 from the *IR* drop across

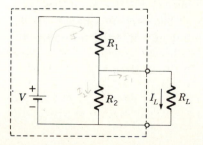

Figure 1-16

V_{eq} = open circuit voltage = voltage when output current is zero

replace battery by short circuit

then:
$(R_3 \| R_1)$ in
series with
$(R_2 \| R_4)$

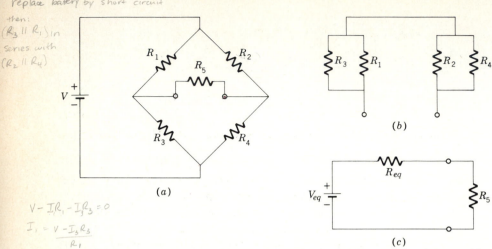

$V - IR_1 - IR_3 = 0$

$I_1 = \dfrac{V - I_3 R_3}{R_1}$

(a)

(b)

R_{eq}

(c)

Figure 1-17 (*a*) Conventional Wheatstone bridge circuit; (*b*) after replacing battery with a short circuit in order to calculate R_{eq}; (*c*) Thévenin equivalent.

R_1. Therefore the equivalent battery is

$V_{at\ 1,3} = R_1' I_1 - R_2 I_2$
junction
$= V$

$$V_{eq} = \frac{V R_1}{R_1 + R_3} - \frac{V R_2}{R_2 + R_4} \tag{1-51}$$

Finally, according to the equivalent circuit, Fig. 1-17c, the current through I_5 is

$$I_5 = \frac{V_{eq}}{R_{eq} + R_5} \tag{1-52}$$

The ease and rapidity with which this result is obtained should be compared with that necessary using Kirchhoff's rules. Note that the balance condition, Eq. (1-46), follows immediately from Eqs. (1-51) and (1-52), since $I_5 = 0$ at balance.

Norton's Theorem

A second form of equivalent circuit useful in situations as, for example, transistor circuits where current sources, rather than emfs, are of major interest, is one given by *Norton's theorem*. Norton's theorem states that any network of batteries and resistors having two output terminals can be replaced by the parallel combination of a current source I_{eq} and a resistance R_{eq}. The current source I_{eq} is the short-circuit current in the output terminals of a circuit, while the resistance R_{eq} is the same as for Thévenin's theorem.

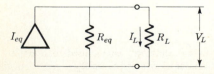

Figure 1-18 Norton equivalent circuit.

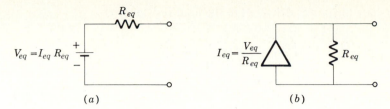

$$V_{eq} = I_{eq} R_{eq}$$ R_{eq}

$$I_{eq} = \frac{V_{eq}}{R_{eq}}$$ R_{eq}

(a)   (b)

Figure 1-19 Relationship between (a) Thévenin equivalent circuit and (b) Norton equivalent circuit.

Norton's equivalent circuit is shown in Fig. 1-18, where the triangle represents the current source I_{eq}. No simple electric component acts as a current source the way a battery acts as a voltage source. Nevertheless, the idea of a current source is conceptually very useful in circuit analysis.

Since it is possible to represent any network by either the Thévenin or the Norton equivalent circuit, it must be possible to convert from one equivalent circuit to the other. Referring to Fig. 1-19a and b, the short-circuit current through the load is V_{eq}/R_{eq} for the Thévenin equivalent circuit, and is equal to I_{eq} for the Norton equivalent circuit. For both circuits to represent the same network, it must be true that

$$I_{eq} = \frac{V_{eq}}{R_{eq}} \tag{1-53}$$

Thus, it is a simple matter to convert from one circuit to another. Which circuit is used to represent any given network is entirely a matter of choice and convenience.

Maximum Power Transfer

In many electronic circuits, such as a radio transmitter or hifi amplifier, it is important to transfer efficiently the maximum amount of electric power from the source to the load, which may be an antenna or loudspeaker. Therefore, it is of interest to determine circuit conditions for which it is possible to achieve the maximum power transfer. Suppose the network is represented by its Thévenin equivalent circuit shown in Fig. 1-20 and that the load connected to the output terminals is represented by the resistance R_L. The subscripts on the equivalent battery and equivalent resistance have been eliminated for convenience.

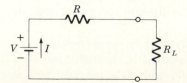

R

V I R_L

Figure 1-20 Thévenin's equivalent circuit used to examine maximum power transfer to load resistance R_L.

Following Joule's law the power delivered to the load resistance is

$$P = I^2 R_L = \left(\frac{V}{R + R_L}\right)^2 R_L$$

$$= \frac{V^2/R_L}{(1 + R/R_L)^2} \tag{1-54}$$

According to Eq. (1-54), the power in the load is zero if the load resistance is very small and is also zero when the load resistance is very large. Thus, there must be an optimum load resistance for which the power in R_L is a maximum.

To find the condition for *maximum power transfer*, differentiate Eq. (1-54) with respect to R_L and equate the result to zero,

$$\frac{dP_L}{dR_L} = \frac{V^2}{R_L}\frac{2R/R_L^2}{(1 + R/R_L)^3} + \frac{V^2}{(1 + R/R_L)^2}\frac{-1}{R_L^2} = 0 \tag{1-55}$$

$$\frac{2R}{R_L} = 1 + \frac{R}{R_L} \tag{1-56}$$

so that
$$R_L = R \tag{1-57}$$

This means that maximum power is delivered to the load when the load resistance is equal to the internal resistance of the network delivering the power. When the load resistance is equal to the internal resistance of the network, the load is said to be *matched* to the circuit.

Since the equivalent circuit of Fig. 1-20 represents any network, the result applies equally well to all circuits. The use of the equivalent-circuit concept in this connection has thus made it possible to prove a very general result quite easily.

ELECTRICAL MEASUREMENTS

D'Arsonval Meter

By far the most common design for electric current–measuring instruments is the *d'Arsonval meter*, named after its inventor. A multiturn coil of fine wire wound on an aluminum frame is pivoted between the poles of a horseshoe permanent magnet (Fig. 1-21). Two fine spiral springs serve to position the coil and to carry the current to be measured. A pointer attached to the coil indicates the current on a scale as the coil rotates in response to the interaction between the current in the coil and the magnetic field of the magnet. A soft-iron pole piece is fitted between the poles of the magnet so that the sides of the coil move in a radially directed field.

With this design, the deflection of the pointer is directly proportional to the current through the coil. The sensitivity of the meter, that is, the deflection for a given current, is improved by increasing the magnetic field of the magnet, the coil

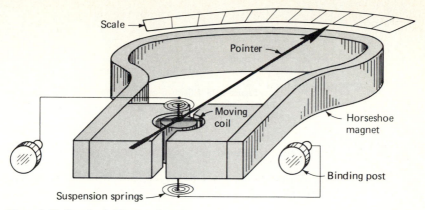

Scale

Pointer

Moving
coil

Horseshoe
magnet

Binding post

Suspension springs

Figure 1-21 Sketch of d'Arsonval meter.

area, and the number of turns on the coil, or by decreasing the torque constant of the springs.

The size of the coil and springs is dictated by mechanical ruggedness. Also, it is not desirable to increase the number of turns of wire inordinately in an effort to improve sensitivity, since the resistance of the coil is increased thereby. This may adversely affect the operation of the meter, as explained in the following section. The magnetic field is limited to that available from conventional permanent magnets. In spite of these practical limitations, common d'Arsonval meters have full-scale deflections for currents as small as 10^{-3} A (1 *milliampere*, abbreviated mA), or even 50×10^{-6} A (50 *microamperes*, abbreviated μA). Laboratory instruments, which may be shock-mounted and therefore designed for maximum sensitivity, are capable of measuring 10×10^{-12} A (10 *picoamperes*, abbreviated pA).

Ammeters and Voltmeters

The d'Arsonval meter is a current-sensitive device, or *ammeter*. It is often convenient to change the current required for full-scale deflection in order to increase the range of currents over which the meter is useful. This is accomplished by *shunting* a portion of the current around the ammeter with a parallel resistance, as diagrammed in Fig. 1-22. Note that the internal resistance of the ammeter's coil R_m is indicated explicitly. Following Kirchhoff's rules, $I = I_m + I_s$ and

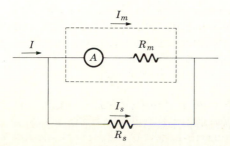

I_m

R_m

I

A

I_s

R_s

Figure 1-22 Increasing range of ammeter by using shunt resistor in parallel with meter resistance.

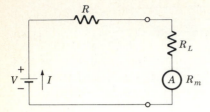

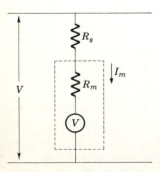

Figure 1-23 Effect of ammeter resistance on current in circuit.

$I_m R_m = I_s R_s$, so that the current to be determined is

$$I = I_m + \frac{I_m R_m}{R_s} = I_m\left(1 + \frac{R_m}{R_s}\right) \tag{1-58}$$

If, for example, the shunt resistance is one-ninth of the meter resistance, $1 + R_m/R_s = 10$, and the full-scale deflection is extended to ten times the inherent sensitivity of the meter.

It is always necessary to consider the effect of the meter resistance on the circuit. Suppose it is desired to measure the current in R_L of the Thévenin equivalent circuit of Fig. 1-23 using an ammeter having an internal resistance of R_m. With the ammeter connected into the circuit the current is

$$I = \frac{V}{R + R_L + R_m} \tag{1-59}$$

Unless $R_m \ll R + R_L$, the current indicated by the ammeter is different from the true current. For this reason it is always desirable that the internal resistance of an ammeter be small. On the other hand, in circumstances where the internal resistance may not be small compared with circuit resistances, it is possible to correct for the disturbing influence of the meter resistance and thus determine the true current.

Since the full-scale deflection of an ammeter may be attributed to the voltage $V_m = R_m I$ across the meter resistance R_m, a d'Arsonval meter also is a *voltmeter*. Here again it is useful to change the range of any given voltmeter by introducing a resistance this time in series with the meter. Referring to the circuit of Fig. 1-24,

Figure 1-24 Using d'Arsonval meter as voltmeter with series resistor multiplier.

the voltage to be measured is

$$V = I_m(R_m + R_s) \tag{1-60}$$

and it is obvious that the series resistance R_s increases the maximum full-scale voltage of a given meter. It is common practice to provide several such series resistance *multipliers* to allow a given meter to be used over a wide range of voltages.

The effect of a voltmeter on the circuit to which it is attached must be considered, just as in the case of an ammeter. This is so because the voltmeter requires a small current to deflect the pointer and this current must be supplied by the circuit. If the meter current is not negligible compared with the normal currents in the circuit, the voltmeter is said to *load* the circuit and a correction must be applied to the indicated meter reading to determine the true voltage in the absence of the disturbing meter influence.

Sensitive d'Arsonval meters are useful as voltmeters since they require only a very small current to achieve full-scale deflection. It is common practice to specify the sensitivity of a voltmeter in terms of the ratio of its internal resistance to the full-scale voltage in units of *ohms per volt*. Note that, according to Eq. (1-60), the ratio of $R_m + R_s$ to the voltage required for full-scale deflection is simply the reciprocal of the current sensitivity of the meter, and therefore the two specifications are quite equivalent. For example, a voltmeter which uses a d'Arsonval meter having a full-scale sensitivity of 1 mA is rated at $1/10^{-3} = 1000 \ \Omega/V$. This means that the voltmeter has an internal resistance of 100,000 Ω on the 100-V scale, etc. Similarly, a 20,000-Ω/V voltmeter (employing a 50-μA meter) has a resistance of 2 megohms (MΩ) on its 100-V scale.

A convenient technique for determining the resistance of a portion of a network is to measure both the current and the voltage and then apply Ohm's law. There are two distinct ways that the meters can be connected in this *voltmeter-ammeter method* (Fig. 1-25a and b). The choice between the two possibilities depends upon the relative values of the meter resistances and the circuit

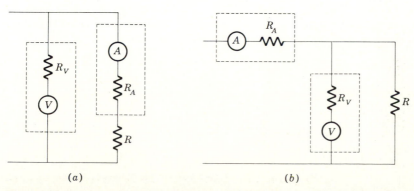

(a)   (b)

Figure 1-25 Two ways of connecting voltmeter and ammeter to measure resistance R.

resistances, as can be shown in the following way. Consider first the circuit in Fig. 1-25a. From Kirchhoff's rules.

$$V = AR + AR_A \tag{1-61}$$

where V and A are the meter readings. The unknown resistance is given by

$$R = \frac{V}{A} - R_A \tag{1-62}$$

which shows that the true resistance is smaller than the indicated V/A ratio.

Similarly, in the circuit of Fig. 1-25b, the current A divides between the parallel paths R and R_V so that

$$V = \frac{R_V R}{R_V + R} A \tag{1-63}$$

Solving for the unknown resistance, the result is, after some rearrangement,

$$R = \frac{V}{A} \frac{1}{1 - (V/A)/R_V} \tag{1-64}$$

According to Eqs. (1-62) and (1-64), the first circuit is most useful when the ammeter resistance is small compared with the unknown (or to V/A), while the second circuit applies when the voltmeter resistance is large compared with the unknown. In either case, the unknown resistance is then given simply by the ratio V/A.

Ohmmeters and Multimeters

A simple extension of the voltmeter-ammeter circuit can be used as an *ohmmeter* in which the meter scale is calibrated directly in ohms. In a typical circuit (Fig. 1-26) the same meter is used successively to measure first the voltage across the unknown resistor and then the current in it. The way the meter scale is calibrated directly in ohms can be understood from the following analysis. First, suppose the terminals a and b in Fig. 1-26 (which are usually connected to *test leads* to facilitate connecting the ohmmeter to the unknown resistor) are shorted together. The voltmeter then measures the battery voltage V.

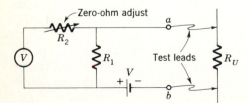

Figure 1-26 Simple ohmmeter circuit.

Next, the test leads are connected to the unknown resistance. If the voltage across R_1, as measured by the meter, is now V_R, Ohm's law gives

$$R_U + R_1 = \frac{V}{V_R/R_1} \tag{1-65}$$

Solving for R_U

$$R_U = R_1\left(\frac{V}{V_R} - 1\right) \tag{1-66}$$

According to Eq. (1-66) the unknown resistance can be calculated directly from the two meter readings, but it is more useful to calibrate the meter scale directly in ohms in the following way.

The variable resistance R_2 is used to adjust the meter reading to full-scale when the test leads are first shorted together. This point on the scale is then "zero ohms," and is so marked. Suppose that when the test leads are connected to the unknown resistor, the meter deflects to half-scale. This means $V_R = V/2$ and, according to Eq. (1-66), $R_U = R_1$. Thus, the midpoint on the scale can be marked with the resistance corresponding to R_1. Similarly, note that the quarter-scale reading, $V_R = V/4$, corresponds to $3R_1$, while a zero reading indicates an open circuit, or infinite ohms. The scale of an ohmmeter is clearly nonlinear although it is not difficult to use since it is direct-reading in ohms.

According to Eq. (1-66) the midscale reading of an ohmmeter depends upon R_1, so that by selecting different values for R_1 it is possible to encompass a wide range of unknown resistances. Equation (1-66) assumes that the meter current is negligible, which may not be true on high-resistance ranges where R_1 is large. Therefore, practical ohmmeter circuits are slightly more complicated than the elementary one illustrated by Fig. 1-26, but the principle of operation is identical. Note that an ohmmeter cannot be used to determine resistance values in a circuit which is in operation, for erroneous readings result due to IR drops in the circuit itself.

It is convenient to include the functions of a voltmeter, ammeter, and ohmmeter within one instrument since all three employ the same basic d'Arsonval meter. In such an instrument, commonly termed a *multimeter* or *VOM* (volt-ohm-milliammeter), switches or a number of terminals select the function and range to be used. The circuit in Fig. 1-27 is an example of an elementary instrument which has four voltage ranges, two current ranges, and a single ohms scale. By carefully tracing through the connections on this circuit diagram it is possible to draw the simple functional circuits for each use and in this way subdivide the analysis of the circuit into easy stages.

The VOM circuit of Fig. 1-27 employs a multitap switch with three wipers which are mechanically linked to move together. More elaborate multimeters have considerably more complicated switching arrangements in order to accommodate additional ranges and other functions.

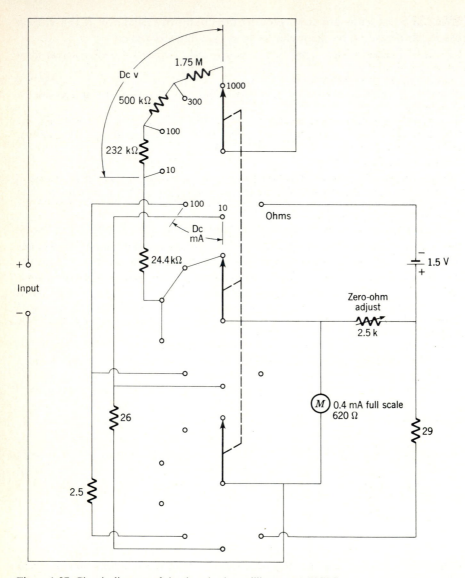

Figure 1-27 Circuit diagram of simple volt-ohm-milliammeter (VOM).

SUGGESTIONS FOR FURTHER READING

Bernard Grob: "Direct and Alternating Current Circuits," McGraw-Hill Book Company, New York, 1986.

Leigh Page and Norman Ilsley Adams: "Principles of Electricity," D. Van Nostrand, Inc., Princeton, N. J., 1931.

Robert E. Simpson: "Introductory Electronics for Scientists and Engineers," Allyn and Bacon, Inc., Boston, 1987.

W. F. Stubbins: "Essential Electronics," John Wiley & Sons, Inc., New York, 1986

EXERCISES

1-1 Calculate the resistance of a copper wire 1 m long and 0.5 mm in diameter. Repeat for a similar-sized nichrome wire.

Answer: $8.65 \times 10^{-2} \Omega$; 5.1Ω

1-2 What is the maximum current that can be in a 1-W 1-MΩ resistor? In a $\frac{1}{2}$-W 10,000-Ω resistor?

Answer: 10^{-3} A; 7.07×10^{-3} A

1-3 Given that the maximum-current capability of a flashlight dry cell is 0.5 A, what is the internal resistance of the cell? Compare with the internal resistance of a storage battery, if the maximum current in this case is 500 A.

Answer: 3Ω; $4.2 \times 10^{-3} \Omega$

1-4 Assume that Fig. 1-7a represents a battery power source connected to a load R_3 by means of copper wires of resistance R_1 and R_2. If $V = 10$ V, the load is 5 Ω, and the wires are 1.0 mm in diameter and 100 m long, calculate the current, the power delivered to the load, the power lost in the wires, and the voltage across the load resistor.

Answer: 1.07 A; 5.7 W; 4.9 W; 5.4 V

1-5 Design a voltage divider (Fig. 1-8) in which the output voltages possible are 1.0, 2.0, 5.0, and 10.0 V, if the battery voltage is 10 V and no current is taken by the output terminals.

1-6 How many identical 1-W resistors, and of what resistance value, are needed to yield an equivalent 1000-Ω 10-W resistor? Obtain two different solutions.

Answer: Ten 10-kΩ resistors in parallel; ten 2.5-kΩ resistors in series parallel

1-7 Determine the current through each resistor of Fig. 1-10a. Verify that the total I^2R loss in the resistors equals the power delivered by the battery.

Answer: 2 A in R_1 and R_2; 1 A in R_3 and R_4; 0.5 A in R_5 and R_6

1-8 In Fig. 1-28, if $R_1 = 2 \Omega$, $R_2 = 5 \Omega$, $R_3 = 2 \Omega$, $R_4 = 5 \Omega$, $R_5 = 10 \Omega$, and $V = 10$ V, find the total current supplied by the battery.

Answer: 2.18 A

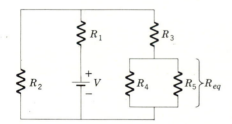

Figure 1-28

1-9 Find the resistance of the network in Fig. 1-29 between the input terminals. What voltage applied to the input results in a 1-A current in the 4-Ω resistor?

Answer: 8 Ω; 72 V

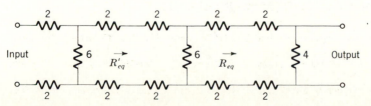

Figure 1-29

1-10 Using Kirchhoff's rules, find the current in the 4-Ω resistor in the network in Fig. 1-30.

   Answer: 0.12 A

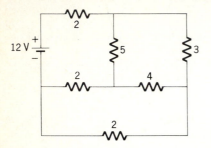

Figure 1-30

1-11 Find the current in each resistor of the circuit in Fig. 1-31.

   Answer: 1.15 A; 0.883 A; 0.267 A

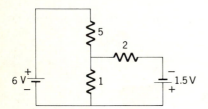

Figure 1-31

1-12 Note that in the Wheatstone bridge (Fig. 1-13) at balance $I_5 = 0$. Using this condition, compare the voltage drops in corresponding arms of the bridge and thus derive the balance condition, Eq. (1-46).

1-13 Solve the Wheatstone bridge network of Fig. 1-14 by drawing loop currents such that there is only one current in R_5. Use the expression for this current to derive the balance condition.

1-14* In the Wheatstone bridge circuit of Fig. 1-13 take $R_1 = R_2 = 100\ \Omega$, $R_4 = 100\ \Omega$, $V = 12\ V$, and R_3 to be a calibrated variable resistor. Plot the voltage across the detector, $R_5 = 100\ \Omega$, as a function of R_3 for the region near balance. Repeat for $R_1 = R_2 = 1000\ \Omega$. Which configuration is more sensitive?

   Answer: The first.

1-15* Experimental values of load current and terminal voltage of the unknown network in Fig. 1-32 are listed in Table 1-2. Determine the Thévenin equivalent network from the data if the voltmeter is

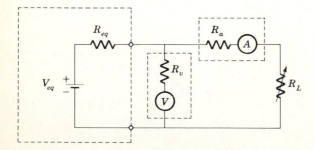

Figure 1-32

1000 Ω/V used on the 1-V scale. Plot the load power IV and the measured resistance V/I as a function of V and compare the load resistance at the maximum power point with R_{eq}.

$\quad$ *Answer:* 67 Ω, 0.5 V, 65 Ω, 55 Ω

Table 1-2

I, $-$mA	V, volts
0 $(R_L = \infty)$	0.468
1.0	0.405
2.0	0.342
3.0	0.279
4.0	0.216
5.0	0.153
6.0	0.090
6.4 $(R_L = 0)$	0.064

1-16 Design an *Ayrton shunt* as in Fig. 1-33 for a 50-μA meter which has an internal resistance of 1000 Ω, if the desired current ranges are 10 mA, 100 mA, 1 A, and 10 A.

$\quad$ *Answer:* 4.523 Ω; 0.4523 Ω; 4.523 $\times$ 10^{-2} Ω; 5.025 $\times$ 10^{-3} Ω

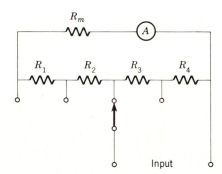

$\quad$ Input $\quad$ **Figure 1-33**

1-17 Suppose, in the circuit of Fig. 1-23, $R = 1000$ Ω, $R_L = 5000$ Ω, and $R_m = 1000$ Ω. If the indicated current is 1.5 mA, what is the true current when the ammeter is not present?

$\quad$ *Answer:* 1.75 mA

1-18 The voltage in a circuit as measured with a 20,000-Ω/V meter on the 500-V scale is 200 V. On the 100-V scale the reading is 95 V. What is the true voltage?

$\quad$ *Answer:* 278 V

1-19 What are the meter readings in the two versions of the voltmeter-ammeter method of Fig. 1-25 if the voltmeter is 1000 Ω/V, the internal resistance of the ammeter is 100 Ω, the "unknown" resistance is 1000 Ω, and the applied voltage is 10 V? Which of the two versions is more satisfactory?

$\quad$ *Answer:* 10 V, 9.1 $\times$ 10^{-3} A; 9.008 V, 9.92 $\times$ 10^{-3} A; circuit b is better

1-20 Draw the scale of an ohmmeter, assuming $R_1 = 10,000$ Ω in Eq. (1-66).

TWO

ALTERNATING CURRENTS

The currents and voltages in most practical electronic circuits are not steady but vary with time. For example, when a circuit is used to measure some physical quantity, such as the temperature of a chemical reaction, the voltage or current in the circuit representing the temperature may vary significantly. Similarly, detection of nuclear disintegrations results in a series of rapid voltage pulses of very short duration. In order to understand such effects it is necessary to study the properties of time-varying currents.

The simplest time-varying current alternates direction periodically and accordingly is called an alternating current, abbreviated ac. Obviously an ac circuit is one in which alternating currents are active, but direct currents may be present as well. Most of the concepts developed for dc circuits in the previous chapter carry over to ac circuits. Two new elements in addition to resistance are important in ac circuits and these are treated in this chapter.

SINUSOIDAL SIGNALS

Frequency, Amplitude, and Phase

The simplest alternating waveform is *sine-wave* voltage or current, which varies sinusoidally with time. A sinusoidal waveform is generated by the variation of the vertical component of a vector rotating counterclockwise with a uniform angular velocity ω as in Fig. 2-1. One complete revolution is termed a *cycle* and the time interval required for one cycle is called the *period T*. The number of cycles per second is the *frequency f*, which is just the reciprocal of the period. The scope of frequencies encountered in electronic circuits is often as small as a few cycles per second, which is called a *hertz* (abbreviated Hz) in honor of the German scientist Heinrich Hertz, discoverer of radio waves. The range of frequency extends to *kilohertz* (kHz, 10^3 Hz) and *megahertz* (MHz, 10^6 Hz), on up to the *gigahertz* (GHz, 10^9 Hz) range.

Since there are 2π radians in one complete revolution and this requires T seconds, the *angular frequency* ω is $2\pi f$. If the length of the rotating vector is V_p, the instantaneous value at any time t is just $V_p \sin \omega t$, where V_p is the *maximum* or *peak amplitude* of the sine wave.

Two sinusoidal waveforms that have the same frequency but pass through zero at different times are said to be out of *phase*, and the angle between the two rotating vectors is called the *phase angle*. In Fig. 2-2 the voltage v_2 is *leading* the sine-wave voltage v_1 because it passes through zero first, and the phase difference is the angle ϕ. Note that it is only possible to specify the phase angle between two sine waves if they have the same frequency. A voltage sine wave is completely described by its frequency and amplitude unless it is compared with another signal of the same frequency. In this case the most general equation for the voltage must include the phase angle

$$v = V_p \sin (\omega t + \phi) \tag{2-1}$$

Note that lowercase symbols are used to indicate time-varying voltages (and currents), while capital letters refer to constant values or to dc quantities.

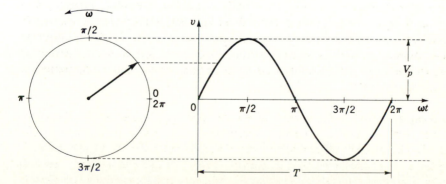

Figure 2-1 Generation of sine wave by vertical component of rotating vector.

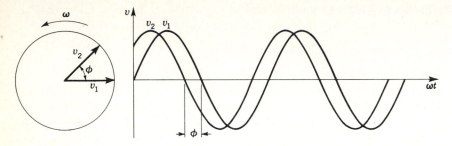

Figure 2-2 Illustrating phase angle between two sinusoidal voltages.

RMS Value

It is often necessary to compare the magnitude of a sine-wave current with a direct current. This is accomplished by comparing the joule heat produced in a resistor by the two currents. That is, the *effective value* of a sinusoidal current is equal to the direct current which produces the same joule heating as the alternating current. To determine this value, the heating effect of an alternating current is calculated by averaging the I^2R losses over a complete cycle. Thus, the average power is given by

$$P = \frac{1}{T}\int_0^T i^2 R \; dt = \frac{I_p^2 R}{T}\int_0^T \sin^2 \omega t \; dt = \frac{I_p^2 R}{T}\left[\frac{t}{2} - \frac{\sin 2\omega t}{4\omega}\right]_0^T = \frac{I_p^2 R}{2} \quad (2\text{-}2)$$

Since the joule heating in a resistor caused by a direct current is equal to I^2R, the effective value of an alternating current I_e is simply

$$I_e^2 R = \frac{I_p^2 R}{2} \quad (2\text{-}3)$$

or

$$I_e = \frac{I_p}{\sqrt{2}} \quad (2\text{-}4)$$

According to Eq. (2-4) the effective value of a sine wave is simply its peak value divided by the square root of two. The effective value is commonly referred to as the *root-mean-square* or *rms* value. Voltmeters and ammeters capable of measuring ac signals are almost universally calibrated in terms of rms readings to facilitate comparison with dc meter readings. It is generally understood that ac currents and voltages are characterized by their rms values, unless stated otherwise.

Power Factor

Suppose that the current and voltage in some portion of a circuit are given by

$$i = I_p \sin \omega t$$

$$v = V_p \sin (\omega t + \phi) \quad (2\text{-}5)$$

where the phase angle ϕ is introduced to account for the possibility that the current and voltage are not in phase. The instantaneous power p is then

$$p = vi = V_p I_p \sin \omega t \sin (\omega t + \phi) \qquad (2\text{-}6)$$

According to Eq. (2-6) the instantaneous power in this part of the circuit varies with time and may even become negative, as illustrated by the waveforms in Fig. 2-3. The interpretation of negative power in Eq. (2-6) is that during some portion of a cycle this part of the circuit gives up electrical power to the rest of the circuit. For the remainder of the cycle the circuit delivers power to the part under investigation.

The average power is found by averaging Eq. (2-6) over a complete cycle,

$$P = \frac{1}{T} \int_0^T vi \, dt = \frac{V_p I_p}{T} \int_0^T \sin \omega t \sin (\omega t + \phi) \, dt \qquad (2\text{-}7)$$

The second factor under the integral is expanded using a standard trigonometric identity

$$P = \frac{V_p I_p}{T} \left(\cos \phi \int_0^T \sin^2 \omega t \, dt + \sin \phi \int_0^T \cos \omega t \sin \omega t \, dt \right) \qquad (2\text{-}8)$$

Both integrals are standard forms and may be evaluated directly to give

$$P = \frac{V_p I_p}{2} \cos \phi \qquad (2\text{-}9)$$

$$P = VI \cos \phi \qquad (2\text{-}10)$$

where V and I are rms values.

The meaning of Eq. (2-10) is that the useful power in an ac circuit element depends not only upon the current and voltage in the element but also upon the phase difference. The term $\cos \phi$ is called the *power factor* of the circuit. Note that when the phase angle is 90° the power factor is zero and no useful electric power is delivered. Therefore, it is possible for the current and voltage to be very large

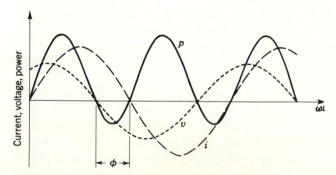

Figure 2-3 Instantaneous power in ac circuit.

and, consequently, the instantaneous power large, yet the average power to be zero. On the other hand, when the current and voltage are in phase, the power factor is unity and the power is equal to the current times the voltage, as in a dc circuit.

CAPACITANCE AND INDUCTANCE

Capacitive Reactance

Consider two parallel metal plates separated by a narrow air gap and connected to the terminals of a battery. The potential difference of the battery exists between the plates and the positive electric charge on the plate connected to the positive terminal of the battery is attracted to the negative electric charge on the plate connected to the negative terminal. The charge Q on each of the plates is proportional to the voltage, so that

$$Q = CV \qquad (2\text{-}11)$$

where C is a constant of proportionality called the *capacitance*. Capacitance is a geometric factor that depends upon the size, shape, and separation between any two conductors such as the two plates.

Capacitance is important in ac circuits because a voltage which changes with time gives rise to a time-varying charge, and this is equivalent to a current. For example, differentiating both sides of Eq. (2-11) with respect to time and using the definition of current in Eq. (1-1) yields

$$i = \frac{dq}{dt} = C\frac{dv}{dt} \qquad (2\text{-}12)$$

In particular, if the voltage is sinusoidal the current in the capacitance,

$$i = C\frac{d}{dt}\,(V_p \sin \omega t) = \omega C V_p \cos \omega t = \omega C V_p \sin\left(\omega t + \frac{\pi}{2}\right) \qquad (2\text{-}13)$$

is also sinusoidal and leads the voltage with a phase angle of $\pi/2$.

In terms of rms values, Eq. (2-13) can be written

$$V = \frac{1}{\omega C}\,I \qquad (2\text{-}14)$$

which shows that the ac current in a capacitance increases linearly with the ac voltage applied quite analogously to the current in a resistance. Indeed, the form of Eq. (2-14) is quite similar to Ohm's law, Eq. (1-4). The proportionality factor between current and voltage, $1/\omega C$, is called the *capacitive reactance*. It is analogous to resistance in dc circuits except, of course, that it decreases with respect to frequency. Note also that capacitive reactance is measured in ohms since it is the ratio of the voltage to the current.

Capacitors

Circuit elements having specific values of capacitance are known as *capacitors*. Most of the capacitors used in electronic circuits consist of two conducting plates separated by a small air gap or thin insulator. The capacitance of such a parallel-plate capacitor is increased by making the area of the plates large and by making the separation between them small. The unit of capacitance is the *farad*, named in honor of the Englishman Michael Faraday. Actually, values of practical capacitors used in electronic circuits are 10^{-6} farad (1 *microfarad*, abbreviated μF), or even 10^{-12} farad (1 *picofarad*, abbreviated pF).

It turns out that insulating material between the parallel plates increases the capacitance of a capacitor because the insulator, in effect, permits a greater charge on the plates for a given applied voltage. The increased capacitance is accounted for by the *dielectric constant* of the insulator. For example, the dielectric constant of mica is about 6 and that of paper is about 2, so that the capacitances of capacitors made from these materials are greater by factors of 6 and 2, respectively, than a parallel-plate capacitor with air between the plates.

Conventional capacitors are made of two thin metal foils separated by a thin insulator or *dielectric*, such as paper or mica. This sandwich is then rolled or folded into a compact size and covered with an insulating coating. One axial wire lead is attached to each foil. In order to increase the capacitance, it is desirable for the insulator to be as thin as possible. This can only be done at the expense of limiting the maximum voltage that can be applied before the insulator ruptures because of the intense electric field. Another important factor is the resistivity of the insulator. Thin, large-area shapes increase the *leakage* resistance between the plates and thus degrade the capacitor. Mica and paper dielectric capacitors are available in capacitances ranging from 0.001 to 1 μF and can be used in circuits where the maximum voltage is of the order of hundreds of volts.

Ceramic and plastic-film capacitors are also used, generally with metal-film plates deposited directly on the dielectric. Plastic dielectrics have very high resistivity, which means that the leakage resistance is extremely small. The large dielectric constant of many ceramic materials provides large capacitance values in a small package.

In several applications, most notably transistor circuits, very large capacitance values are desirable and leakage resistances are of secondary concern. *Electrolytic* capacitors made of an oxidized metal foil in a conducting paste or solution are used to achieve high capacitance values. The thin oxide film is the dielectric between the metal foil and the solution. Because the film is extremely thin, the capacitance is quite large. Several metals, such as tantalum and aluminum, can be used in electrolytic capacitors, and capacitance values range from 1 to 10^5 μF. The largest capacitances are useful in circuits where the applied voltage does not exceed a few volts, because the oxide dielectric is so thin. Electrolytic capacitors can only be used in circuit situations where the metal foil never becomes negative with respect to the solution. If the foil becomes negative, electrolytic action destroys the film and the capacitor becomes useless. Some typical capacitors are shown in Fig. 2-4.

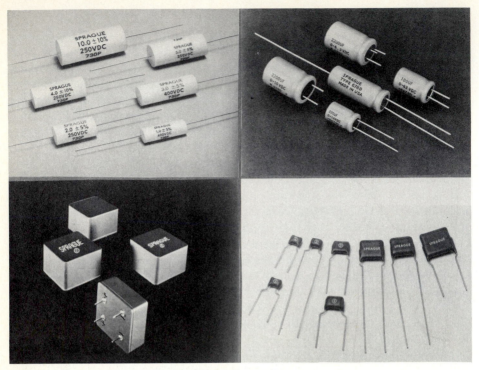

Figure 2-4 Typical capacitors. (*Sprague Electric Company.*)

It is often convenient to vary the capacitance in a circuit without removing the capacitor. The components described above are *fixed* capacitors, since it is clearly not easy to vary their capacitance. The common *variable* capacitor is made of two sets of interleaved plates, one immobile and the other set attached to a shaft. Rotating the shaft effectively changes the area of the plates, thus changing the capacitance. Because the dielectric is air and it is necessary to make the plate separation relatively large to assure that they do not touch, maximum capacitance values are limited to about 500 pF. In the fully unmeshed state a 500-pF variable capacitor may have a minimum capacitance of 10 pF or so. Mica *trimmer* capacitors use a mica dielectric. The separation between plates is adjusted with a screwdriver. They are commonly used where variation in capacitance is only occasionally necessary. The range of capacitance values is about the same as for air dielectric capacitors.

Conventional symbols for capacitors in circuit diagrams are reminiscent of the parallel-plate construction, as illustrated in Fig. 2-5. The slightly curved plate is the negative terminal in the case of electrolytic capacitors, and the outside plate on foil capacitors. This is sometimes significant in electronic circuits in order to connect the capacitor properly, even though foil capacitors are not sensitive to voltage polarity.

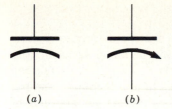

(a)   (b)

Figure 2-5 Conventional circuit symbols for (a) fixed and (b) variable capacitors.

Inductive Reactance

The current in any electric circuit is accompanied by a magnetic field that exists in the region surrounding the current. Changes in the magnetic field arising from a varying current induce an emf in the circuit which is given by

$$v = L\frac{di}{dt} \tag{2-15}$$

where the factor L is called the *inductance*. Inductance is a geometric constant that depends upon the size and shape of the circuit and upon magnetic properties of nearby materials. The induced emf acts to oppose the change in current and increases with the rate of change of current, according to Eq. (2-15). This means that the circuit inductance prevents current in the circuit from changing instantaneously.

The inductance of simple circuits is small enough so that the induced emf may usually be ignored. This is true except at the very highest frequencies of interest in electronic circuits, where the rate of change of current becomes exceedingly large. Circuits used at such frequencies are kept as small as possible to minimize inductive effects. By contrast, it is possible to produce electric components that have appreciable inductance, and these prove to be very useful in ac circuits.

Consider the case of a sinusoidal current in such an inductance. The induced emf in the inductance is also sinusoidal, for, according to Eq. (2-15),

$$v = L\frac{d}{dt}(I_p \sin \omega t) = \omega L I_p \cos \omega t = \omega L I_p \sin\left(\omega t + \frac{\pi}{2}\right) \tag{2-16}$$

and in this case the current lags the voltage by a phase angle of $\pi/2$. Expressing Eq. (2-16) in terms of rms values

$$V = \omega L I \tag{2-17}$$

which shows that the ac current in an inductance increases linearly with applied ac voltage just as in the case of a resistor and in the case of a capacitor. The quantity ωL is called the *inductive reactance* and is measured in ohms. Note that the magnitude of the inductive reactance increases with frequency in contrast to the case of capacitive reactance.

Inductors

Electric components with appreciable inductance are called *inductors*, or *induct-ances*, and in some applications, *chokes*. They consist of many turns of wire wound adjacent to one another on the same support. In this way every single coil of wire links the magnetic flux produced by the current in all other coils, and the total flux intercepted by all the coils together can be made large. The unit of inductance is called the *henry* after Joseph Henry, an early American investigator of inductive effects.

High-frequency electronic circuits frequently employ inductances of the order of 10^{-6} henry (*microhenry*, or μH) which may be a helical coil of a few turns on, say, a 1-cm-diameter support. A few tens of turns produce inductance values in the 10^{-3}-henry (*millihenry*, or mH) range. Large inductances for use at low frequencies are obtained by winding many hundreds of turns of wire on a core of a ferromagnetic material such as iron. The magnetic properties of these materials are such that the magnetic flux is increased appreciably. In this fashion induct-ances of several hundred henrys are attained. Several typical inductors are pictured in Fig. 2-6.

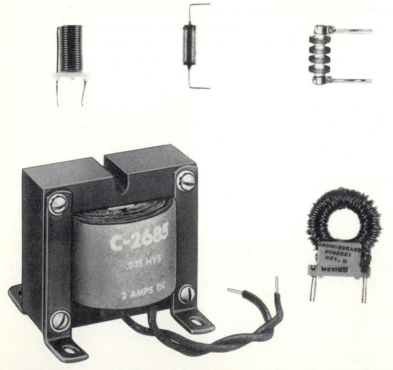

Figure 2-6 Typical inductors used in electronic circuits. (*J. W. Miller Co. and Hamilton Standard Controls, Inc.*)

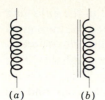

(a)   (b)   **Figure 2-7** Circuit symbols for (a) inductor and (b) iron-core inductor.

The cores of iron-core inductors are laminated in order to interrupt currents induced in the metal core by the changing magnetic flux. This reduces the I^2R losses of these so-called *eddy currents* in the core. Individual laminations are stacked on top of each other and separated with insulating varnish to make a core of desired size. The circuit symbol for an inductor is a helical coil, as shown in Fig. 2-7. Parallel lines along the helix, also illustrated in Fig. 2-7, signify a magnetic core.

Variable inductances can be achieved by moving the core relative to the windings, but such components are not widely used and most inductors are fixed. In many applications it is necessary to take account of the resistance of the wire in the windings and the capacitance between layers of the winding in determining the effect of a choke in a given ac circuit.

The resistive component in a practical inductor may also include the power lost in the magnetic material as a result of the rapidly changing magnetic field in the core. As mentioned earlier, metallic cores must be laminated to reduce eddy-current losses and the thickness of the lamination determines the maximum useful frequency of the inductor. This technique reaches its practical limit in powdered iron cores in which the iron is present as fine particles. *Ferrite* cores, made of high-resistivity ferromagnetic materials, are used at high frequencies because their high resistivity makes eddy-current losses negligible. Such materials are not as useful as iron at low frequencies because magnetic-saturation effects limit the maximum power levels of the inductor.

SIMPLE CIRCUITS

RL Filter

The series combination of a resistor and an inductor is a simple but useful ac circuit. Suppose this *RL filter* is connected to a source of sinusoidal voltage symbolized on the circuit diagram, Fig. 2-8, by a circle containing a one-cycle

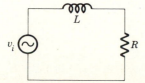

Figure 2-8

sine wave. The current and voltage in the circuit are determined in the following way. According to Kirchhoff's rules the sum of the voltages around the loop must be equal to zero at every instant. This means that the source voltage must equal the induced voltage across the inductor plus the IR drop across the resistor, or

$$v_i = Ri + L\frac{di}{dt} \tag{2-18}$$

where i is the current. Assuming the current is sinusoidal,

$$i = I_p \sin \omega t \tag{2-19}$$

the derivative is

$$\frac{di}{dt} = \omega I_p \cos \omega t \tag{2-20}$$

Introducing Eqs. (2-19) and (2-20) into the circuit equation, Eq. (2-18), the voltage is given by

$$v_i = RI_p \sin \omega t + \omega L I_p \cos \omega t \tag{2-21}$$

This expression may be put into a more illustrative form by introducing the phase angle between the voltage and the current with the aid of Fig. 2-9. Expressing the coefficients of $\sin \omega t$ and $\cos \omega t$ in terms of $\cos \phi$ and $\sin \phi$, Eq. (2-21) becomes

$$v_i = I_p\sqrt{R^2 + (\omega L)^2}\,(\cos \phi \sin \omega t + \sin \phi \cos \omega t) \tag{2-22}$$

The expression in parentheses is just the trigonometric identity for $\sin (\omega t + \phi)$, so that

$$v_i = I_p\sqrt{R^2 + (\omega L)^2}\,\sin (\omega t + \phi) \tag{2-23}$$

where

$$\phi = \arctan \frac{\omega L}{R} \tag{2-24}$$

Note that, according to Eq. (2-24), the current lags the voltage, as previously observed in connection with Eq. (2-16). Note also that the inductive reactance and resistance both are important in determining the current in this ac circuit.

Suppose that the voltage drop across R is considered to be an output voltage of the circuit. Then, using Eq. (2-19),

$$v_o = RI_p \sin \omega t \tag{2-25}$$

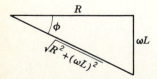

Figure 2-9

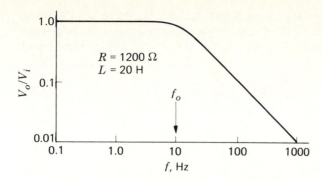

Figure 2-10 Frequency characteristic of *RL* low-pass filter.

The ratio of the rms output voltage to the rms input voltage is, from Eqs. (2-23) and (2-25)

$$\frac{V_o}{V_i} = \frac{RI_p}{I_p\sqrt{R^2 + (\omega L)^2}} \tag{2-26}$$

or

$$\frac{V_o}{V_i} = \frac{1}{\sqrt{1 + (\omega L/R)^2}} \tag{2-27}$$

According to Eq. (2-27), at low frequencies where $\omega L/R \rightarrow 0$, the output voltage is equal to the input voltage. At high frequencies the output voltage is smaller than the input, as a plot of Eq. (2-27) in Fig. 2-10 illustrates, and the circuit is known as a *low-pass* filter. Consider the frequency f_0, where

$$\frac{2\pi f_0 L}{R} = \frac{\omega_0 L}{R} = 1 \tag{2-28}$$

According to Eq. (2-27) this is where

$$\frac{V_o^2}{V_i^2} = \frac{1}{2} \tag{2-29}$$

Since the output power in R is proportional to the voltage squared, f_0 is called the *half-power* frequency. In effect, input signals above this frequency are choked off by the inductance; this is the origin of the term *choke* introduced earlier.

RC Filter

An elementary but very useful circuit employing a capacitor and resistor connected in series, is shown in Fig. 2-11. This *RC filter* is connected to a source of sinusoidal voltage

$$v_i = V_p \sin \omega t \tag{2-30}$$

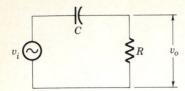

Figure 2-11 *RC* filter.

Voltage drops around the loop give

$$v_i = \frac{Q}{C} + Ri \qquad (2\text{-}31)$$

where i is the current. Differentiating each term with respect to time and putting $i = dQ/dt$, Eq. (2-31) becomes, after rearranging,

$$R\frac{di}{dt} + \frac{1}{C}i = \omega V_p \cos \omega t \qquad (2\text{-}32)$$

To solve this circuit differential equation assume that the current is given by

$$i = I_p \sin (\omega t + \phi) \qquad (2\text{-}33)$$

where I_p and ϕ are to be determined. Differentiating Eq. (2-33)

$$\frac{di}{dt} = \omega I_p \cos (\omega t + \phi) \qquad (2\text{-}34)$$

Now Eqs. (2-33) and (2-34) are substituted into the circuit differential equation, Eq. (2-32). The result is

$$R\omega I_p \cos (\omega t + \phi) + \frac{I_p}{C} \sin (\omega t + \phi) = V_p \omega \cos \omega t \qquad (2\text{-}35)$$

and the equation is solved by choosing values for I_p and ϕ that make Eq. (2-35) true for all values of t. This substitution has changed the differential equation into a trigonometric equation.

The values of I_p and ϕ which satisfy Eq. (2-35) are found by first expanding each term using a trigonometric identity. This gives, after dividing through by ωI_p,

$$R(\cos \omega t \cos \phi - \sin \omega t \sin \phi)$$

$$+ \frac{1}{\omega C}(\sin \omega t \cos \phi + \cos \omega t \sin \phi) = \frac{V_p}{I_p} \cos \omega t \quad (2\text{-}36)$$

Collecting terms in $\sin \omega t$ and $\cos \omega t$,

$$\left(R \cos \phi + \frac{1}{\omega C} \sin \phi - \frac{V_p}{I_p}\right) \cos \omega t + \left(\frac{1}{\omega C} \cos \phi - R \sin \phi\right) \sin \omega t = 0$$

$$(2\text{-}37)$$

Now consider that $t = 0$; then $\sin \omega t = 0$ and

$$R \cos \phi + \frac{1}{\omega C} \sin \phi - \frac{V_p}{I_p} = 0 \tag{2-38}$$

Similarly, suppose $\omega t = \pi/2$, so that $\cos \omega t = 0$. Then

$$\frac{1}{\omega C} \cos \phi - R \sin \phi = 0 \tag{2-39}$$

In order that Eq. (2-37) be satisfied for all values of t, both Eqs. (2-38) and (2-39) must be true. Equation (2-39) may be solved immediately for ϕ,

$$\frac{1}{\omega C} \cos \phi = R \sin \phi$$

$$\tan \phi = \frac{\sin \phi}{\cos \phi} = \frac{1}{R\omega C}$$

so that

$$\phi = \arctan \frac{1}{R\omega C} \tag{2-40}$$

The solution for ϕ is used in Eq. (2-38) to solve for I_p. This is done exactly as in the case of the RL filter, Eq. (2-21). Equation (2-38) becomes

$$R \frac{R}{\sqrt{R^2 + (1/\omega C)^2}} + \frac{1}{\omega C} \frac{1/\omega C}{\sqrt{R^2 + (1/\omega C)^2}} - \frac{V_p}{I_p} = 0 \tag{2-41}$$

So finally

$$I_p = \frac{V_p}{\sqrt{R^2 + (1/\omega C)^2}} \tag{2-42}$$

Thus the current in the circuit is

$$i = \frac{V_p}{\sqrt{R^2 + (1/\omega C)^2}} \sin(\omega t + \phi) \tag{2-43}$$

where

$$\phi = \arctan \frac{1}{\omega RC}$$

Note that according to Eq. (2-40), the phase angle is positive. This means that the current leads the voltage, which is characteristic of a capacitive circuit. When $1/\omega RC \to 0$ at high frequencies, the phase angle is zero and the current is in phase with the voltage. At very low frequencies, the phase angle approaches $\pi/2$.

The output voltage across the resistance is

$$v_o = RI_p \sin(\omega t + \phi) \tag{2-44}$$

which means that the ratio of the rms output voltage to the input voltage, found from Eqs. (2-30), (2-42), and (2-44), is then

$$\frac{V_o}{V_i} = \frac{1}{\sqrt{1 + (1/\omega RC)^2}} \tag{2-45}$$

A plot of Eq. (2-45), Fig. 2-12, shows that the output voltage V_o is very small at low frequencies and is equal to the input voltage at high frequencies. Since low frequencies are attenuated while high frequencies are not, this circuit is called an *RC high-pass filter*. The frequency f_0 where

$$2\pi f_0 RC = 1 \tag{2-46}$$

is called the *half-power* frequency, as in the *RL* filter.

Differentiating and Integrating Circuits

Suppose that the series resistance and capacitance in the simple *RC* filter are small enough to make $\omega RC \ll 1$ over a given frequency range. Under this condition the output voltage is, from Eqs. (2-44) and (2-45),

$$v_o = V_p \omega RC \sin(\omega t + \pi/2)$$
$$= V_p \omega RC \cos \omega t \tag{2-47}$$

But, note that the time derivative of the input signal is

$$\frac{dv_i}{dt} = \omega V_p \cos \omega t \tag{2-48}$$

Therefore, combining Eqs. (2-47) and (2-48),

$$v_o = RC \frac{dv_i}{dt} \tag{2-49}$$

The interpretation of Eq. (2-49) is that when $\omega RC \ll 1$ the *RC* filter circuit performs the operation of differentiation. That is, the output voltage signal is the time derivative of the input voltage. This useful property is applied extensively in electronic circuits.

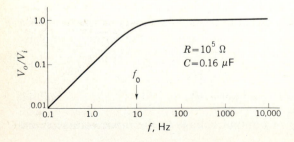

Figure 2-12 Frequency characteristic of *RC* high-pass filter.

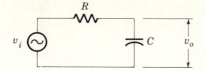

Figure 2-13 *RC* low-pass filter.

Correspondingly, the voltage across the capacitor is, by Eq. (2-12), the integral of the current,

$$v_c = \frac{1}{C} \int I_p \sin{(\omega t + \phi)} \, dt = -\frac{I_p}{\omega C} \cos{(\omega t + \phi)} \tag{2-50}$$

Suppose now $\omega RC \gg 1$, which may be accomplished by making R and C very large. Then, $I_p = V_p/R$, $\phi = 0$, and Eq. (2-50) becomes

$$v_c = -\frac{V_p}{RC\omega} \cos{\omega t} \tag{2-51}$$

Introducing the integral of the input voltage results in

$$v_c = \frac{1}{RC} \int v_i \, dt \tag{2-52}$$

which means that under these conditions the *RC* circuit performs the operation of integration. The output voltage is only a small fraction of the input voltage when the circuit is used to differentiate or integrate, according to Eqs. (2-49) and (2-52). This is not a serious disadvantage in practical applications, however, since the output signal may be subsequently increased in magnitude using amplifier circuits described in later chapters.

It is permissible to ignore the limits of integration in deriving Eq. (2-52), because we are concerned with only steady-state conditions after any initial transient voltages, which may accompany turning on the input voltage, have died away. The possibility of transient effects in *RC* circuits is considered in the next section. The *RC* circuit is a *low-pass* filter when the output voltage signal is taken from the capacitor, Fig. 2-13. The transition frequency from one domain to the other is marked approximately by the half-power frequency, where $\omega_0 RC = 1$, as in the case of the high-pass filter.

RC low pass — output across capacitor
RC high pass — output across Resistor

TRANSIENT CURRENTS

Time Constant

The study of ac circuits to this point has implicitly assumed that the rms values of the sinusoidal emfs are constant. Transient effects are associated with sudden changes in the voltages applied to ac networks, as, for example, when an emf is first applied. These transient currents dissipate rapidly, leaving steady-state currents in the network which persist as long as the ac emf is applied. Although most often the steady-state currents are of primary concern, in many cases transient effects are important as well. This is particularly so in determining the response of a network to isolated voltage pulses.

Transient currents in ac networks are found by solving the circuit differential equation, taking into account the magnitude of the current at the time the applied voltage is changed. Actually, the sum of the transient current and the steady-state current is the complete solution of the circuit differential equation. It is generally satisfactory to consider the two aspects of ac-circuit analysis separately because the response of circuits is most often of interest in connection either with steady emfs alone or with transient effects alone.

It is convenient to investigate first the transient currents in a simple RC series circuit incorporating a battery and a two-position switch, as shown in Fig. 2-14. Suppose the switch is suddenly connected to terminal 1. The circuit voltage equation is, from Kirchhoff's rule,

$$V = Ri + \frac{q}{C} \tag{2-53}$$

Differentiating with respect to t and rearranging,

$$\frac{di}{dt} + \frac{1}{RC} i = 0 \tag{2-54}$$

To solve this circuit differential equation, rewrite it as

$$\frac{di}{i} = -\frac{1}{RC} dt \tag{2-55}$$

Both the left side and the right side of Eq. (2-55) are standard integrals, so that upon integrating,

$$\ln i = -\frac{1}{RC} t + K \tag{2-56}$$

where K is a constant to be evaluated from the initial conditions. Another way of writing Eq. (2-56) is

$$i = \exp\left(-\frac{t}{RC} + K\right) = Ae^{-t/RC} \tag{2-57}$$

where A replaces $\exp K$ for simplicity. In order to evaluate A, note that the current is equal to V/R when the switch is closed at $t = 0$ since the initial voltage

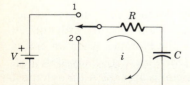

Figure 2-14

across C is zero. Therefore,

$$i = \frac{V}{R} e^{-t/RC} \qquad (2\text{-}58)$$

According to Eq. (2-58), the current in this simple RC circuit decays exponentially when the switch is put in position 1. The quantity RC, which has the dimension of seconds, is called the *time constant*. It determines how rapidly the current decreases.

The voltage across the capacitor at any time is found by subtracting the voltage drop across the resistor from the battery voltage,

$$v_c = V - Ri$$

$$= V(1 - e^{-t/RC}) \qquad (2\text{-}59)$$

The decay of charging current and the growth of voltage across the capacitor are symmetrical according to Eqs. (2-58) and (2-59), as shown in Fig. 2-15. Note that the voltage starts from zero and increases exponentially to the battery voltage. After a time equal to one time constant the voltage is equal to $1 - e^{-1} = 1 - 0.356 = 63$ percent of the final value. The actual time at which this voltage is attained depends upon the magnitude of the resistance and the capacitance. A long time is required for large values of R and C, and vice versa.

The decay of current upon closing the switch to position 2 as the charge on the capacitor discharges through the resistor is the same as that given by Eq. (2-58). This can be seen by writing the circuit voltage equation for this case

$$\frac{q}{C} + Ri = 0 \qquad (2\text{-}60)$$

Differentiating and rearranging,

$$\frac{di}{dt} + \frac{1}{RC} i = 0 \qquad (2\text{-}61)$$

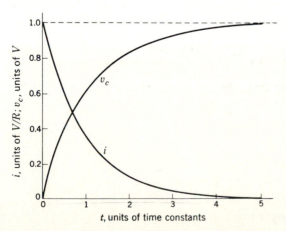

Figure 2-15 Charging current and capacitor voltage in RC circuit.

which is identical to Eq. (2-54). Furthermore, the potential across the capacitor is equal to V when the capacitor is fully charged and this means that the initial current is V/R, just as in the charging current case. The voltages across both the capacitor and the resistor decay exponentially to zero with a time constant of RC.

The simple inductive circuit is analogous to the RC case, as is illustrated by considering the circuit in Fig. 2-16. The circuit voltage equation with the switch in position 1 is

$$V = Ri + L\frac{di}{dt} \tag{2-62}$$

which upon rearranging becomes

$$\frac{di}{dt} + \frac{R}{L}i - \frac{V}{L} = 0 \tag{2-63}$$

This equation is solved exactly like Eq. (2-54) with the result that the current is given by

$$i = \frac{V}{R}(1 - e^{-tL/R}) \tag{2-64}$$

Evidently, the current in an inductive circuit grows exponentially with a time constant L/R. That is, small R and large values of inductance result in long transient currents. Note that, according to Eq. (2-64), the steady-state value is just the dc current V/R.

The decay of current when the switch is put in position 2 is also exponential since the circuit differential equation in this case is simply

$$L\frac{di}{dt} + Ri = 0 \tag{2-65}$$

By analogy with Eq. (2-54), the current is given by

$$i = \frac{V}{R}e^{-tL/R} \tag{2-66}$$

The growth and decay of current in an inductive circuit are symmetrical, according to Eqs. (2-64) and (2-66), exactly as in the corresponding case of the RC circuit.

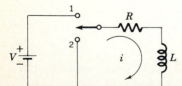

Figure 2-16

AC Transients

Transient currents also accompany the turning on of ac voltage sources. For example, consider the circuit in Fig. 2-17, for which the circuit differential equation is

$$R\frac{di}{dt} + \frac{1}{C}i = \frac{dv}{dt} \tag{2-67}$$

It is convenient to express the applied voltage as

$$v = V_p \sin(\omega t + \theta) \tag{2-68}$$

where the phase angle θ is introduced in order to choose the instantaneous input voltage amplitude at $t = 0$. Since the full solution of Eq. (2-67) involves a transient current and a steady-state current of frequency ω, the total current is

$$i = i_1 + i_2 \tag{2-69}$$

where i_1 is the transient current and i_2 is the steady-state current.

Substituting Eq. (2-69) into the differential equation (2-67) and rearranging, the result is

$$\left(R\frac{di_1}{dt} + \frac{1}{C}i_1\right) + \left(R\frac{di_2}{dt} + \frac{1}{C}i_2 - \frac{dv}{dt}\right) = 0 \tag{2-70}$$

This equation is solved if both parentheses are equal to zero. The first one is just Eq. (2-54), for which the solution previously found is Eq. (2-58). Similarly, the second parentheses is just Eq. (2-32), and the solution is Eq. (2-43). Using these expressions, the total current is written

$$i = Ae^{-t/\tau} + \frac{V_p}{\sqrt{R^2 + (1/\omega C)^2}} \sin(\omega t + \theta + \phi) \tag{2-71}$$

where $\tau = RC$
$\phi = \arctan 1/R\omega C$

and the constant A is determined from the voltage of the source at the instant the switch is closed. Note that Eq. (2-71) consists of a transient part together with a steady-state term, as discussed above.

Suppose the applied voltage is at its peak, that is, $v = V_p$, when the switch is closed at $t = 0$. This means we choose $\theta = \pi/2$, according to Eq. (2-68), and the

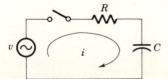

Figure 2-17

initial current is V_p/R. Thus, Eq. (2-71) for $t = 0$ becomes

$$\frac{V_p}{R} = A + \frac{V_p}{\sqrt{R^2 + (1/\omega C)^2}} \sin\left(\frac{\pi}{2} + \phi\right) \tag{2-72}$$

This is solved for the constant A, which is then substituted into Eq. (2-71). The result is the final solution

$$i = \frac{V_p}{R} \frac{1}{1 + (\omega RC)^2} e^{-t/\tau} + \frac{V_p}{\sqrt{R^2 + (1/\omega C)^2}} \sin\left(\omega t + \frac{\pi}{2} + \phi\right) \tag{2-73}$$

In order for the transient term of Eq. (2-73) to have an appreciable magnitude compared with the steady-state term, the denominator of its coefficient must not be too large. This means that $(\omega RC)^2 < 1$, or

$$\tau < \frac{1}{\omega} \tag{2-74}$$

According to this expression, the time constant of the transient current is smaller than one period of the sine-wave current. Therefore, any transient current is negligible after only a fraction of a cycle of the ac wave, and the transient may therefore be ignored. If, on the other hand, the time constant is much greater than one period, the transient persists over many cycles, but, according to Eq. (2-73), the amplitude is small. This reasoning demonstrates why the steady-state solution alone is sufficiently accurate for most purposes.

Ringing

Sudden changes of voltage in a *RLC* circuit may produce *ringing*, in which a decaying ac current is generated at a frequency characteristic of the circuit, called the *resonant* frequency. The amplitude of the ac current dies away exponentially as in nonresonant circuits. For example, after the switch is closed in the circuit of Fig. 2-18, an ac current may be generated which decreases relatively slowly. The phenomenon is analogous to striking a bell a sharp mechanical blow with a hammer, which is the origin of the term ringing.

Resonant circuits are examined in greater detail in Chap. 5. To examine ringing in the circuit of Fig. 2-18 we start by applying Kirchhoff's rules, which becomes,

$$V = Ri + \frac{q}{C} + L\frac{di}{dt} \tag{2-75}$$

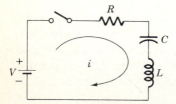

Figure 2-18

after differentiating,

$$\frac{d^2i}{dt^2} + \frac{R}{L}\frac{di}{dt} + \frac{1}{LC}i = 0 \tag{2-76}$$

To solve this circuit differential equation it is assumed, on the basis of previous analyses, that an ac current is present which has an amplitude that decays exponentially with time. Therefore, the current is written as

$$i = Ae^{-at} \sin(\omega t + \phi) \tag{2-77}$$

where A, a, ω, and ϕ are constants. Equation (2-77) is differentiated once, twice, and all three expressions introduced into the circuit differential equation. After some algebraic manipulation, (2-76) reduces to

$$\left[\left(a^2 - \omega^2 - \frac{Ra}{L} + \frac{1}{LC}\right)\sin(\omega t + \phi) + \left(\frac{R\omega}{L} - 2a\omega\right)\cos(\omega t + \phi)\right]Ae^{-at} = 0 \tag{2-78}$$

In order for (2-77) to be a solution of (2-76) the coefficients of both terms in (2-78) must separately be equal to zero. From the coefficient of the cosine term,

$$a = \frac{R}{2L} \tag{2-79}$$

while the coefficient of the sine term gives an expression for the frequency

$$\omega^2 = \frac{1}{LC} - \frac{R^2}{4L^2} \tag{2-80}$$

Therefore, the current in this circuit is sinusoidal at a frequency given by (2-80) and with an amplitude that decreases with a time constant determined from (2-79). A sketch of the current behavior is illustrated in Fig. 2-19.

It is conceivable that values of the circuit elements are such that

$$\frac{R^2}{4L^2} > \frac{1}{LC} \tag{2-81}$$

in which case the ac frequency calculated from (2-80) is imaginary. The interpretation of this situation may be developed from the mathematical expression for the sine of an imaginary angle,

$$\sin j\omega t = \frac{j}{2}(e^{\omega t} - e^{-\omega t}) \tag{2-82}$$

Therefore, Eq. (2-77) reduces to a simple exponential decay with a time constant determined by the R, C, and L values in the circuit. In effect, according to (2-81), when the circuit resistance is very large, the ac current is damped out by the joule losses in the resistance, and the circuit is said to be *overdamped*. The point of *critical damping* is when the frequency calculated from (2-80) vanishes and the exponential decay is governed solely by (2-79).

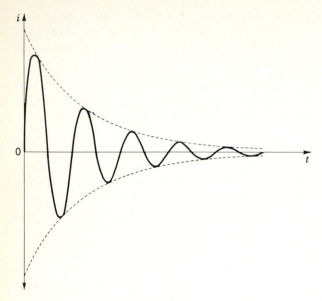

Figure 2-19 Transient current in *RLC* circuit can be decaying oscillation.

COMPLEX WAVEFORMS

Fourier Series

Most steady ac signals of practical interest are not simple sine waves, but often have more complex waveforms. Nevertheless, the analysis based on sine waves can apply because complex waveforms may be represented by summing sine waves of various amplitudes and frequencies. Consider, for example, the sum of the two sine waves shown in Fig. 2-20, one of which is twice the frequency of the

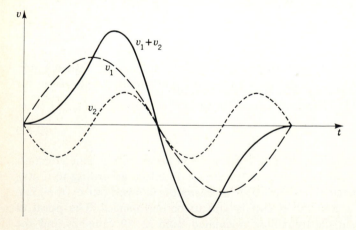

Figure 2-20 Sum of two sine waves is complex waveform.

other. The resulting waveform is more complex than either constituent. If this periodic voltage is applied to a circuit, the resulting current may be determined by calculating the current caused by each sinusoidal voltage component separately. In this way the effect of any complex waveshape can be investigated.

Any single-valued periodic waveform $f(t')$ may be represented by a *Fourier series* which is a summation of a dc term and a *fundamental* (or lowest) frequency together with its *harmonics* (integral multiples of the fundamental frequency),

$$f(t') = \frac{a_0}{2} + \sum_{n=1}^{\infty} (a_n \cos nt' + b_n \sin nt') \tag{2-83}$$

where a_0, a_n, and b_n are constants and $t' = \omega t$ is in radians. If $f(t')$ is known in the interval $t' = -\pi$ to $t' = \pi$, the values of the a_n coefficients are determined by multiplying both sides of Eq. (2-83) by $\cos mt'$ (where m is an integer) and integrating from $= -\pi$ to π. Since

$$\int_{-\pi}^{\pi} \cos mt' \cos nt' \, dt' = \int_{-\pi}^{\pi} \sin mt' \sin nt' \, dt' = \begin{cases} 0, & m = n \\ \pi, & m = n \neq 0 \end{cases}$$

and

$$\int_{-\pi}^{\pi} \sin mt' \cos nt' \, dt' = 0$$

the result is

$$a_n = \frac{1}{\pi} \int_{-\pi}^{\pi} f(t') \cos nt' \, dt' \tag{2-84}$$

Similarly, using $\sin mt'$ as the multiplier,

$$b_n = \frac{1}{\pi} \int_{-\pi}^{\pi} f(t') \sin nt' \, dt' \tag{2-85}$$

Note that in the special case $n = 0$, $\cos nt' = 1$ and

$$a_0 = \frac{1}{\pi} \int_{-\pi}^{\pi} f(t') \, dt' \tag{2-86}$$

which is twice the average, or dc, value of the waveform.

The Fourier series for a square wave of peak voltage V_p and frequency ω is given by

$$v = \frac{4V_p}{\pi} \left(\sin \omega t + \frac{1}{3} \sin 3\omega t + \frac{1}{5} \sin 5\omega t + \frac{1}{7} \sin 7\omega t \cdots \right) \tag{2-87}$$

The faithfulness of the series representation improves as the number of terms included in Eq. (2-87) is increased, as illustrated graphically in Fig. 2-21. This means that the higher frequencies are necessary in order to reproduce the sharp corners of the square wave. A second example is the sawtooth wave (Fig. 2-22)

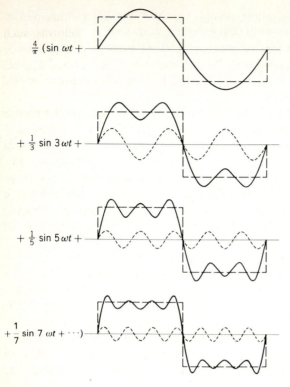

$\frac{4}{\pi}$ (sin ωt +

+ $\frac{1}{3}$ sin 3ωt +

+ $\frac{1}{5}$ sin 5ωt +

+ $\frac{1}{7}$ sin 7 ωt + $\cdots$)

Figure 2-21 Many frequency components are present in a square wave. Faithfulness of representation improves as number of harmonics is increased.

which is represented by

$$v = \frac{2V_p}{\pi}\left(\sin \omega t - \frac{1}{2}\sin 2\omega t + \frac{1}{3}\sin 3\omega t - \frac{1}{4}\sin 4\omega t + \cdots\right) \qquad (2\text{-}88)$$

Here, again, the high-frequency terms must be included in order to obtain the sharp corner of the sawtooth waveform.

Note that the a_n and b_n coefficients in Eq. (2-83) represent the magnitude of the frequency component, n, in the waveform. Thus, the variation of a_n's and b_n's

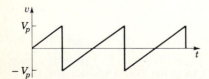

Figure 2-22 Sawtooth wave.

with frequency is an equally valid description of the waveform as is the time depen- dence, $f(t')$. In fact, it proves possible to represent nonperiodic waveforms, such as a single pulse, using the Fourier technique, in which case the summation in Eq. (2-83) is replaced by an integral and the distribution of the a_n's and b_n's be- comes continuous in the frequency domain. This often proves to be a more useful description than the time variation.

According to Eqs. (2-87) and (2-88), complex waveforms, such as the square wave and the sawtooth wave, are composed of a wide spectrum of frequencies. If the waveform of a given input signal applied to any network is to be preserved at the output, the attenuation of the network must be independent of frequency. For example, a square wave applied to an RC high-pass filter can result in a square- wave output signal only if the fundamental frequency of the square wave is greater than the cutoff frequency of the filter. It is equally important that the relative phases of the harmonics remain constant in traversing the network if the waveform is to be preserved.

On the other hand, the harmonic composition of a wave is often purposefully altered by a network in order to change one waveform into another. Suppose a square wave is applied to an RC differentiating circuit. The output waveform is a series of alternating positive and negative sharp pulses, as sketched in Fig. 2-23b. In order for the circuit to differentiate a square wave adequately, it is necessary to include all terms in the Fourier series representation that satisfy the approxi- mation $\omega RC \ll 1$ leading to Eq. (2-49). This is usually easily accomplished in practice, and the simple differentiating circuit is widely used to generate pulses from a square wave.

A second example is the integration of a square wave using a simple RC circuit. The result is a triangular wave, Fig. 2-23c. Here the fundamental fre- quency of the square wave must be large enough that the approximation $\omega RC \gg 1$ leading to Eq. (2-52) is satisfied.

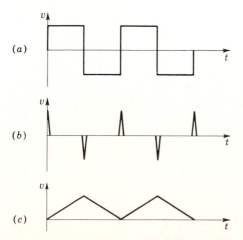

Figure 2-23 (a) Square wave, (b) differentiated square wave, (c) integrated square wave.

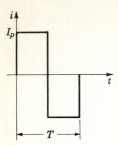

Figure 2-24 Square wave.

Effective Value

The effective value of a complex waveform is defined in exactly the same way as in the case of the sinusoidal wave discussed earlier. Consider, for example, the square wave of period T and maximum value I_p illustrated in Fig. 2-24. The effective current is just

$$I_e^2 = \frac{1}{T} \int_0^T I_p^2 \, dt = \frac{I_p^2}{T} \int_0^T dt = I_p^2 \qquad (2\text{-}89)$$

This means that the effective value of a square wave is equal to the maximum value. Since the joule heat in a resistor depends upon the square of the current, the current direction is immaterial and Eq. (2-89) is just what is expected.

Consider next the simple triangular sawtooth waveform illustrated in Fig. 2-25. The current varies as

$$i = \frac{2I_p}{T} t \qquad (2\text{-}90)$$

over the interval from $-T/2$ to $T/2$. Therefore, the rms value is

$$I_e^2 = \frac{1}{T} \int_{-T/2}^{T/2} \left(\frac{2T_p}{T} t \right)^2 dt = \frac{4I_p^2}{T^3} \int_{-T/2}^{T/2} t^2 \, dt = \frac{I_p^2}{3}$$

$$I_e = \frac{I_p}{\sqrt{3}} \qquad (2\text{-}91)$$

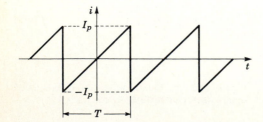

Figure 2-25 Sawtooth wave.

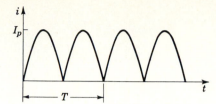

Figure 2-26 Waveform of half-sinusoids.

Another waveform of interest is a series of half-sine waves, Fig. 2-26. The effective value can be found by the procedure used above. This waveform has, however, a dc component, as can be shown by computing the average value

$$I_{dc} = \frac{2}{T} \int_0^{T/2} I_p \sin \omega t \, dt = \frac{4I_p}{\omega T} = \frac{4I_p}{2\pi}$$

$$I_{dc} = \frac{2}{\pi} I_p \tag{2-92}$$

This dc component contributes to the heating effect of the complete waveform.

For most purposes the rms value of the ac components alone is of interest. It is convenient to consider the Fourier series of complex waveforms. Since the heating effect of each frequency component is independent of the others, the effective value of any complex waveform is given by

$$I_e^2 = I_{dc}^2 + I_1^2 + I_2^2 + I_3^2 + \cdots \tag{2-93}$$

where I_{dc} is the average value and $I_1, I_2, I_3, \ldots$ are the rms currents of each frequency component. It is often sufficient to consider only the fundamental frequency, since it predominates.

The Fourier series of the waveform in Fig. 2-26 is

$$i = \frac{2}{\pi} I_p - \frac{4I_p}{3\pi} \cos 2\omega t - \frac{4I_p}{15\pi} \cos 4\omega t + \cdots \tag{2-94}$$

Note that the first term corresponds to Eq. (2-92), in conformity with the definition of the dc component. The effective value of the ac components alone is, using Eqs. (2-4) and (2-89),

$$I_{rms}^2 = \frac{I_p^2}{2} \left[\left(\frac{4}{3\pi} \right)^2 + \left(\frac{4}{15\pi} \right)^2 + \cdots \right] \cong \frac{1}{2} \left(\frac{4I_p}{3\pi} \right)^2 \tag{2-95}$$

$$I_{rms} = \frac{2\sqrt{2}I_p}{3\pi} \tag{2-96}$$

The preceding results are used in Chap. 3 in connection with further applications of inductance and capacitance filters.

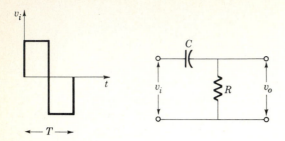

Figure 2-27 Square-wave input signal applied to high-pass filter.

Square-Wave Response

It is useful to investigate the response of circuits to square-wave voltages by the transient methods of the previous section. This is possible since a square wave can be considered to be a dc voltage which is switched on and off at the frequency of the wave. Transient-current analysis complements the Fourier-series method and is often simpler. The transient technique yields the output waveform directly and makes it unnecessary to determine the response of the circuit to each of the harmonics.

Consider first the high-pass RC filter of Fig. 2-27 to which is applied the square wave of period T as indicated on the diagram. According to Eq. (2-58) the

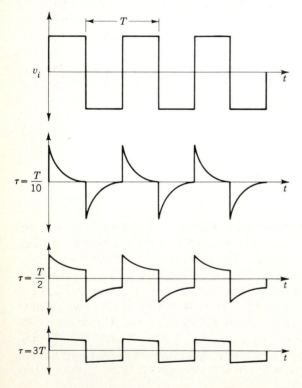

Figure 2-28 Square-wave input to a high-pass filter produces different output waveforms depending upon ratio of period to circuit time constant.

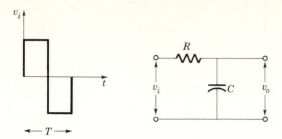

Figure 2-29 Square-wave input signal applied to low-pass filter.

transient current accompanying each half-cycle of the square wave has the form

$$i = Ae^{-t/\tau} \tag{2-97}$$

where the constant A depends upon the current in the circuit resulting from the previous transient. For example, if $\tau = T/2$, the current has decayed 63 percent of the way at the time the square-wave voltage reverses. This means that the voltage across the capacitor causes a transient current in the reverse direction, and so forth. The output voltage $(=Ri)$ is a succession of exponential transients, as shown in Fig. 2-28. These waveforms correspond to different values of the RC time constant compared with the period of the square wave.

Note that when the circuit time constant is very small compared with the period of the square wave, the output waveform approaches the differential of the square wave. This is an alternative approach to the study of differentiating circuits discussed in the previous section.

The low-pass filter circuit, Fig. 2-29, may be analyzed in the same fashion. The voltage across the capacitor is an exponential, as discussed earlier, and the output waveform, Fig. 2-30, again depends upon the relative magnitude of the circuit time constant compared with the period. In particular, note that when the circuit time constant is large the capacitor voltage rise is essentially linear during the pulse time. The output voltage is thus just the integral of the input signal, again in conformity with harmonic-circuit analysis.

Oscilloscope

Undoubtedly the most useful instrument for studying complex waveforms is the *cathode-ray oscilloscope*, which is capable of displaying voltage waveforms visually. The heart of the oscilloscope is the cathode-ray tube, in which the position of a thin beam of electrons is electrically controlled to "paint" the waveform on a fluorescent screen. In the cathode-ray tube (CRT) an *electron gun* directs a high-velocity focused beam of electrons onto a glass face-plate covered with fluorescent material that emits light when bombarded by electrons. The beam passes between a pair of horizontal deflecting plates and a pair of vertical deflecting plates on its way to the screen, as sketched in Fig. 2-31. The beam produces a spot of light in the center of the screen when the voltage on the deflection plates

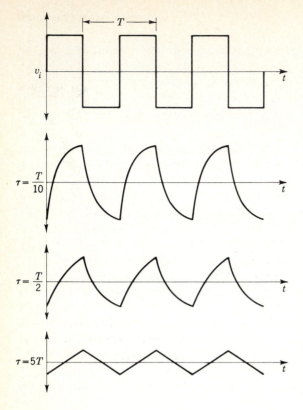

Figure 2-30 Square-wave input to low-pass filter produces different output waveforms depending upon ratio of period to circuit time constant.

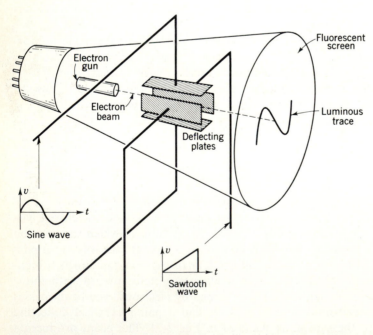

Figure 2-31 Sketch of cathode-ray tube.

is equal to zero. The position of the spot on the screen is easily changed by deflecting the electron beam with voltages applied to either pair of deflecting plates. The beam is deflected as a result of the sidewise force on the electrons caused by the electric field between each pair of deflecting plates. Practical oscilloscopes include a number of auxiliary circuits to increase the versatility of the instrument, as discussed in a subsequent chapter.

The way in which the oscilloscope displays a waveshape can be understood by referring to Fig. 2-32. Suppose a sine-wave voltage of sufficient amplitude to cause a peak-to-peak deflection of about one-half the screen diameter is applied to the vertical pair of deflection plates. A sawtooth voltage waveform of the same frequency is applied to the horizontal deflecting plates with sufficient amplitude to cause the beam to sweep one-half the screen diameter in this direction. If both waveforms start from zero at the same instant, the spot begins in the center of the screen. The sawtooth wave on the horizontal plates causes the spot to move uniformly to the right while vertical motion depends upon the sine-wave voltage. The result is that the spot traces out a sine-wave pattern on the screen, as the point-by-point plot in Fig. 2-32 illustrates.

When the sawtooth voltage on the horizontal plates drops to zero at the end of the cycle, the spot rapidly returns to the starting point. The same pattern is traced out on subsequent cycles. At most frequencies of interest the spot moves too rapidly to be seen and the pattern on the screen appears to be a stationary sine wave because of the persistence of vision and of the phosphor. The sawtooth

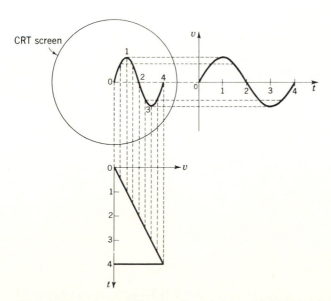

Figure 2-32 Oscilloscope having sawtooth horizontal-deflection voltage displays waveform of voltage applied to vertical plates.

voltage and the sine wave must start from the beginning at exactly the same instant on each cycle, for otherwise the spot does not trace out the same pattern every time. Therefore, the sawtooth *sweep* voltage is *synchronized* with the signal on the vertical plates by auxiliary circuits associated with the oscilloscope. It is also common practice to turn off the beam during the fast return of the spot to the beginning, in order to eliminate the *retrace* line from the pattern. The sweep voltage may conveniently be made a submultiple of the signal frequency in order to display more than one cycle of the waveform.

The oscilloscope is also useful in measuring the phase angle between sine waves. This is done by applying one sine wave to the vertical plates and the other to the horizontal plates. The pattern which results can be a straight line, a circle, or an ellipse, depending upon the phase angle, as illustrated in Fig. 2-33 for the corresponding phase angles of 0, 45, and 90°. The manner in which the phase

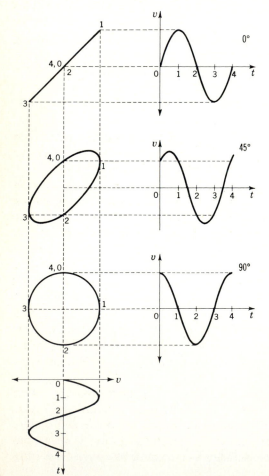

Figure 2-33 Lissajous figures for phase differences of 0°, 45°, and 90° between horizontal- and vertical-deflection voltages.

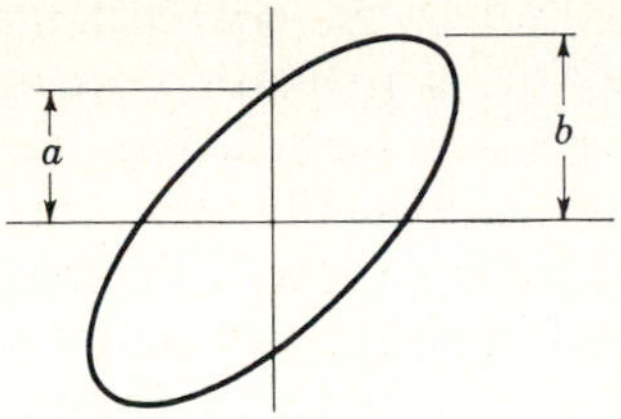

Figure 2-34 Phase angle between sine-wave voltages applied to horizontal and vertical plates is determined from ratio a/b.

angle is calculated is as follows. Suppose the horizontal voltage is

$$v_H = V_p \sin \omega t \tag{2-98}$$

and the vertical sine wave is given by

$$v_V = b \sin (\omega t + \phi) \tag{2-99}$$

When $t = 0$, $v_H = 0$, which means the horizontal deflection is zero. The vertical deflection at this point may be labeled a,

$$v_V = b \sin \phi = a \tag{2-100}$$

Solving for ϕ,

$$\phi = \arcsin \frac{a}{b} \tag{2-101}$$

The ratio a/b can be determined directly from the dimensions of the pattern on the screen, as shown in Fig. 2-34. Note that the pattern must be centered on the horizontal and vertical reference lines in order for Eq. (2-101) to be valid.

The patterns in Fig. 2-33 are specific examples of so-called *Lissajous figures* which are used to determine the ratio between the frequencies of two sine waves. One sine wave is applied to the vertical deflection plates and the other is applied to the horizontal deflection plates. If the ratio of their frequencies is some integral fraction, such as $\frac{1}{2}$, $\frac{1}{4}$, and $\frac{2}{3}$, the pattern is stationary. The frequency ratio is determined by the number of loops of the pattern touching a vertical line at the edge of the pattern compared with the number of loops touching a horizontal line at the edge of the pattern. The reason for this is that an integral number of sine waves on the horizontal deflection plates are compared in the same time that an integral number of sine waves are completed on the vertical plates. Some examples of common Lissajous figures are given in Fig. 2-35 for different frequency ratios. These patterns are derived in the same fashion used in connection with Figs. 2-32 and 2-33.

Several different patterns can represent the same frequency ratio, depending upon the interval between the times the two sine waves pass through zero. Figure 2-33 is an example of this for a $1 : 1$ frequency ratio. Although in principle Lissajous figures can be used to determine the frequency ratio between non-sinusoidal waveforms, in practice the resulting patterns are confusing and difficult to interpret.

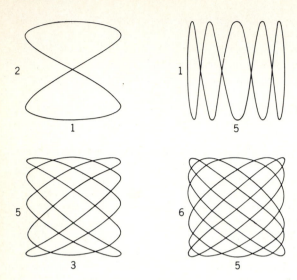

Figure 2-35 Lissajous figures for frequency ratios of 2 : 1, 1 : 5, 5 : 3, and 6 : 5.

SUGGESTIONS FOR FURTHER READING

Bernard Grob: "Basic Electronics," McGraw-Hill Book Company, New York, 1984.

R. M. Kirchner and G. F. Corcoran: "Alternating Current Circuits," 4th ed., John Wiley & Sons, Inc., New York, 1960.

A. Mottershead: "Introduction to Electricity and Electronics," John Wiley & Sons, Inc., New York, 1982.

Leigh Page and Norman Isley Adams: "Principles of Electricity," D. Van Nostrand Company, Inc., Princeton, N. J., 1931.

EXERCISES

2-1 Calculate the capacitive reactance in ohms of a 0.01-μF capacitor at (*a*) 100 Hz, (*b*) 1 kHz, (*c*) 100 kHz, (*d*) 1 MHz.

    Answer: 0.16 MΩ, 16 kΩ, 159 Ω, 15.9 Ω.

2-2 Calculate the inductive reactance in ohms of a 2.5-mH choke at (*a*) 100 Hz, (*b*) 1 kHz, (*c*) 100 kHz, (*d*) 1 MHz.

    Answer: 1.57 Ω, 15.7 Ω, 1.57 kΩ, 157 kΩ.

2-3 Three capacitors are connected in series across a 100-V battery. If the capacitances are 1.0, 0.1, and 0.01 μF, respectively, calculate the potential difference across each capacitor.

    Answer: 0.90 V, 9.0 V, 90.1 V

2-4 Verify that Eq. (2-43) is the solution of the circuit differential equation for a series RC circuit, Eq. (2-32), by direct substitution.

2-5 Show that the stored energy in a charged capacitor is equal to $\frac{1}{2}CV^2$ by integrating the expression $dW = V\,dq$. Substitute for dq using Eq. (2-11).

2-6 Plot the frequency-response characteristic of an RC high-pass filter using Eq. (2-45) for $R = 10^6\ \Omega$ and $C = 0.01\ \mu$F. Cover the frequency interval from two decades below the half-power frequency to two decades above it.

2-7 Develop an expression analogous to Eq. (2-45) for an *RC* low-pass filter. Plot the frequency-response characteristic for $R = 10^6 \ \Omega$ and $C = 100$ pF.

2-8* Plot the frequency-response characteristic of an *RL* low-pass filter, Eq. (2-27), if $L = 10$ H and $R = 100 \ \Omega$. Design an *RC* circuit having the same frequency characteristic.

2-9 Show that the Fourier series for a square wave is given by Eq. (2-87). Repeat for the sawtooth waveform in Fig. 2-22.

2-10 Using the Fourier series representation of a square wave, determine the result of integrating a square wave. Do this by integrating Eq. (2-87) term by term and plotting the resulting series. Compare the waveform with the sketch in Fig. 2-23.

2-11 Using the Fourier series representation of a sawtooth wave, determine the result of differentiating a sawtooth wave. Do this by differentiating Eq. (2-88) term by term and plotting the result.

2-12* Show how an *RC* low-pass filter integrates a square wave by computing the amplitude and phase angle of the first three terms in the Fourier series representation of a square wave, Eq. (2-87), at the output of the filter if the fundamental frequency is equal to 10 times the half-power frequency of the filter. Determine the output waveform graphically as in Fig. 2-21.

2-13* Repeat Exercise 2-12 if the fundamental frequency of the square wave is equal to the half-power frequency of the filter.

2-14 By a procedure analogous to that used in preparing Fig. 2-32, determine the observed waveform on an oscilloscope screen if the frequency of the horizontal sawtooth is one-half of the vertical sine-wave frequency.

2-15 By a procedure analogous to that used in preparing Fig. 2-33, determine the observed waveform on an oscilloscope screen if a sine wave is applied to both the horizontal and vertical deflection plates and the phase angle between them is 30°. Use the resulting pattern to confirm Eq. (2-101).

Voltage across a capacitor can't change instantaneously
Current thru "   "   can   "   "   "
Voltage across an inductor "   "   "   "
Current thru "   "   "   "

resistor can do both.

After a long time inductor becomes a wire
   "   "   "   "   capacitor   "   an open
                                        switch

THREE

DIODE CIRCUITS

Resistors, capacitors, and inductors are called linear components because the current increases in direct proportion to the applied voltage in accordance with Ohm's law. Components for which this proportionality does not hold are termed nonlinear devices, and they are the basis for all practical electronic circuits. This chapter examines the properties of an important nonlinear device, the diode rectifier. The term "diode" comes from the fact that rectifiers have two terminals, or electrodes.

A rectifier is nonlinear in that it passes a greater current for one polarity of applied voltage than for the other. If a rectifier is included in an ac circuit, the current is negligible whenever the polarity of the voltage across the rectifier is in the reverse direction. Therefore, only a unidirectional current exists and the alternating current is said to have been rectified. A major application of rectifiers is in power-supply circuits which convert conventional 115-V, 60-Hz ac line voltages to dc potentials suitable for use in place of batteries with transistors and integrated circuits.

NONLINEAR COMPONENTS

Current-Voltage Characteristics

The most useful description of the electrical properties of a nonlinear component is the relationship between the current in the unit and the applied voltage. Indeed, this is analogous to the importance of Ohm's law for linear components since Ohm's law forms the basis for all linear-circuit operation and analysis. The relationship between current and voltage is conventionally displayed graphically in the *current-voltage* (or *IV*) *characteristic* of the device. The performance of a nonlinear circuit element in any circuit may be analyzed with the aid of the appropriate current-voltage characteristic.

There exists a rich variety of current-voltage characteristics associated with various practical electrical components, rectifiers included, which are discussed in this and subsequent chapters. In many cases it is possible to understand the origin of these nonlinear relationships in terms of basic physical principles. For example, the motions of electrons in the device may lead directly to the variation of current with applied voltage. This is the case with Ohm's law discussed in Chap. 1, where the drift of electrons in a conductor results in a linear I vs. V, or IV characteristic. While an understanding of the physical principles involved is often important in the development of new device designs having favorable current-voltage properties, such an understanding is not necessary for the analysis of device performance and circuit operation.

Actually, the IV characteristics of practical nonlinear devices are universally determined by empirical measurement anyway, rather than by theoretical analysis. This is so because of the complexities inherent in practical units. Characteristic current-voltage curves of commercial devices are presented in manufacturer's data sheets and are widely available. All the principles of circuit analysis, in particular Kirchhoff's rules, are applicable to circuits with nonlinear components. Most often it is convenient to employ graphical techniques in the analysis because the IV characteristic is not expressible in a manageable mathematical form.

The Ideal Rectifier

An elementary but very important nonlinear current-voltage characteristic is that of the *ideal rectifier*, Fig. 3-1. This shows that an ideal rectifier has zero resistance for one polarity of applied voltage and infinite resistance for the reversed polarity. It is, in effect, a voltage-sensitive switch which is closed when the voltage polarity is in the *forward* direction and open for *reverse* polarity. This ideal characteristic is approximated by several practical types of diodes and is the basis for many important electronic circuits.

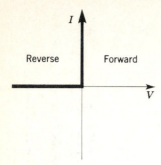

Figure 3-1 Current-voltage characteristic of ideal rectifier.

The Junction Diode

As shown in Chap. 4, the addition of certain foreign atoms to otherwise pure semiconductor materials such as germanium or silicon produces free electrons which carry electric current. A semiconductor crystal containing such foreign atoms is called an *n-type* crystal because of the negative charge of the current-carrying electrons. Similarly, by incorporating certain other kinds of foreign atoms, the semiconductor appears to conduct current by positive current carriers. Such a crystal is called a *p-type* conductor because the current carriers have a positive charge. It is possible to change from *n*-type to *p*-type conductivity in the same crystal by introducing an abrupt change from one impurity type to the other across a given region of the semiconductor. The junction between an *n*-type region and a *p*-type region in a semiconductor crystal, called a *pn junction*, exhibits distinctive nonlinear electrical characteristics.

In fact the *pn junction diode* is very nearly an ideal rectifier for the voltages commonly encountered in practical applications. This is illustrated by the current-voltage characteristics of typical germanium and silicon *pn* junctions shown in Fig. 3-2. In both units the forward current increases very rapidly for applied potentials greater than a few tenths of a volt, and even at appreciable reverse voltages the reverse currents are negligible by comparison. The forward voltage required for conduction is about 0.2 V for germanium diodes at room temperature compared to approximately 0.6 V in the case of silicon junctions. This difference can be related to the difference in electrical properties of the two semiconductor materials and is examined further in the next chapter.

The reverse current is several microamperes at room temperature in germanium *pn* junctions and only a few picoamperes in silicon diodes, a difference also related to the basic properties of the two semiconductors. These values are sufficiently minute to approximate the current-voltage characteristic of an ideal diode. Of greater concern is the fact that the reverse current increases exponentially with temperature. This rapid increase means that only a modest temperature rise can be tolerated and silicon diodes, for example, become nearly inoperative above about 200°C. Temperature changes also influence the forward characteristic, as can be seen in Fig. 3-3 where the potential for forward conduc-

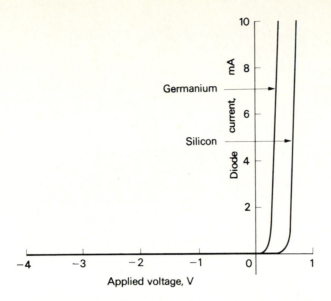

Figure 3-2 Current-voltage characteristics of germanium and silicon junction diodes.

tion in a silicon diode is reduced more than 0.1 V by only a 20°C temperature rise.

Junction diodes used in circuits at power levels above a few watts or so are cooled in order to dissipate joule heat caused by the current in the device. It is common practice to attach junction diodes firmly to metal heat sinks having radiator fins to conduct the heat away from the diode.

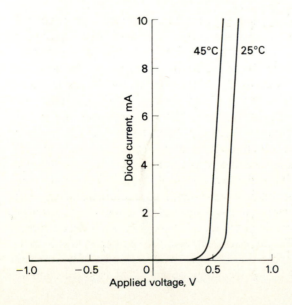

Figure 3-3 Effect of temperature on current-voltage characteristics of silicon junction diode.

Figure 3-4 Circuit symbol for semiconductor diode.

In spite of their temperature sensitivity, a characteristic common to most semiconductor devices, junction diodes are extremely good rectifiers. Commercial junction diodes are made of either silicon or germanium by processes similar to those described in Chap. 4, and practical devices are completely encapsulated to protect the semiconductor surface from contamination. This is accomplished by placing the diode in a small metal can filled with an inert atmosphere or by encasing the diode in plastic. Because of their superior electrical characteristics, silicon and germanium junction diodes have supplanted older forms of semiconductor rectifiers made of selenium or copper oxide. It is interesting to note that the latter rectifiers were the first commercially useful semiconductor devices. The circuit symbol for all semiconductor diodes, given in Fig. 3-4, has an arrow to indicate the direction of conventional forward current in the device.

The excellent forward conductivity of a *pn* junction means that practical diodes can be quite small. Therefore the stray capacitive reactances are correspondingly small and junction diodes are useful at high frequencies. This feature is enhanced in *point-contact* diodes, in which a metal probe is placed in contact with a semiconductor crystal. During processing, a minute *pn* junction is formed immediately under the point. Such tiny devices can operate at frequencies corresponding to millimeter wavelengths and are used, for example, in radar and high-speed computer circuits. An early type of point-contact diode which has now been replaced by tiny *pn* junction diodes is shown in Fig. 3-5.

Figure 3-5 Early point-contact diode. (*Ohmite Manufacturing Company.*)

RECTIFIER CIRCUITS

Half-Wave Rectifier

An elementary rectifier circuit, Fig. 3-6, has a junction diode in series with an ac source and a resistive load. When the polarity of the source is in the forward direction, the diode conducts and produces a current in the load. On the reverse half-cycle the diode does not conduct and the current is zero. The output current is therefore a succession of half sine waves and the circuit is called a *half-wave* rectifier. The average value of the half sine waves is clearly not zero, so that the output current has a dc component. That is, the input sine wave has been rectified.

The circuit is analyzed in the usual way except that, since the current-voltage characteristic of the diode is nonlinear, the solution must be carried out graphically. The voltage equation around the circuit is

$$v - v_d - iR_L = 0 \tag{3-1}$$

which may be put in the form

$$i = \frac{v}{R_L} - \frac{1}{R_L}v_d \tag{3-2}$$

Equation (3-2) is a straight line with slope $-1/R_L$ and intercept at $i = 0, v_d = v$. It may be looked upon as the current-voltage characteristic of R_L and is called the *load line*.

The load line together with the current-voltage characteristic of the diode represent two relations in the two unknowns v_d and i. By plotting the load line and the current-voltage characteristic of the diode together as in Fig. 3-7 we obtain the circuit current corresponding to every instantaneous value of the applied voltage from the intersection of the load line with the diode characteristic. Thus the output-current waveform can be determined by point-by-point graphical solution of Eq. (3-2) as shown in Fig. 3-7.

The output-current waveform departs from half-sinusoids because of curvature in the diode characteristic. For most practical applications, however, the applied voltage is much greater than the forward voltage drop across the diode so that the second term in Eq. (3-2) is negligible and the diode acts essentially as an ideal rectifier.

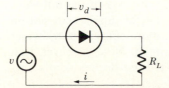

Figure 3-6 Elementary half-wave rectifier circuit.

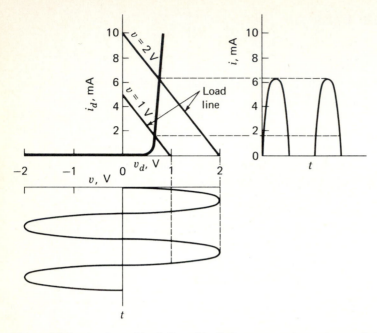

Figure 3-7 Intersection of load line with diode current-voltage characteristic gives instantaneous circuit current.

Full-Wave Rectifier

The half-wave rectifier is inactive during one-half of the input cycle and is therefore less efficient than is possible. By arranging two diodes as in Fig. 3-8a so that each diode conducts on alternate half-cycles, *full-wave* rectification results. This is accomplished by using a center-tapped transformer winding. Then, diode D_1 conducts on one half-cycle while diode D_2 is reverse-biased. On the alternate half-cycle, the conditions are interchanged so that the output-current waveform, Fig. 3-8b, has only momentary zero values, in contrast to the half-wave circuit.

Note that each diode in the full-wave rectifier must withstand the full end-to-end voltage of the transformer winding. Therefore the *peak inverse voltage* rating of the diodes must be at least twice the peak output voltage. This is usually not a serious drawback except for circuits designed to operate at the highest voltages. Comparing the output waveforms of half-wave and full-wave rectifiers, Figs. 3-7 and 3-8, reveals that the fundamental frequency equals the supply voltage in the half-wave rectifier but is twice the supply frequency in the case of the full-wave circuit. The center-tapped transformer in the full-wave circuit supplies current on both half-cycles of the input voltage, which permits a more efficient transformer design than is possible in the case of the half-wave circuit. Note, however, that the output voltage is only one-half the total voltage of the transformer secondary.

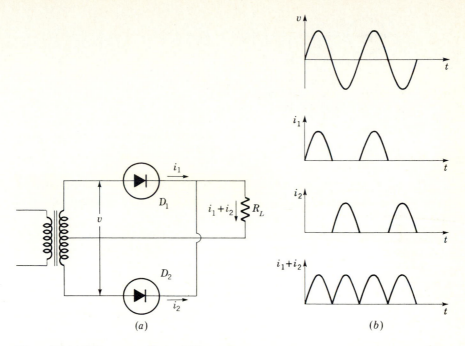

Figure 3-8 (*a*) Full-wave rectifier and (*b*) waveforms.

Bridge Rectifier

Full-wave rectification without a center-tapped transformer is possible with the *bridge rectifier*, Fig. 3-9. The operation of this circuit may be described by tracing the current on alternate half-cycles of the input voltage. Suppose, for example, that the upper terminal of the transformer is positive. This means that diode D_2 conducts, as does diode D_3, and current is present in the load resistor. On the alternate half-cycle diodes D_4 and D_1 conduct and the current direction in the

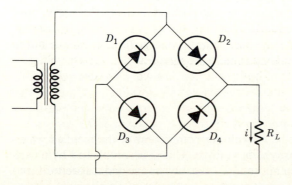

Figure 3-9 Bridge rectifier is full-wave circuit without center-tapped transformer.

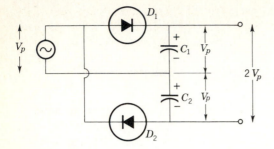

Figure 3-10 Voltage-doubler rectifier yields dc output voltage equal to twice peak input voltage.

load resistance is the same as before. The voltage across R_L corresponds to full-wave rectification and the peak voltage is equal to the peak transformer voltage less the potential drops across the diodes.

Since two diodes are in series with the load, the output voltage is reduced by twice the diode drop. On the other hand, the peak inverse voltage rating of the diodes need only be equal to the transformer voltage, in contrast to the previous full-wave circuit.

Voltage Doubler

Consider the circuit of Fig. 3-10, in which two diodes are connected to the same voltage source but in the opposite sense. On the half-cycle during which the upper terminal of the source is positive, diode D_1 conducts and capacitor C_1 charges to the peak value of the input voltage. On the reverse half-cycle D_2 conducts and C_2 also charges to the full input voltage. Meanwhile, the charge on C_1 is retained, since the potential across D_1 is in the reverse direction. Thus both C_1 and C_2 charge to the peak supply voltage and the dc output voltage is equal to twice the peak input voltage. Accordingly, this circuit is called a *voltage doubler*.

The above analysis applies to the case when no current is delivered to the load. When a load resistance is connected, current is supplied by the discharge of the capacitors. On alternate half-cycles the capacitors are subsequently recharged. This implies, however, that the output voltage under load is no longer dc, but has an ac component. It is necessary to make the capacitance of C_1 and C_2 large enough to minimize this variation in output voltage, taking into account the current drain and the supply frequency. The filtering action of the capacitors in this circuit is treated in greater detail in the next section.

FILTERS

It is usually desirable to reduce the alternating component of the rectified waveform so that the output is primarily a dc voltage. This is accomplished by means of low-pass filters which are composed of suitably connected capacitors and inductances. A power-supply filter reduces the amplitudes of all alternating components in the rectified waveform and passes the dc component. A measure of the effectiveness of a filter is given by the *ripple factor r*, which is defined as the ratio of the rms value of the ac component to the dc or average value. That is,

$$r = \frac{V_{rms}}{V_{dc}} \tag{3-3}$$

According to this definition it is desirable to make the ripple factor as small as possible. Although rectified waveforms contain many harmonics of the input voltage frequency, it is generally satisfactory to determine the ripple factor at the fundamental frequency only. This is so because the fundamental frequency is predominant and also because higher harmonics are attenuated more than the fundamental by the low-pass filter characteristic.

Capacitor Filter

The simplest filter circuit consists of a capacitor connected in parallel with the load resistance, as in the typical low-voltage bridge-rectifier power supply shown in Fig. 3-11. If the reactance of the capacitor at the power-line frequency is small compared with the load resistance R_L, the ac component is shorted out and only dc current remains in the load resistor.

Characteristics of the *capacitor filter* are determined by examining waveforms in the circuit, Fig. 3-12. The capacitor is charged to the peak value of the rectified voltage V_p and begins to discharge through R_L after the rectified voltage decreases from the peak value. The decrease in capacitor voltage between charging pulses depends upon the relative values of the $R_L C$ time constant and the period of the input voltage. A small time constant means the decrease is large and the ripple voltage is also large. On the other hand, a large time constant results in a small ripple component. The diodes conduct only during the portion of the cycle that the capacitor is charging, because only during this interval is the sum of the supply voltage and the capacitor voltage such that the potential across the diodes is in the forward direction.

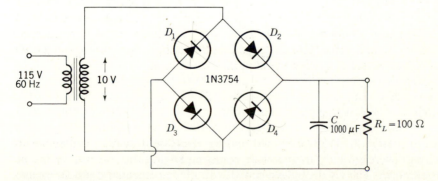

Figure 3-11 Practical low-voltage power supply using capacitor filter.

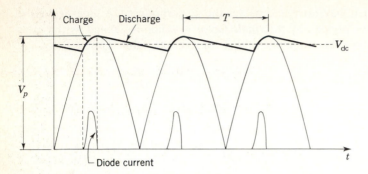

Figure 3-12 Output voltage of capacitor filter is dc voltage and small triangular ripple voltage.

The ripple voltage is approximately a triangular wave if the $R_L C$ time constant is long compared with the period, $R_L C \gg T$. In this case the exponential decrease of capacitor voltage during one period is given approximately by $V_p \times T/R_L C$. The rms value of a triangular wave is calculated in Chap. 2, Eq. (2-91), so that the ripple factor of the capacitor filter is

$$r = \frac{1}{V_p} \frac{1}{\sqrt{3}} \frac{V_p T}{2 R_L C} = \frac{1}{2\sqrt{3}} \frac{1}{f R_L C} \tag{3-4}$$

where the dc output voltage has been taken to be equal to V_p for simplicity.

This result shows that ripple is reduced by increasing the value of the filter capacitor. When the load current is equal to zero ($R_L \to \infty$), the ripple factor becomes zero, which means that the output voltage is pure dc. As the load current is increased (smaller values of R_L), the ripple factor increases. Inserting component values given on the circuit diagram of Fig. 3-11, the ripple factor is found to be 0.05. Thus the ripple voltage is about 5 percent of the dc output. Note that the ripple voltage of a full-wave rectifier is approximately one-half that of the half-wave circuit, because the frequency of the rectified component is twice as great.

The actual dc output voltage from the capacitor filter is equal to the peak capacitor voltage less the average ripple component,

$$V_{dc} = V_p - \frac{V_p T}{2 R_L C} = V_p - \frac{1}{2 f C} I_{dc} \tag{3-5}$$

where, again, the approximation $I_{dc} \approx V_p/R_L$ is used for simplicity. According to Eq. (3-5), the dc output voltage decreases linearly as the dc current drawn by the load increases. The constancy in output voltage with current is called the *regulation* of the power supply; Eq. (3-5) shows that a large value of filter capacitance improves regulation. It should be noted that the decrease in output voltage given by this equation refers only to the change accompanying the increase in the ac ripple component. The IR drops associated with diode resistances and the resistance of the transformer winding further reduce the output voltage as the load current increases.

The simple capacitor filter provides very good filtering action at low currents and is often used in high-voltage low-current power supplies. Because of its simplicity, the circuit also is found in those higher current supplies where ripple is relatively less important. The dc output voltage is high, equal to the peak value of the supply voltage. The disadvantages of the capacitor filter are poor regulation and increased ripple at large loads.

L-Section Filter

It is useful to add a series inductor to the capacitance filter, as in the circuit of Fig. 3-13. The series inductance in this *L-section* or *choke-input* filter opposes rapid variations in the current and so contributes to the filtering action. The ripple factor is calculated by noting that the ac voltage components of the recti-fied waveform divide between the inductance and the capacitance (assuming R_L is greater than the reactance of the capacitor), while the dc output voltage is simply equal to the dc component of the rectified waveform. Therefore, the ripple factor of the L-section filter is written as

$$r = \left(\frac{V_{rms}}{V_{dc}}\right)_{rect} \frac{1/\omega C}{1/\omega C + \omega L} \tag{3-6}$$

inserting the rms and dc values appropriate for a full-wave rectified waveform, Eqs. (2-96) and (2-92),

$$r = \frac{\sqrt{2}}{3}\frac{1}{1 + \omega^2 LC} \tag{3-7}$$

Note that the ripple factor is independent of the load, in contrast to the case of the capacitor filter, and that large values of L and C improve the filtering action.

The inductance chokes off alternating components of the rectified waveform and the dc output voltage is simply the average, or dc, value of the rectified wave. For a series of half-sinusoids this means that the dc output voltage is $2V_p/\pi$, or approximately $0.6V_p$. Therefore, the output voltage of the choke-input filter is considerably less than that of the capacitor filter.

According to Eq. (3-7) the ripple is independent of the load current. This means that there is no decrease in output voltage due to a decrease in filtering action at high currents, such as in Eq. (3-5). Therefore, the output voltage is inde-pendent of the load, except for IR drops in the diodes and the inductor and trans-former windings. For this reason the L-section filter is used in applications where wide variations in the load current are expected. The advantage of good regulation

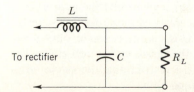

Figure 3-13 L-section filter.

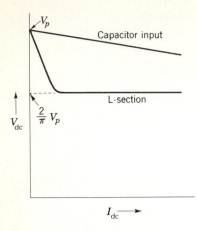

V_p

Capacitor input

L-section

V_{dc} $\frac{2}{\pi} V_p$

I_{dc} ⟶

Figure 3-14 Voltage-regulation characteristics of capacitor-input filter and L-section filter.

must be balanced against the comparatively low output voltage in considering any given application.

When the load current is zero the capacitor charges to the peak value of the input waveform as in the capacitor filter. The output voltage is then equal to V_p and regulation is poor. As discussed in connection with the capacitor filter, in this circumstance the diodes conduct for only a part of the applied voltage cycle. The transition between this performance and that of the true L-section filter occurs at the load current for which the diodes just conduct over the entire cycle, and the voltage across the capacitor remains at the dc component of the rectified waveform.

A comparison of the voltage-regulation curves of a capacitor filter and a choke-input filter, Fig. 3-14, shows the poor regulation but high output of the former and the good regulation but lower voltage of the latter. The minimum current necessary to ensure good regulation in the case of the L-section filter is clearly evident.

π-Section Filter

The combination of a capacitor-input filter with an L-section filter, as shown in the power-supply diagram of Fig. 3-15, is a very popular circuit. The output voltage of this *π-section filter* is nearly equal to that of the capacitor-input filter, and the regulation characteristics are about the same. The ripple is very much reduced by the double filtering action, however. In fact, the overall ripple factor is essentially the product of the ripple factor of the capacitor filter times the impedance ratio of the L-section filter. In spite of the inferior regulation properties of the π-section filter, it is widely used because of its excellent filtering action.

A useful variant of the π-section filter particularly suitable for low-current circuits replaces the inductance with a resistor. This circuit is useful only if the current is low enough so that the IR drop across the resistor is not excessive and if regulation is of secondary concern. Within these restrictions, the circuit provides better filtering action than a simple capacitance filter.

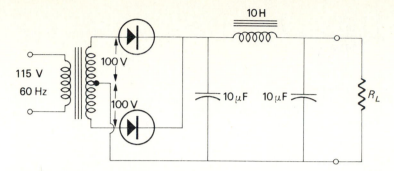

Figure 3-15 Practical 100-V power supply.

DIODE REGULATORS

It is often desirable that the voltage of a power supply remain fixed independent of the load current. Although the regulation of the choke-input filter is good, the output voltage decreases with increasing current even in this circuit because of the *IR* drops across the transformer and filter choke winding resistances. Other variables, such as changes in the line voltage and component aging, may also contribute to variations in the power-supply voltage. In order to maintain the output voltage constant, *voltage regulator* devices are made part of the power supply. These are fairly elaborate electronic circuits if very precise regulation is necessary, as discussed in Chap. 10. In other cases, special components that maintain modest voltage stability are used.

Zener Diodes

At some particular reverse-bias voltage the reverse current in a *pn* junction increases very rapidly. This happens when electrons are accelerated to high velocities by the electric field at the junction and produce other free electrons by ionization collisions with atoms. These electrons are similarly accelerated by the field and in turn cause other ionizations. The avalanche process leads to a very large current and the junction is said to have suffered *breakdown*. The breakdown is not destructive, however, unless the power dissipation is allowed to increase the temperature to the point where local melting destroys the semiconductor. The voltage across the junction remains quite constant over a wide current range in the breakdown region. This effect can be used to maintain the output of a power supply at the breakdown voltage.

These *pn* junctions are called *zener diodes*, because Clarence Zener first suggested an explanation for the rapid current increase at breakdown. The current-voltage curve of a zener diode, Fig. 3-16, has a sharp transition potential and a flat current plateau above breakdown. Zener diodes may be obtained with breakdown voltages ranging from about two to several hundred volts and with current ratings of a few milliamperes to many amperes.

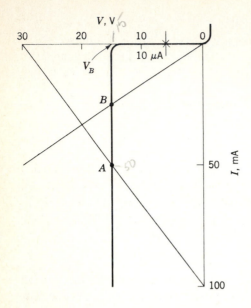

Figure 3-16 Current-voltage characteristic of zener diode.

The way in which a zener diode is used to regulate the output voltage of a power supply is illustrated in Fig. 3-17. The unregulated output voltage of the supply V_s must be greater than the zener breakdown voltage of the diode. Then the voltage drop across R_s caused by the diode current plus the diode breakdown potential adds up to the power-supply potential. As the load current increases, the diode current decreases, so the drop across R_s always is the difference between the breakdown potential and the power-supply voltage. Even if the power-supply voltage varies under load, the regulated output voltage remains constant at the diode breakdown potential.

The zener diode voltage regulator is analyzed by plotting the current-voltage characteristic pertaining to R_s on the same axes as the current-voltage characteristic of the diode. Kirchhoff's rule applied to the circuit of Fig. 3-17 is

$$V_s - IR_s - V = 0 \qquad (3\text{-}8)$$

Solving for I,

$$I = \frac{V_s}{R_s} - \frac{1}{R_s} V \qquad (3\text{-}9)$$

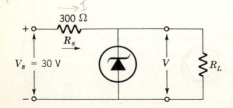

Figure 3-17 Zener diode voltage regulator.

which is the equation of a straight line of slope $-1/R_s$ and intercept $I = V_s/R_s$. Using numerical values given on the circuit diagram, Eq. (3-9) intersects the current-voltage characteristic of the diode at point A on Fig. 3-16. This intersection gives the current in R_s. The load current is just V_B/R_L, where V_B is the breakdown voltage of the diode.

It is useful to plot the IV characteristic of the load,

$$I_L = \frac{1}{R_L} V \tag{3-10}$$

on Fig. 3-16. The load current is given by the intersection of Eq. (3-10) with the diode characteristic, point B. Note that the load current must be less than the current in R_s, for otherwise the voltage drop across R_s is too large to maintain V_B across the diode and regulating action ceases. So long as intercept B is at a smaller current than point A, the circuit performs properly.

Controlled Rectifiers

It is often necessary to control the power delivered to some load, such as an electric motor or the heating element of a furnace. Series resistances or potentiometers waste power, a serious drawback in high-power circuits. *Controlled rectifiers* have been developed which are capable of adjusting the transmitted power with little waste.

The most satisfactory unit of this kind is the *silicon controlled rectifier*, or *SCR*. This semiconductor device contains four parallel *pn* junctions; it is described in detail in Chap. 4. For present purposes it is sufficient to note that it is similar to a junction rectifier in which forward conduction is controlled by the current in a control electrode, called the *gate*. The inclusion of the gate electrode is shown on the SCR symbol, as in Fig. 3-18.

The current-voltage characteristic of a typical SCR, Fig. 3-19, is identical to a junction rectifier in the reverse direction. The forward direction has both an "on" state which is equivalent to normal forward conduction in a junction rectifier, and a low-current "off" state. So long as the forward anode-cathode potential remains below a certain critical value (actually it is the corresponding current that is critical), the forward current is small. Above this critical value the SCR is in the high-current low-voltage on condition. The critical current may be supplied by the gate electrode, which means that the SCR may be triggered into the

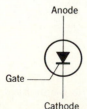

Anode

Gate

Cathode   **Figure 3-18** Circuit symbol for silicon controlled rectifier.

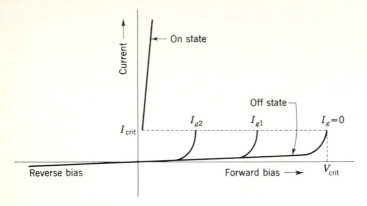

Figure 3-19 Current-voltage characteristic of SCR. Unit can be triggered into on state by gate current. Here $I_{g2} > I_{g1} > 0$.

on position by a small current (as low as 100 μA) supplied to the gate terminal. It is not necessary to maintain the gate current, for once the SCR is triggered into the high-conduction state it remains in this condition until the anode potential is reduced to zero.

Consider the simple SCR circuit of Fig. 3-20, in which the motor speed is controlled by the power delivered to it. The peak value of the transformer voltage V_p is less than the critical value, so that, unless the SCR is triggered by a gate current, the load current is zero. Suppose the gate is supplied with a current pulse each cycle that lags the transformer voltage by a phase angle α, as illustrated in Fig. 3-21. Forward conduction is delayed until this point in each cycle and the rectified output waveform is similar to that of a half-wave rectifier, except that the first part of each cycle is missing. The average, or dc, value of the current in the motor is found by integrating the output current over a complete cycle, or

$$I_{dc} = \frac{I_p}{2\pi} \int_{\alpha}^{\pi} \sin \omega t \; d\omega t \qquad (3\text{-}11)$$

where I_p is the peak value of the current. Integrating,

$$I_{dc} = \frac{I_p}{2\pi} (1 + \cos \alpha) \qquad (3\text{-}12)$$

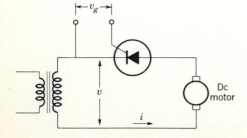

Figure 3-20 Simple SCR circuit to control speed of dc motor.

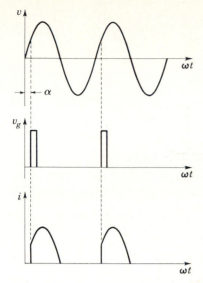

Figure 3-21 Output current waveform depends upon phase angle between transformer voltage and gate-current pulse.

According to Eq. (3-12) the motor current can be adjusted from a maximum ($\alpha = 0$) to zero ($\alpha = \pi$) simply by changing the phase angle of the gate pulses.

DIODE CIRCUITS

Diodes prove to be useful in circuits other than power supplies. A major application is in circuits designed to operate with square-wave pulses. In addition, the rectification characteristics of diodes are put to work in circuits dealing with sine-wave signals. In these applications the nonlinear diode modifies the sine-wave signals in a specified manner.

Clippers

Consider the diode *clipper* circuit, Fig. 3-22, in which two diodes are connected in parallel with an ac voltage source. Batteries V_1 and V_2 bias each diode in the reverse direction. Whenever the input voltage signal is greater than V_1, diode D_1 conducts and causes a voltage drop across the series resistor R. Similarly, every

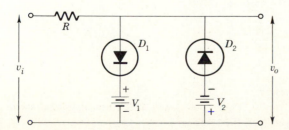

Figure 3-22 Diode clipper.

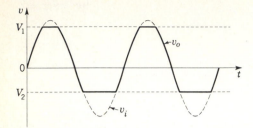

Figure 3-23 Maximum amplitudes in output waveform of diode clipper are limited to values of bias voltage.

time the input voltage becomes more negative than V_2, diode D_2 conducts. Thus, the output waveform is *clipped* or *limited* to the voltages set by the reverse biases V_1 and V_2. The clipping action is most efficient when the series resistance is much greater than the load impedance.

Clipper circuits produce square waves from a sine-wave generator, as illustrated by the waveforms in Fig. 3-23. If $V_1 = V_2$ and the amplitude of the input signal is considerably greater than the bias voltages, the output waveform is very nearly a square wave. Note that if, say, $V_2 = 0$, the output can never swing negative, so the waveform is a train of positive-going square pulses. The clipping action operates on any input waveform; it can be adjusted by altering the reverse-bias voltages V_1 and V_2.

The clipper circuit is also a useful protective device which limits input voltages to a safe value set by the bias voltage. In radio-receiver circuits limiters are often used to reduce the effect of strong noise pulses by limiting their amplitude to that of the desired signal. The signal waveform is transmitted undistorted so long as the instantaneous amplitude remains smaller than the bias voltages.

Clamps

The diode *clamp* circuit is shown in Fig. 3-24. Consider first the situation in which the bias voltage V is set equal to zero. The diode conducts on each negative cycle of input voltage and charges the capacitor to a voltage equal to the negative peak value of the input signal. If the load current is zero, the capacitor retains its charge on the positive half-cycle, since the diode voltage is in the reverse direction. Thus, the output voltage is

$$v_o = v_i + V_p \tag{3-13}$$

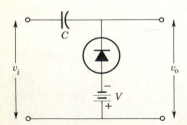

Figure 3-24 Diode clamp.

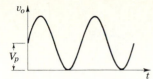

Figure 3-25 Negative peak of output waveform is clamped at zero when $V = 0$ in diode clamp circuit of Fig. 3-24.

where V_p is the negative peak value of the input voltage. According to Eq. (3-13) the output voltage waveform replicates the input signal except that it is shifted by an amount equal to the dc voltage across the capacitor.

The output waveform corresponding to a sinusoidal input, Fig. 3-25, has the negative peaks of the sine wave *clamped* at zero voltage. This is always the case, independent of the amplitude of the input voltage. Furthermore, the negative peaks are always clamped at zero volts no matter what form the input waveshape takes. When the terminals of the diode are interchanged, the same circuit analysis applies; the positive peaks of the output wave are clamped at zero.

If the bias voltage V in Fig. 3-24 is set at a potential other than zero, the capacitor charges to a voltage equal to $V_p + V$. Therefore, the negative peaks are clamped at the voltage V. Similarly, when the bias is $-V$, the negative peaks are clamped at this potential. Reversing the diode polarity makes it possible to clamp the positive peaks of the input wave at the voltage equal to the bias potential.

The diode clamp is used in circuits which require that the voltages at certain points have fixed peak values. This is important, for example, if the waveform is subsequently clipped at a given voltage level. In most diode clamp circuits it is useful to connect a large resistance across the output terminals so that the charge on the capacitor can eventually drain away. This also makes it possible for the circuit to adjust to changes in amplitude of the input voltage.

AC Voltmeters

The rectifying action of diodes makes it possible to measure ac voltages with a dc voltmeter, such as the d'Arsonval meter discussed in the first chapter. Two ways in which this is accomplished in VOM meters are illustrated in Fig. 3-26. In Fig. 3-26a diode D_1 is a half-wave rectifier and the dc component of the rectified waveform registers on the dc voltmeter. Diode D_2 is included to rectify the negative cycle of the input waveform. Although no meter current results from the action of diode D_2, it is included so that both halves of the ac voltage are rectified. This ensures that the voltmeter loads the circuit equally on both half-cycles and avoids possible waveform distortion. The bridge circuit, Fig. 3-26b, has greater sensitivity than the half-wave circuit because it is a full-wave rectifier.

The deflection of a d'Arsonval meter is proportional to the average current, so that the voltmeters of Fig. 3-26 measure the average value of the ac voltage. The meter scale is commonly calibrated in terms of rms readings, however, in order to facilitate comparison with dc readings. This calibration assumes that the ac waveform is sinusoidal, and if this is not the case, the meter readings must be

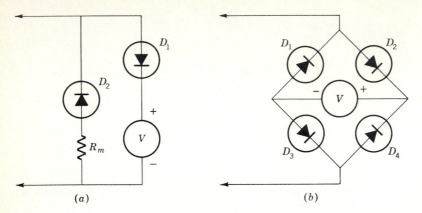

Figure 3-26 (*a*) Half-wave and (*b*) full-wave ac voltmeters.

interpreted in terms of the actual waveform measured. Series resistor multipliers are used with ac voltmeters to increase the range, just as in the case of dc meters.

The half-wave rectifier, Fig. 3-27, makes a useful *peak-reading* voltmeter if the current drain through the meter circuit is small. In this case the capacitor charges to the positive peak value of the input waveform on each half-cycle, and the dc meter deflection corresponds to this value. Here again, it is common practice to calibrate the meter scale in terms of rms readings, assuming that the unknown voltage is sinusoidal. This is accomplished by designing the scale to indicate the peak value divided by $\sqrt{2}$. In case the unknown voltage is not sinusoidal, the peak value is obtained by multiplying the scale reading by $\sqrt{2}$. A difficulty with this circuit is that the system being measured must contain a dc path. If this is not so and a series capacitor is present, the voltage across C depends upon the relative values of the two capacitors, since they are effectively in series during the forward cycle. This results in erroneous voltage readings.

The diode clamp is a peak-reading circuit which does not have this difficulty. It is customary to include an RC filter, Fig. 3-28, to remove the ac component of the clamped wave. Referring to the waveform in Fig. 3-25, the average value is equal to the negative peak of the input voltage, so the dc meter reading corresponds to the negative voltage. Note that a series capacitor in the circuit to be measured in effect becomes part of the clamping action and does not result in a false meter indication. If the filter in Fig. 3-28 is replaced with a diode peak

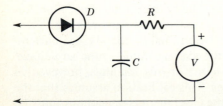

Figure 3-27 Peak-reading ac voltmeter.

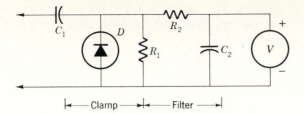

Figure 3-28 Diode clamp used as peak-reading voltmeter.

rectifier, Fig. 3-27, the combination reads the peak-to-peak value of the input wave (Exercise 3-13). Although usually calibrated in terms of rms readings for a sine wave, a peak-to-peak voltmeter is very useful in measuring complex waveforms.

Detectors

Nonlinear diode properties are useful in other ways than those corresponding to actual rectification. Suppose that the amplitude of the ac voltages applied is very small, so that the diode current-voltage characteristic can be represented by the first two terms of a series expansion,

$$i = a_1 v + a_2 v^2 \tag{3-14}$$

where a_1 and a_2 are constants. According to Eq. (3-14), one part of the current is proportional to the square of the input voltage.

Consider the circuit shown in Fig. 3-29, in which two sinusoidal signals of somewhat different frequencies, ω_1 and ω_2, are supplied to a diode. The total diode voltage is the sum of the individual waves,

$$v = V_1 \sin \omega_1 t + V_2 \sin \omega_2 t \tag{3-15}$$

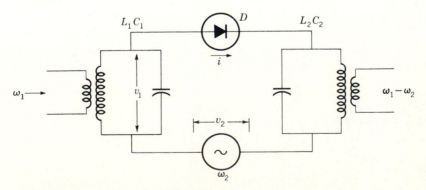

Figure 3-29 Diode mixer uses nonlinear properties of diode.

where V_1 and V_2 are the peak amplitudes. The current is found by substituting Eq. (3-15) into Eq. (3-14), assuming for the moment that the diode impedance is greater than other impedances in the circuit.

$$i = a_1 V_1 \sin \omega_1 t + a_1 V_2 \sin \omega_2 t + a_2 V_1^2 \sin^2 \omega_1 t$$

$$+ a_2 V_2^2 \sin^2 \omega_2 t + 2a_2 V_1 V_2 \sin \omega_1 t \sin \omega_2 t \quad (3\text{-}16)$$

It is instructive to express Eq. (3-16) in the form of a Fourier series type expression in which each term represents a given frequency in the total waveform. This is accomplished by introducing trigonometric substitutions for $\sin^2 \omega t$ and $\sin \omega_1 t \sin \omega_2 t$ and rearranging,

$$i = \frac{a_2}{2} (V_1^2 + V_2^2) + a_1(V_1 \sin \omega_1 t + V_2 \sin \omega_2 t)$$

$$- \frac{a_2}{2} (V_1^2 \cos 2\omega_1 t + V_2^2 \cos 2\omega_2 t)$$

$$+ a_2 V_1 V_2 \cos (\omega_1 - \omega_2)t - a_2 V_1 V_2 \cos (\omega_1 + \omega_2)t \quad (3\text{-}17)$$

The first term in Eq. (3-17) is a direct current, while the second corresponds to the input voltages. The last two terms are at frequencies not present in the original inputs. Suppose ω_1 and ω_2 are not too different; then the term involving $\omega_1 - \omega_2$ is a comparatively low frequency.

In the *diode mixer* circuit of Fig. 3-29 the output circuit $L_2 C_2$ is tuned (see Chap. 5) to the frequency $\omega_1 - \omega_2$, so only this component has an appreciable amplitude at the output terminals. In effect, an input signal of frequency ω_1 is mixed with a constant-amplitude sine wave at ω_2 and is converted to an output frequency $\omega_1 - \omega_2$. This is called *heterodyning* and is of considerable importance in, for example, radio receivers. The incoming high-frequency signal is converted into a lower-frequency signal which is easier to amplify. Furthermore, changing the frequency ω_2 means that the receiver is tuned to a different ω_1, such that $\omega_1 - \omega_2$ is a constant. This is important because the amplification of the $\omega_1 - \omega_2$ signal is accomplished by fixed tuned circuits that are inexpensive and easy to keep in adjustment, yet the receiver may be tuned to a different input frequency. In this application Fig. 3-29 is called a *first detector* because the signal frequency is modified in the first stages of the receiver.

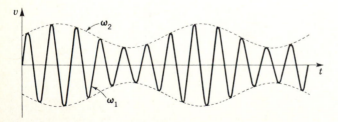

Figure 3-30 Modulated waveform.

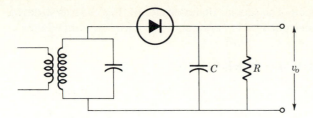

Figure 3-31 Diode peak rectifier used to demodulate modulated waveform.

This circuit has another important use, which may be illustrated by an alternate rearrangement of Eq. (3-17), using the substitution $2 \sin a \sin b = \cos (a - b) - \cos (a + b)$:

$$i = \frac{a_2}{2} (V_1^2 + V_2^2) + a_1 V_2 \sin \omega_2 t - \frac{1}{2} (V_1^2 \cos 2\omega_1 t + V_2^2 \cos 2\omega_2 t)$$

$$+ a_1 V_1 \sin \omega_1 t + 2a_2 V_1(V_2 \sin \omega_2 t) \sin \omega_1 t \quad (3\text{-}18)$$

Suppose ω_2 is much smaller than ω_1 and that the output circuit $L_2 C_2$ is tuned to the frequency ω_1. Then only the last two terms in Eq. (3-18) have an appreciable output amplitude. The first of these is simply the current resulting from the input voltage at the frequency ω_1. Careful inspection of the last term in Eq. (3-18) shows that it is a sinusoidal waveform at a frequency ω_1 but with an amplitude that varies sinusoidally at a frequency corresponding to ω_2.

The waveform of the last two terms in Eq. (3-18), illustrated in Fig. 3-30, shows how the amplitude varies. The frequency ω_1 is said to be *modulated* at the frequency of ω_2. This is the basis for radio and television communication wherein a high-frequency *carrier* ω_1 is modulated in accordance with a low-frequency signal ω_2 corresponding to the sound or picture to be transmitted. The high-frequency voltage is radiated from the sending station to a receiver where the waveform of the signal is recovered.

The signal is recovered by *demodulating* the modulated waveform with a diode peak rectifier, Fig. 3-31. Since the output of the peak rectifier is equal to the peak value of the input voltage, the waveform across the output capacitor corresponds to that of modulating signal, as in Fig. 3-32. In this application the peak rectifier is called a *second detector*, because the waveform is changed for the

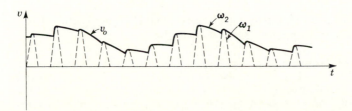

Figure 3-32 Modulating waveform recovered at output of second detector.

second time in the receiver circuit. The second detector circuit, Fig. 3-31, is widely used in radio and television receivers.

SUGGESTIONS FOR FURTHER READING

William H. Hayt, Jr., and Jack E. Kemmerly: "Engineering Circuit Analysis," McGraw-Hill Book Company, New York, 1986.
W. Ryland Hill: "Electronics in Engineering," McGraw-Hill Book Company, New York, 1961.
Jacob Millman and Samuel Seely: "Electronics," McGraw-Hill Book Company, New York, 1951.
Julien C. Sprott: "Introduction to Modern Electronics," John Wiley & Sons, Inc., New York, 1981.

EXERCISES

3-1 What is the slope of the load line in Fig. 3-7 for a very small load resistance? For a very large load resistance?
   Answer: Nearly vertical, nearly horizontal.

3-2 Determine the dc output current from the half-wave rectifier circuit in Fig. 3-6, if $v = 10$ V, $R_L = 100$ Ω, and the rectifier is a silicon diode. Repeat for the full-wave rectifier in Fig. 3-8.
   Answer: 0.45 A, 0.90 A

3-3 In the voltage-doubler circuit, Fig. 3-10, the input voltage is 115 V rms at a frequency of 60 Hz. If the load resistance is 10,000 Ω, what is the minimum value of the capacitors necessary to be sure the voltage drop between charging pulses is less than 10 percent of the output voltage?
   Answer: 4.2×10^{-6} F

3-4 Calculate the output voltage and ripple factor of the circuit of Fig. 3-11 after diodes D_1 and D_4 are removed.
   Answer: 12.9 V; 0.048

3-5* Plot the dc output and the ripple voltage of the full-wave rectifier circuit given in Fig. 3-15 up to a current of 200 mA.

3-6 Repeat Exercise 3-5 if the output capacitor is defective and is open-circuited.

3-7 Determine the ripple factor of an L-section filter comprising a 10-H choke and a 8-μF capacitor used with a full-wave rectifier. Compare with a simple 8-μF capacitor filter at a load current of 50 mA and also 150 mA, assuming an output of 50 V.
   Answer: 0.0104; 0.300; 0.900

3-8 In the zener diode voltage regulator of Fig. 3-17 determine the range of load resistances over which the circuit is useful if $R_s = 1500$ Ω and the supply voltage $V = 150$ V; the diode breakdown voltage is 100 V and maximum rated current is 100 mA. For a fixed $R_L = 10,000$ Ω, over what range of input voltages does the circuit regulate?
   Answer: Greater than 3 kΩ; 115 to 265 V

3-9 In Exercise 3-8, what is the proper value of R_s if the maximum diode current is 10 mA?
   Answer: 5 kΩ

3-10* Plot the dc output voltage of the zener diode voltage regulator in Fig. 3-17 as a function of load current from 0 to 75 mA. Over what current range is regulation effective?
   Answer: 0 to 50 mA

3-11 Determine the phase angle between the gate voltage v_g and the anode voltage (the same as v_1) for the *phase-shift controlled rectifier*, Fig. 3-33. Plot the dc load current as a function of control resistance R if the transformer secondary voltage is 100 V rms, the load resistance is 10 Ω, and $C = 0.1$ μF. Assume the SCR turns on whenever the gate voltage is positive with respect to the cathode.
   Answer: $-2 \arctan R\omega C$

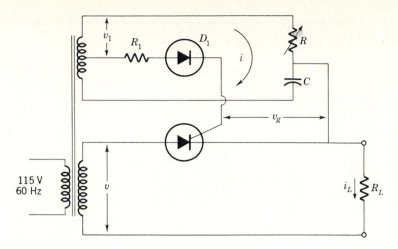

Figure 3-33 Phase-shift control of SCR.

3-12 Sketch the waveforms of the sinusoidal input voltage, the diode voltage, and the output voltage for the diode clamp, Fig. 3-24, with a positive bias voltage V. Repeat for $-V$.

3-13* Analyze the peak-to-peak voltmeter circuit, Fig. 3-34, by sketching the voltage waveforms at points A, B, and C, assuming the input voltage A is sinusoidal. Note that the circuit is a diode clamp followed by a peak rectifier.

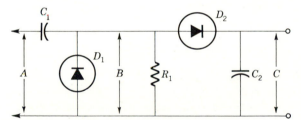

Figure 3-34

3-14 The output reading of the peak-to-peak voltmeter in Fig. 3-34 measuring a sawtooth waveform is 1.4 V. What is the voltage of the waveform?

Answer: 1.15 V (4 V peak-to-peak)

3-15 Determine the expression corresponding to Eq. (3-14) for the rectifier equation given in the next chapter, Eq. (4-7).

Answer: $i/I_0 = (e/kT)v + \frac{1}{2}(e/kT)^2 v^2$

FOUR

SEMICONDUCTOR DEVICES

The versatility inherent in nonlinear electronic components is enhanced immeasurably by the capability to influence current in the device in accordance with signals introduced at a control electrode. These are considered to be active devices, rather than passive components, because the control electrode permits active interaction with currents in the device.

Electrical properties of active devices are described by current-voltage characteristics as for other nonlinear components. It is often useful to develop an appreciation for the origin of these current-voltage characteristics in terms of basic physical processes. In this way it is possible to optimize performance of a nonlinear element in any application, and also to design improved components having desirable electrical features. It turns out that the current-voltage characteristics of electronic devices depend primarily upon the motions of free electrons in them.

The electronic functions of transistors take place within solid materials called semiconductors. Accordingly, the properties of transistors and other semiconductor devices, such as the junction diode, stem directly from the behavior of electrons in semiconductor crystals. The great variety of semiconductor devices that have been developed since the discovery of the transistor are possible because of the versatility of semiconductor materials.

SEMICONDUCTORS

Energy Bands

The properties of any solid material, including semiconductors, depend upon the nature of the constituent atoms and upon the way in which the atoms are grouped together. That is, the properties are a function of both the atomic structure of the atoms and the crystal structure of the solid. Experiments have shown that an atom consists of a positively charged nucleus surrounded by electrons located in discrete orbits. Actually electrons can exist in stable orbits near the nucleus only for certain discrete values of energy called *energy levels* of the atom.

The allowed energies of electrons in an atom are depicted by horizontal lines on an *energy-level diagram*, Fig. 4-1a. The curved lines in the diagram represent the potential energy of an electron near the nucleus. As a consequence of the *Pauli exclusion principle*, only a certain maximum number of electrons can occupy a given energy level and the result is that in any atom, electrons fill up the lowest possible levels first.

When atoms come close together to form a solid crystal, electrons in the upper levels of adjacent atoms interact to bind the atoms together. Because of the

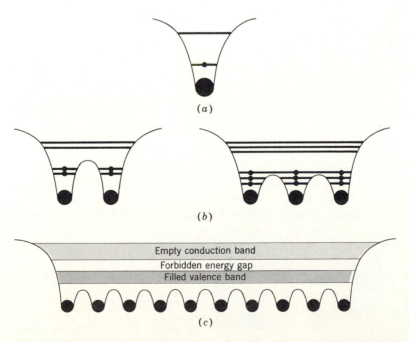

(a)

(b)

Empty conduction band
Forbidden energy gap
Filled valence band

(c)

Figure 4-1 Energy-level diagram for (a) isolated atom, (b) two and three atoms close together, and (c) solid crystal. In the crystal, energy levels are broadened into bands.

strong interaction between these outer, or *valence*, electrons, the upper energy levels are drastically altered. This can be illustrated by an energy-level diagram for the entire crystal. Consider first two isolated atoms, each with an energy-level diagram pertaining to the outer electrons as in Fig. 4-1a. When these are brought close together, Fig. 4-1b, the valence electrons in both atoms are attracted by both nuclei. The result is that the energy required to remove an electron from one nucleus and place it on the other is reduced. This means that an outer electron is equally likely to be located near either nucleus. The appropriate energy-level diagram for the combination of two atoms has two energy levels near each atom core. The higher, unoccupied, levels are similarly split, indicating that these levels, too, can each contain two electrons. When three atoms are brought together, Fig. 4-1b, the outer electrons of all three atoms can be associated with any of the three nuclei. Consequently, three energy levels are available.

Even the tiniest crystal contains many hundreds of millions of atoms, so that very many energy levels are associated with each nucleus and the energy-level diagram appropriate for the entire crystal is a band of levels. The lowest energy band, called the *valence band* (Fig. 4-1c), is completely filled with electrons for there is one electron for each of the available energy levels. Conversely, the upper energy band is empty of electrons because it corresponds to the unoccupied higher levels in the isolated atom. It is called the *conduction band*, for reasons explained below. The energy region between the valence band and the conduction band is called the *forbidden energy gap* since no electrons with such energies exist in the crystal. The forbidden energy gap corresponds to the energy region between energy levels in the isolated atom, as can be seen by comparing the energy-level diagrams in Fig. 4-1.

This picture of the electronic energy levels in a crystal is known as the *energy-band model* of a crystal. It is very useful in determining the electrical properties of any solid, since it shows how electrons can move in the crystal. While the general features of the band model for any solid are as described, many important details depend upon the specific atomic and crystal structure. In particular, the differences between metals, semiconductors, and insulators are reflected in their energy-band models.

The atomic and crystal structures of metals are such that the valence and conduction bands overlap, as indicated in the conventional energy-band model for a metal, Fig. 4-2a. Since there is no forbidden energy gap in a metal crystal, any of the many valence electrons are free to roam throughout the solid and to

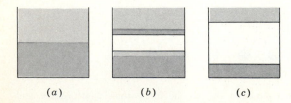

(a)   (b)   (c)

Figure 4-2 Energy-band models for (a) metal, (b) semiconductor, and (c) insulator.

move in response to an electric field. Therefore, metals are excellent electric conductors.

An insulating crystal has a wide forbidden energy gap, Fig. 4-2c. The valence band is completely filled with electrons and the conduction band is completely empty. Obviously the upper band cannot contribute to electric conductivity since no electrons are present to act as carriers. It may seem paradoxical at first, but electrons in the completely filled valence band also cannot conduct electricity. When an electron moves in response to an electric field, it must move slightly faster than before. Consequently, it has greater energy and must find an empty level at a slightly higher energy. Every nearby level is filled, however, so that it is impossible for any electron in the filled valence band to be accelerated by the electric field and the crystal is therefore an insulator.

The energy-band model of a semiconductor, Fig. 4-2b, is similar to that of an insulator except that the forbidden energy gap is comparatively narrow. A few electrons can be excited from the valence band to the conduction band across the forbidden energy gap by virtue of the thermal energy of the crystal at room temperature. Electrons promoted to the conduction band can conduct electricity. The corresponding electron vacancies in the valence band make it possible for electrons in this band to contribute to conductivity as well. Since the number of carriers is much fewer than in the case of a metal, semiconductors are poorer conductors than metals but better than insulators.

The width of the forbidden energy gap of semiconductors is of the order of 1 electronvolt, as shown in Table 4-1 for several typical semiconductor crystals. The electronvolt, abbreviated eV, is equal to the kinetic energy gained by an electron in traversing a potential difference of 1 V. It is a convenient energy unit in semiconductor studies. In general, materials with wider forbidden energy gaps are desirable for semiconductor devices. The number of electrons promoted to the conduction band at high temperatures is smaller, and the change in device characteristics with temperature is less severe when the forbidden energy gap is wide. For this reason, silicon crystals are more widely used than germanium crystals.

Table 4-1 Forbidden energy gaps of typical semiconductors

Name	Chemical symbol	Forbidden energy gap, eV
Indium antimonide	InSb	0.18
Germanium	Ge	0.72
Silicon	Si	1.1
Gallium arsenide	GaAs	1.34
Cadmium sulfide	CdS	2.45
Silicon carbide	SiC	2.8
Zinc oxide	ZnO	3.3

Electrons and Holes

According to the preceding discussion, the net current resulting from electrons in a filled valence band is zero. Electrons in the valence band of a semiconductor at room temperature can conduct current, however, because of the few vacant levels left behind by electrons excited to the conduction band. The vacant levels in the valence band are called *holes*. It turns out that holes in the valence band can be treated as positively charged carriers fully analogous to the negatively charged electrons in the conduction band.

The energy-band model, Fig. 4-2b, refers to a perfect crystal structure which contains no chemical impurities and in which no atoms are displaced from their proper sites. The properties of the solid are therefore characteristic of an ideal structure, and the crystal is called an *intrinsic semiconductor*. Although it is not possible to achieve perfect structures in real crystals, this ideal may be approached, and intrinsic behavior is observed experimentally.

Extrinsic Semiconductors

The electrical properties of a semiconductor are drastically altered when foreign, or *impurity*, atoms are incorporated into the crystal. Since the properties now depend strongly upon the impurity content, the solid is called an *extrinsic* semiconductor. Consider, for example, a single crystal of the important semiconductor material silicon. Each silicon atom has four valence electrons which are part of the filled valence band of a silicon crystal. Suppose now a pentavalent atom, such as phosphorus, arsenic, or antimony, substitutes for a silicon atom in the crystal. Four of the impurity atom's electrons play the same role as the four valence electrons of the replaced silicon atom and become part of the valence band. The fifth valence electron is easily detached by thermal energy and moves freely in the conduction band.

Phosphorus, arsenic, or antimony impurity atoms in silicon donate electrons to the conduction band and are called *donor* impurities. One electron is present in the conduction band for each donor atom in the crystal; note that there is not an equivalent number of holes in the valence band. The crystal conducts electricity mainly by virtue of electrons in the conduction band, and such a crystal is called an *n-type* semiconductor because of the negative charge on the current carriers.

By comparison, a trivalent atom like boron, aluminum, gallium, or indium substituted for a silicon atom produces a hole in the valence band. Since only three electrons are available, an electron from an adjacent silicon atom transfers to the impurity atom. The foreign atom is said to have accepted a valence electron, and such impurities are termed *acceptors*. Acceptors in a crystal create holes in the valence band and produce a *p-type* semiconductor because of the effective positive charge on each hole.

A crystal containing both donor and acceptor impurities is either *n*-type or *p*-type, depending upon which impurity concentration is greater, because electrons from donor atoms fill up all available acceptor levels. Intrinsic crystals can be produced by including equal concentrations of donor and acceptor impurities, and such crystals are said to be *compensated*.

Note that a few holes are present in the valence band of an *n*-type crystal since some electrons are thermally excited across the forbidden energy gap. These holes are called *minority* carriers and the electrons are called *majority* carriers because of their relative concentrations. Conversely, holes are majority carriers in a *p*-type semiconductor, while electrons in the conduction band are minority carriers. Intentionally introducing impurity atoms in semiconductors to obtain a desired concentration of majority carriers is called *doping*.

SEMICONDUCTOR DIODES

The *pn* Junction

The junction between a *p*-type region and an *n*-type region in the same semiconductor single crystal, a basic structure in many semiconductor devices, is called a *pn junction*. Several different techniques for producing *pn* junctions are described in a later section of this chapter. All result in a transition from acceptor impurities to donor impurities at a given place in the crystal. Electrons in the *n* region tend to diffuse into the *p* region at the junction. At equilibrium, this is compensated by an equal flow of electrons in the reverse direction. The concentration of electrons is much larger in the *n* material, however, and the electron current from the *n* region would be greater except for a potential rise that develops at the junction that reduces the electron current from the *n* side. An identical argument applies to the hole currents diffusing across the junction. The magnitude and polarity of the internal potential rise at the junction make the two currents equal, and the built-in potential rise leads to excellent rectification in the *pn* junction, as already noted in Chap. 3.

The potential rise V_0 shown in the energy-band model for a *pn* junction, Fig. 4-3, results from donor electrons at the junction transferring to nearby acceptor atoms such that

$$N_d x_1 = N_a x_2 \tag{4-1}$$

where x_1 is the width of the junction in the *n* region, x_2 is the extent of the junction in the *p* region, N_d is the donor concentration, and N_a is the concentration of acceptors. By symmetry, the electric field between such a double layer of charges is uniform and can be shown to be proportional to the charge densities. The electric field is just $V_0/(x_1 + x_2)$, so that,

$$\frac{V_0}{x_1 + x_2} = \frac{1}{K}(N_a x_1 + N_d x_2) \tag{4-2}$$

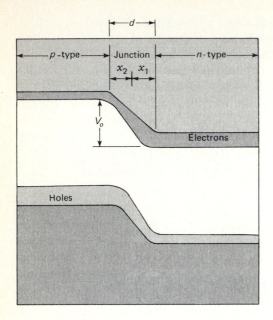

Figure 4-3 Energy-band model of *pn* junction.

where K is the constant of proportionality. Suppose, for simplicity, that the donor and acceptor concentrations are equal, $N_a = N_d = N$, then Eq. (4-2) becomes

$$\frac{V_0}{x_1 + x_2} = \frac{N}{K}(x_1 + x_2) \tag{4-3}$$

and the total width of the junction, $d = x_1 + x_2$, is given by

$$d^2 = K\frac{V_0}{N} \tag{4-4}$$

According to Eq. (4-4), the junction is narrow for large impurity concentrations. This principle is used to set the breakdown voltage of zener diodes at the desired potential. The avalanche breakdown process discussed in Chap. 3 depends upon the electric field in the junction. Therefore, the voltage at which breakdown occurs is small in narrow junctions and large in wide junctions.

When the *pn* junction is in equilibrium the current I_1 resulting from electrons diffusing from the *n* side is equal to the current I_2 which arises from electrons leaving the *p* side, Fig. 4-4*a*. Suppose now an external potential is applied to the junction so as to increase the internal barrier, as in Fig. 4-4*b*. The number of electrons diffusing across the junction from the *n* region is much reduced since very few electrons have sufficient energy to surmount the larger potential barrier. On the other hand, the number moving from the *p* to the *n* side is not affected because these electrons encounter no barrier. Thus a net current exists, but it is

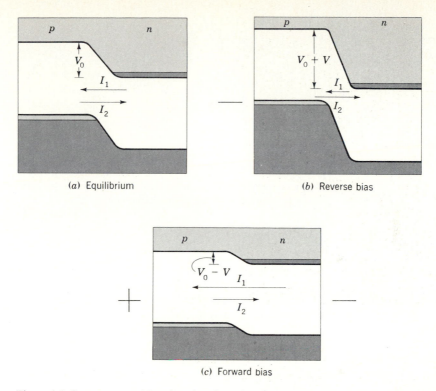

(a) Equilibrium (b) Reverse bias

(c) Forward bias

Figure 4-4 Current across (a) pn junction depends upon height of potential rise at junction, which is (b) increased under reverse bias and (c) decreased by forward bias.

limited by the small number of electrons in the *p* region. If the polarity of the external potential is reversed, Fig. 4-4c, the internal barrier is reduced and I_1 is large because the number of electrons in the *n* region is so great. Again, the electron current I_2 from *p*-type to *n*-type remains unaffected. The net current in this case is large and corresponds to the forward direction.

In determining the current-voltage characteristic of the junction diode, the same considerations apply to the current carried by the holes as to that carried by electrons, and the total current is the sum of the two. Focusing attention on the electrons first, the electron current from the *n* region to the *p* region is controlled by the potential rise at the junction in accordance with Boltzmann's law,

$$I_1 = C_n e^{-e(V_0 - V)/kT} \tag{4-5}$$

where k is Boltzmann's constant, T is the absolute temperature, and C_n is a constant determined by the junction area and properties of the semiconductor. Note that in equilibrium when the applied voltage V is equal to zero, then

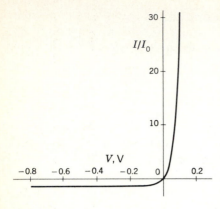

Figure 4-5 Current-voltage characteristic of *pn* junction according to rectifier equation.

$I_1 = I_2$, so that the net electron current can be written as

$$I_n = I_1 - I_2 = C_n e^{-eV_0/kT}(e^{eV/kT} - 1) \tag{4-6}$$

A similar expression applies to the hole currents, which means that the total junction current is just

$$I = I_0(e^{eV/kT} - 1) \tag{4-7}$$

where I_0 is called the *saturation current,*

$$I_0 = Ce^{-eV_0/kT} \tag{4-8}$$

Equation (4-7) is known as the *rectifier equation.*

The polarity of the applied potential is such that the *p* region is positive for *forward bias* because electrons from the *n* region are attracted to the *p* region and vice versa. According to Eq. (4-7) the current increases exponentially in the forward direction. In contrast, the *reverse bias* current is essentially equal to I_0, independent of reverse potentials greater than a few tenths of a volt. A plot of the rectifier equation for small values of applied voltage is shown in Fig. 4-5. Experimental current-voltage characteristics of practical junction diodes, Fig. 3-2, are in good agreement with the rectifier equation.

The double layer of electric charge resulting from transfer of donor electrons onto acceptor atoms at the junction is, in effect, a parallel-plate capacitor. The junction capacitance is reduced by an applied reverse potential because this increases the width of the junction, according to Eq. (4-4), and hence the effective separation between the charge layers. This action makes the *pn* junction an effective electrically variable capacitor.

If, say, the concentration of acceptors on the *p*-type side of the junction is much greater than the concentration of donors in the *n*-type region, $N_a \gg N_d$, then, according to Eq. (4-1), $x_1 \gg x_2$ and $d \cong x_1$. That is, the junction extends predominantly into the *n* region. Obviously, the reverse condition is also possible if the relative concentrations are reversed. This incursion of the junction width into one side of the *pn* junction forms the basis for the field-effect transistor discussed in subsequent sections of this chapter.

The Tunnel Diode

A useful effect occurs in *pn* junctions in which the impurity concentrations on both the *n* and *p* sides are very great. The energy-band model of such a junction is shown in Fig. 4-6a for the case of zero applied voltage.

Very large impurity concentrations mean that the junction is very narrow, according to Eq. (4-4). Widths of the order of 10 to 100 atomic diameters are easily achieved in germanium diodes and similar devices. In this situation it is possible for an electron in the conduction band on the *n* side to jump to the

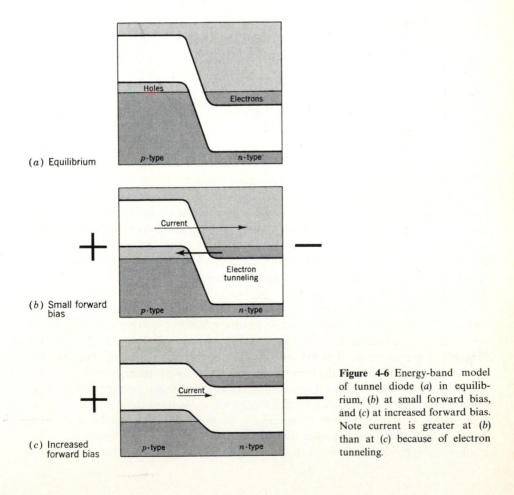

(*a*) Equilibrium

(*b*) Small forward bias

(*c*) Increased forward bias

Figure 4-6 Energy-band model of tunnel diode (*a*) in equilibrium, (*b*) at small forward bias, and (*c*) at increased forward bias. Note current is greater at (*b*) than at (*c*) because of electron tunneling.

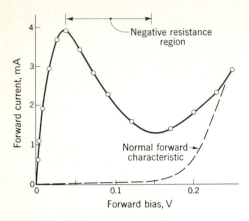

Figure 4-7 Experimental current-voltage characteristic of tunnel diode. Negative resistance region exists between peak and valley currents.

valence band on the *p* side by a process called electron *tunneling*. The tunneling transition takes place with no change in the energy of the electron. The ability of an electron to tunnel through the potential barrier of the junction is a result of the wavelike nature of the electron, and the probability that an electron will penetrate the potential barrier can be calculated from the principles of quantum mechanics. The tunneling current depends upon the junction width, upon the number of electrons capable of tunneling, and upon the number of empty energy levels into which they can transfer. In equilibrium, Fig. 4-6a, the tunneling currents across the junction in the two directions are equal and the net current is zero.

At a small forward-bias potential, electrons tunnel from the conduction band on the *n* side to empty levels in the valence band on the *p* side and a forward current results. The current increases with voltage until the electrons on the *n* side are in line with the holes on the *p* side, Fig. 4-6b. As the bias is increased beyond this point, electrons in the conduction band are raised above the valence-band states and the tunnel current is reduced. In this range of forward bias each increment in voltage causes a decrease in current. Finally, Fig. 4-6c, the electrons and holes are completely out of line and the junction current corresponds to the normal forward *pn*-junction current.

The current-voltage characteristic of a *tunnel diode* (also referred to as an *Esaki diode* after its discoverer), Fig. 4-7, shows how the current rises to a maximum, corresponding to the condition of Fig. 4-6b, and then decreases as the forward bias is increased. The interval between the peak and valley of the characteristic curve exhibits a negative resistance effect since each voltage increase reduces the current. The usefulness of this effect stems from the fact that conventional electronic circuit components have positive resistance and therefore dissipate power. If a tunnel diode is placed in a resonant circuit (Chap. 5) so that the net resistance vanishes, no power loss results. The circuit therefore oscillates at its resonant frequency, as analyzed in Chap. 8.

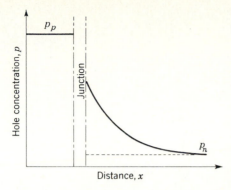

Figure 4-8 Hole concentration near *pn* junction under forward bias.

Minority-Carrier Injection

The internal potential barrier in a *pn* junction is reduced when the junction is biased in the forward direction. The forward current results from holes diffusing across the junction from the *p*-type side and electrons diffusing across from the *n*-type side. The result is that holes are injected into the *n* region and electrons are injected into the *p* region, where in each case they are minority carriers. This *minority-carrier injection* at a *pn* junction is the basis for bipolar transistors, as described in a later section. It is usually desirable that the forward current be carried predominantly by either holes or electrons in order to enhance the injection effect.

If the *n* region is lightly doped and the *p* region is heavily doped, the forward current is carried mostly by holes and, consequently, a large excess hole concentration is injected into the *n* region. If the doping ratio is interchanged, the reverse situation is true and electrons are injected into the *p* region. Injected holes diffuse away from the junction because of their concentration gradient. As they diffuse, they combine with the majority carriers, electrons, so that far from the junction the hole concentration is characteristic of the *n*-type semiconductor. The concentration of holes in the various regions is indicated schematically in Fig. 4-8.

Light-Emitting Diodes

In certain *pn* junctions, most notably combinations of gallium arsenide and gallium phosphide, injected minority carriers recombine quickly in the junction region. Each recombination is accompanied by the emission of light energy of a wavelength corresponding to the width of the forbidden energy gap, as illustrated in the energy-band model, Fig. 4-9. Most often, such a *light-emitting diode*, or *LED*, emits red light, although green and yellow LEDs are also common. LED displays, discussed in Chap. 9, are characterized by an exceptionally long operating lifetime.

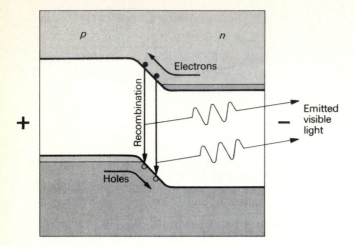

Figure 4-9 Energy-band model of light-emitting diode, or LED.

BIPOLAR TRANSISTORS

Collector Characteristics

A transistor consists of two physically parallel *pn* junctions juxtaposed in the same single crystal. Two distinct types are possible, the *pnp* transistor and the *npn* transistor, depending upon the conductivity type of the common region. The operation of the two is conceptually identical except for the interchange of minority and majority carrier types and the polarity of the bias potentials, so that it suffices to discuss the *pnp* transistor. Since current in *pnp* and *npn* transistors is carried by both minority and majority carriers, these devices are referred to as *bipolar* transistors.

The energy-band model for a *pnp* structure in the absence of applied bias voltages, Fig. 4-10*a*, is simply that of two *pn* junctions placed back to back. In operation, one junction, called the *emitter*, is biased in the forward direction and the other, the *collector*, is biased in the reverse direction, as in Fig. 4-10*b*. Holes injected into the *n*-type *base* region at the emitter junction diffuse across to the collector junction where they are collected by the electric field at the junction.

Variations in the emitter-base bias voltage change the injected current correspondingly, and this signal is observed at the collector junction. The forward-biased emitter represents a small resistance and the reverse-biased collector a large resistance. Since nearly the same current is in both, a large power gain results. For maximum amplification, it is desirable that the collector current be as large a fraction of the emitter current as possible. The current through the emitter junction should be carried primarily by holes, since electrons injected from the base to the emitter cannot influence the collector current. This can be accom-

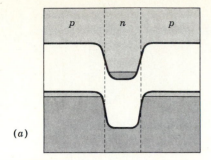

(a)

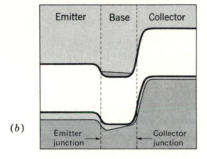

(b)

Figure 4-10 (a) Energy-band diagram of *pnp* transistor in equilibrium, and (b) under operating bias.

plished if the *n*-type base region is lightly doped. Secondly, the base region must be thin so that few holes are lost by recombination before reaching the collector junction. It is convenient to dope the collector region lightly in order to increase the junction width. This reduces the capacitance of the collector junction and also increases the reverse breakdown voltage.

A useful figure of merit for a bipolar transistor is the *current-gain factor* α, which is the ratio of the change in collector current to the change in emitter current for constant collector voltage. Carrier currents in the various regions of a transistor are given in Fig. 4-11, using the emitter current I_e as a starting point. The current I_e in the emitter junction results in a collector current αI_e. The difference between the emitter and collector currents $(1 - \alpha)I_e$ appears at the external connection to the base. Obviously, when the current gain is unity, all the emitter current appears in the collector circuit and the base current is zero.

Actually, α is very nearly unity for most devices so that the base-collector current gain β is a more sensitive measure of transistor quality. The relation between the base-collector current gain and α is, from Fig. 4-11,

$$\beta = \frac{I_c}{I_b} = \frac{\alpha}{1 - \alpha} \tag{4-9}$$

The base-collector current gain is large when α approaches unity. Typical values of β range from 20 to 10^3 in practical transistors.

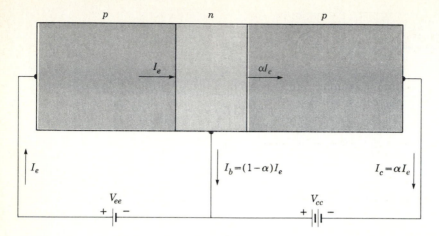

Figure 4-11 Currents in *pnp* transistor expressed in terms of emitter current and current-gain factor.

Minority carriers injected into the base move to the collector by diffusion, and the finite time taken by the carriers to cross the base limits the high-frequency usefulness of transistors. By ingenious fabrication techniques it is possible to achieve very narrow base widths ($\sim 5 \times 10^{-7}$m) and obtain useful amplification at frequencies of 10^9 Hz. It is also necessary to reduce the area of the junctions in high-frequency transistors in order to minimize adverse effects of the junction capacitances.

A convenient way to represent the current-voltage characteristics of a transistor is the collector characteristic, Fig. 4-12, for different values of emitter current. When $I_e = 0$ the characteristic is simply the reverse saturation curve of the collector junction. Emitter current translates the curve along the current axis. These characteristics follow directly from the preceding discussion and pertain to the *grounded-base* circuit configuration already considered in Fig. 4-11. In the grounded-base connection the base terminal is common to both input and output circuits. As is considered in greater detail in Chap. 6, the *grounded-emitter* configuration in which the emitter electrode is common to both input and output circuits exhibits more favorable properties.

The quality of practical transistors is so satisfactory that the grounded-base current gain α is very nearly equal to unity and the reverse leakage current at the collector junction is negligible for all practical purposes. Consequently, the grounded-base collector characteristics are very flat, straight, and uniformly spaced. As a matter of fact, so little new information is presented in such curves that they are almost never measured. The grounded-emitter collector characteristics, Fig. 4-13, are much more informative. In particular, a value for the grounded-emitter current gain β may be estimated directly from the curves by noting the collector current for a given base current.

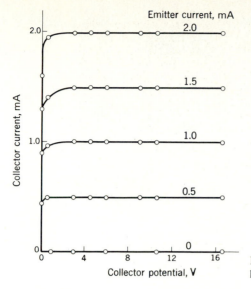

Figure 4-12 Experimental grounded-base collector characteristics of *pnp* transistor.

The operation of a bipolar transistor results from the properties of the emitter junction under forward bias and the collector junction under reverse bias. The properties of both the emitter junction and the collector junction depend sensitively upon temperature as shown by Eqs. (4-7) and (4-8). Comparing collector characteristics at an elevated temperature, Fig. 4-14, with those for the same transistor at room temperature, Fig. 4-13, shows that the currents are increased considerably.

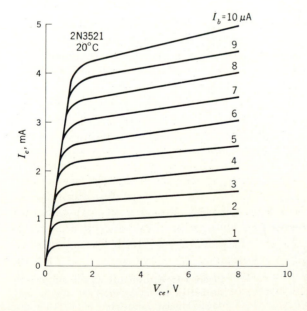

Figure 4-13 Grounded-emitter collector characteristics for type 2N3521 *npn* silicon transistor.

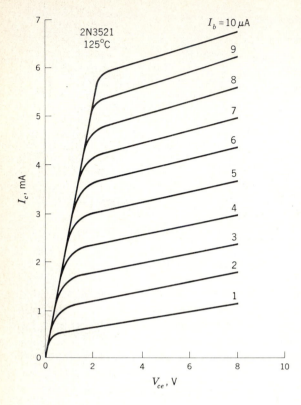

Figure 4-14 Grounded-emitter collector characteristics for type 2N3521 transistor at a temperature of 125°C.

The conventional circuit symbols for *pnp* and *npn* transistors, Fig. 4-15, are derived from the structure of the original invention of the transistor in which two sharp metal probes were placed close together on a small single crystal germanium base. Such *point-contact* transistors have been entirely supplanted by bipolar junction types.

A very important mode of transistor operation is as an electronic switch in which the collector current is either large or equal to zero, depending upon the presence or absence of base current. According to the collector characteristics in Fig. 4-13, for example, if the base current is equal to zero the collector current is very small and the transistor is said to be *cut off*. By contrast, a large base current

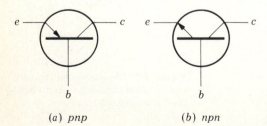

Figure 4-15 Arrowheads indicate direction of current in circuit symbols for (*a*) *pnp* and (*b*) *npn* bipolar transistors.

causes the transistor to conduct fully, and the collector current is said to be *saturated*. The switching mode is employed extensively in digital computer circuits.

Silicon Controlled Rectifier

It is now pertinent to describe in greater detail the silicon controlled rectifier originally discussed in Chap. 3. The useful performance of this device is obtained by adding a third *pn* junction to the transistor structure. The resulting *pnpn* four-layer device, Fig. 4-16a, is also called a *four-layer diode* when connections are made only to the two outer layers. A positive potential applied to the *p*-type terminal puts the center *pn* junction under reverse bias while the two outer junctions are forward-biased. The device may be looked upon as the back-to-back combination of a *pnp* transistor together with an *npn* transistor, Fig. 4-16b, the two transistors having a common collector junction.

Note that I_1 is the emitter current in the *pnp* transistor and that $\alpha_1 I_1$ is the collector current. Similarly, I_2 is the emitter current of the *npn* unit and $\alpha_2 I_2$ is the collector current. Using Kirchhoff's current rule at the collector junction.

$$I_2 = \alpha_1 I_1 + \alpha_2 I_2 \qquad (4\text{-}10)$$

Considering the overall current input to the device

$$I_2 = I_g + I_1 \qquad (4\text{-}11)$$

where I_g is the current into the gate terminal. Substituting for I_1 in Eq. (4-10) and solving the result for I_2 yields

$$I_2 = \frac{-\alpha_1}{1 - (\alpha_1 + \alpha_2)} I_g \qquad (4\text{-}12)$$

According to Eq. (4-12), if the sum of the current gains $\alpha_1 + \alpha_2$ is near unity, the current I_2 can be very large even though the gate current is small. In fact, if

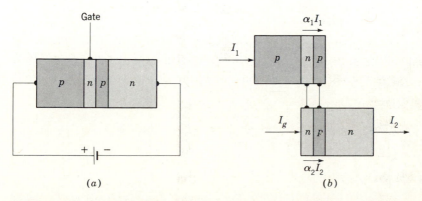

(a)   (b)

Figure 4-16 (a) Sketch of SCR and (b) interpretation in terms of *pnp* transistor coupled to *npn* transistor.

$\alpha_1 + \alpha_2 = 1$, the current is large and limited only by the ohmic resistance of the semiconductor, even if there is no gate current.

The current-voltage characteristics of a SCR, Fig. 3-19, are interpreted as follows. When the gate current I_g is zero, the *pnp* transistor in Fig. 4-16*b* has no base-emitter bias and consequently does not conduct. At a critical applied voltage, avalanche breakdown at the collector junction increases the current and this biases both transistors into conduction. The conducting state is attained at lower values of applied voltage by increasing I_g to bias the *pnp* transistor into conduction. This in turn biases the *npn* transistor into conduction. Once triggered into the conducting state, the gate current may be reduced to zero, since the device current itself maintains bias current in both transistors.

A second gate connection made to the central *p* region may be used to turn the device off by biasing the *npn* transistor off. This double-gate configuration is known as a *silicon controlled switch*, or *SCS*. A further modification of Fig. 4-16*a* has an *n* and a *p* region in parallel at each end. This results in a current-voltage characteristic which is symmetrical about the origin in Fig. 3-19. That is, the device can be switched into conduction for either polarity of applied voltage, since it is basically two reversed SCRs in parallel. Such a *triac* is very useful in controlling ac currents without rectification.

The SCR current-voltage characteristic, Fig. 3-19, indicates negative resistance properties since a voltage decrease results in a current increase when a transition from the off state to the on state occurs. This negative resistance effect is inherently different from that of the tunnel diode, Fig. 4-7. The difference between the two may be identified by noting that a line through the negative resistance region parallel to the voltage axis in Fig. 4-7 intercepts the characteristic curve at another point. By contrast, the line must be parallel to the current axis if it is to intercept the curve again in Fig. 3-19. The shape of the tunnel-diode curve is often called *N-type* negative resistance characteristic and the SCR curve called an *S-type* negative resistance characteristic because of the relative shapes of the two types.

The Unijunction Transistor

An S-type negative resistance characteristic somewhat similar to that of the SCR is produced by the *unijunction* transistor. The circuit symbol for a unijunction transistor, Fig. 4-17, shows schematically that the device consists of a semi-

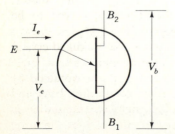

Figure 4-17 Circuit symbol of unijunction transistor.

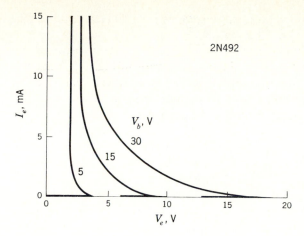

Figure 4-18 Current-voltage characteristics of type 2N492 unijunction transistor.

conductor bar with two ohmic contacts and a single, small-area emitter *pn* junction positioned between them. If no emitter current is present, a fraction of the voltage V_b applied between the base contacts appears at the emitter junction because of the voltage-divider action of the semiconductor bar. Only reverse current is present in the emitter circuit so long as the emitter voltage is less than this fraction and the emitter junction remains under reverse bias. If, however, the junction voltage V_e is increased such that the emitter becomes biased in the forward direction, emitter current increases and carriers are injected into the semiconductor in the region between the emitter and B_1. The resistance of this portion decreases correspondingly and the voltage-divider action reduces the fraction of V_b at the junction. Thus the emitter current increases even if the emitter voltage decreases. This results in a very stable negative resistance characteristic, illustrated in Fig. 4-18. Circuit applications of the unijunction transistor are considered in Chap. 8.

Fabrication of Transistors

Of the several ways to produce a semiconductor single crystal containing two *pn* junctions, four types have been most popular: the grown-junction, alloy-junction, diffused-mesa, and planar transistors, Fig. 4-19. Many modifications of these four types have been devised, but they differ only in details from the ones shown. The grown-junction transistor was historically the first junction type to be fabricated and has now been supplanted. It is prepared during crystal growth from the melt by altering the impurity content of the melt as the solidified crystal is slowly withdrawn. A typical crystal may be 15 cm in diameter, while a single transistor is only about 1 mm square and many transistors can be obtained by cutting up the original crystal. Connections are soldered to the emitter, base, and collector regions. Solder containing a small amount of impurity corresponding to the conductivity type being soldered is used in order to provide good electric

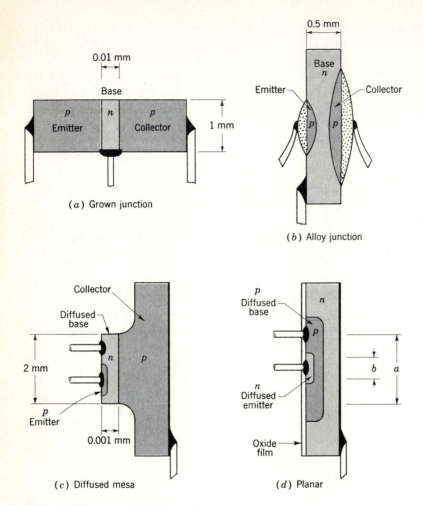

(a) Grown junction

(b) Alloy junction

(c) Diffused mesa

(d) Planar

Figure 4-19 Structure of bipolar transistors.

contacts. Thus, indium is introduced into the solder for the contacts attached to the p regions and antimony is added to the solder for the base. This is particularly important in making the connection to the very narrow base region because the n-type impurity produces a good electric connection to the n-type base and at the same time forms a pn junction with the p regions. This means that the soldered base contact is effectively electrically isolated from the emitter and collector regions, even though it may physically extend beyond the thin base region.

The pnp alloy-junction transistor, now also largely supplanted by other processes, is produced from a thin wafer of n-type single-crystal material by placing indium pellets on opposite surfaces and heating. As the indium melts, it dissolves some of the germanium beneath it. During subsequent cooling the dissolved germanium recrystallizes upon the base crystal and incorporates many indium atoms into its structure. The recrystallized material is therefore p-type, and a transistor structure results. It is convenient to make the collector pellet larger than the emitter pellet, for in this way carriers injected at the emitter junction are collected more efficiently. Thus, the alloy junction is inherently a more satisfactory transistor geometry than the simple grown-junction shape.

As discussed earlier, it is necessary to make the dimensions of the transistor region as small as possible in order to achieve satisfactory high-frequency performance. This is difficult in the case of the alloy-junction transistor, although many ingenious variations of the above process have been developed to accomplish this goal. The diffused-mesa transistor, Fig. 4-19c, however, is much more adaptable to high-frequency transistors because the junction-fabrication process can be precisely controlled. The collector junction is formed by placing a p-type single-crystal wafer in a hot gas of, say, antimony atoms. As the wafer is heated, antimony atoms diffuse into the crystal to a depth of approximately 10^{-3} mm. The crystal is subsequently masked and chemically etched to produce a small elevated region, or mesa, about 2 mm in diameter. The mesa top is then suitably masked so that two regions 0.3 mm in diameter are exposed and appropriate metals are deposited by evaporation in high vacuum. Next, the slab is maintained at an elevated temperature while the two metal deposits diffuse into the n region. This results in the formation of a p-type emitter region by one of the metal deposits while the other one forms a contact to the base region.

The area of the collector junction is defined exactly by the size of the mesa while the emitter junction is set by the area of the vapor-deposited metal. Both processes can be precisely controlled and result in an extremely tiny volume. The width of the base region is determined by the gaseous diffusion process and subsequent diffusion of the deposited metal. Both processes can be adjusted easily by the temperature of the wafer to yield a very thin, yet accurately defined, base region. Therefore, the entire fabrication technique lends itself to precise control of transistor geometry. Very tiny active regions are produced on a wafer which is large enough to handle conveniently.

The mesa transistor, like the grown-junction and alloy-junction transistors, is so constructed that the emitter and collector junctions are exposed at the surface of the semiconductor. The electrical properties of the junction are extremely sensitive to chemical impurities from the surrounding atmosphere because of the high electric fields at the junction. Minute traces of foreign atoms on the surface in the junction region greatly degrade the performance of actual transistors compared with theoretical expectations. For this reason, transistors are often hermetically sealed in carefully cleansed metal enclosures after fabrication.

A more satisfactory solution to the surface-contamination problem is to grow a thin oxide film on the surface of the semiconductor by heating in an oxidizing atmosphere. This technique, applicable primarily to silicon because of the favorable properties of silicon oxide, results in a semiconductor device completely

Figure 4-20 Photomicrograph of planar transistor. (*Raytheon Semiconductor.*)

protected from atmospheric contamination. In addition, the oxide film is a barrier to gaseous diffusion of impurities so the film itself acts as a mask during
fabrication. This technique makes possible the planar transistor, Fig. 4-19*d*, so
called because the entire transistor appears to be a simple plane wafer of silicon.

Construction of a planar *npn* transistor begins with a single-crystal wafer of
n-type silicon. A thin oxide layer is grown on the top surface by heating in an
oxygen atmosphere. The oxide is chemically etched away in a region of size *a*.
The wafer is then exposed to a hot boron gas atmosphere. Boron atoms diffuse
into the exposed silicon, forming a *p*-type base region and the collector junction.
Boron atoms also diffuse laterally under the oxide film so that the base region is
somewhat larger than *a* and the collector junction is protected at the surface by
the oxide layer. Next, the wafer is reoxidized to produce a film covering the entire
area once again. After the oxide is etched away in an area of size *b*, the wafer is
again exposed to a hot impurity gas, this time phosphorus. These *n*-type impurity
atoms diffuse into the silicon, forming the emitter region and the emitter junction.
Lateral diffusion of phosphorus atoms under the oxide film again ensures that the
emitter junction is protected by the oxide layer.

Finally, electric contacts are provided by vapor deposition and alloying as in
the case of the mesa transistor. The result is an *npn* transistor entirely contained
in one plane wafer and completely encased in an impervious oxide film. Geometric definition of the junction positions and the base width is extremely accurate because of the well-controlled gaseous diffusion doping technique. Several
hundred transistors can be fabricated simultaneously from one silicon wafer a few
centimeters or so in diameter. These are subsequently cut into individual units by
sawing up the wafer. A micrograph of a typical planar transistor is shown in Fig.
4-20. Connections to the emitter and base regions are clearly visible.

FIELD-EFFECT TRANSISTORS

Drain Characteristics

Very effective control action is obtained in the *field-effect transistor*, or *FET*, in which the flow of majority carriers is controlled by signal voltages applied to a reverse-biased *pn* junction. Consider the *n*-type semiconductor bar (most often of silicon) in Fig. 4-21 which has an ohmic contact, the *source*, at one end and a similar contact, the *drain*, at the other. Electrons moving from the source to the drain in response to the drain voltage V_{ds} pass through a channel parallel to the junction. This *pn* junction is called the *gate* because the width of the reverse-biased gate junction determines the width of the channel and consequently the magnitude of the current between source and drain. Signal voltages applied to the gate result in corresponding variations in drain current.

In effect the FET is a variable resistor which is controlled by the electric field associated with the *pn* junction, whence the quite descriptive name. Very little power is expended by the applied signal because of the small reverse current in a *pn* junction. Note that the unit described is an *n*-channel FET. Clearly a *p*-channel FET is equally possible. In this case the potentials applied to the drain and gate have the opposite polarity so that the gate junction is again under reverse bias. The circuit symbol for an *n*-channel FET is shown in Fig. 4-22. The corresponding symbol for a *p*-channel unit has the arrow representing the gate junction pointing in the other direction. Note that the direction of the arrow is consistent with the direction of conventional current flow in a junction diode.

The current-voltage characteristic of a FET is derived from analysis of current flow in the channel. As sketched in Fig. 4-21, the width of the channel is narrowest near the drain end of the gate because the reverse bias here is the sum of the gate potential plus the drain potential and the junction is at its widest. The sum of V_{gs} and V_{ds} may be large enough for the junction to completely pinch off the channel. It turns out, in fact, that this is the normal operating condition and the reverse bias required for pinchoff is labelled V_p.

The nonuniform voltage drop along the channel and the nonlinear variation in channel width combine to make the exact analysis complicated. A simplified

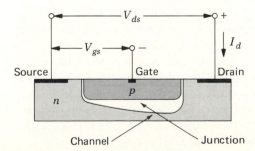

Channel · · · · · · · · Junction   **Figure 4-21** Sketch of *n*-channel FET.

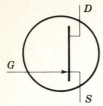

Figure 4-22 Circuit symbol for *n*-channel FET.

study can be carried out by noting that the current in the channel resistance is proportional to V_{ds} in accordance with Ohm's law, and to $V_{gs} - V_p$ because of pinchoff, as discussed above. The drain voltage results in an additional reverse bias varying from zero at the source end to $-V_{ds}$ at the drain end. As an approximation, the added reverse bias is taken to be the simple average, $-V_{ds}/2$. The drain current is then,

$$I_d = K\left(V_{gs} - V_p - \frac{V_{ds}}{2}\right)V_{ds} \qquad (4\text{-}13)$$

where K is a constant.

At pinchoff the sum of the gate voltage and the drain voltage is equal to the

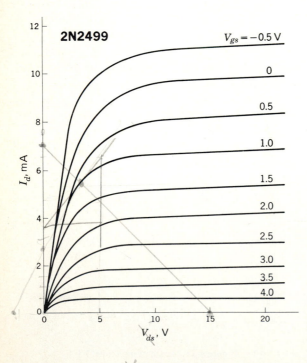

Figure 4-23 Current-voltage characteristics of type 2N2499 *p*-channel FET.

pinchoff voltage, or $V_{gs} + V_{ds} = V_p$. Therefore, Eq. (4-13) becomes

$$I_d = K\left(V_{gs} - V_p - \frac{V_{gs} - V_p}{2}\right)(V_{gs} - V_p) = \frac{K}{2}(V_{gs} - V_p)^2$$

$$= I_{dss}\left(1 - \frac{V_{gs}}{V_p}\right)^2 \tag{4-14}$$

where I_{dss} is a constant that depends primarily upon the geometrical design of the FET. Evidently the drain current does not depend upon the drain voltage above pinchoff. Equations (4-13) and (4-14) are a good representation of the *drain characteristics* of practical FETs below and above pinchoff, respectively, as illustrated in Fig. 4-23, for the case of a *p*-channel device. The maximum source-drain voltage is limited by reverse breakdown at the gate junction.

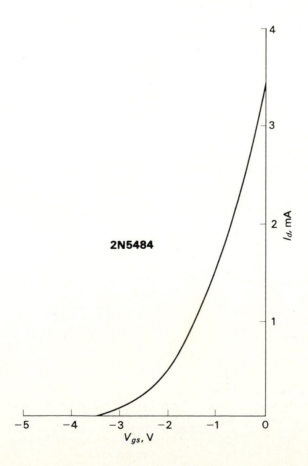

Figure 4-24 Transfer characteristic of type 2N5484 *n*-channel FET.

The Transfer Characteristic

The variations in drain current corresponding to voltage changes at the gate electrode described by Eq. (4-14) are most clearly displayed by a plot of the drain current as a function of gate voltage, which is called the *transfer characteristic* because it relates changes in the output circuit to signals at the input. The typical transfer characteristic in Fig. 4-24 shows the square-law dependence given by Eq. (4-14).

The transfer characteristic is an alternative way of describing the nonlinear electrical properties of the FET. It is complementary to the drain characteristics and shows directly the change in drain current per unit change in gate voltage. This ratio is a measure of the device's control feature, and a large value implies favorable amplification properties. It should be noted that the FET control characteristics are fundamentally different from those of the bipolar transistor in that the input signal is a voltage rather than a current. The consequences of this difference are explored in various ways in succeeding chapters. Actually, both types of control lead to quite effective amplifying and switching actions.

Since current flow in a FET is by majority carriers, in contrast to the case of the bipolar transistor, its electrical properties tend to be less sensitive to temperature. Nevertheless, the change in the transfer characteristic with temperature illustrated in Fig. 4-25 shows that the effect is not negligible. Suitable circuit configurations have been devised to minimize changes in circuit performance with temperature so that both types of transistors are operable over practical temperature ranges.

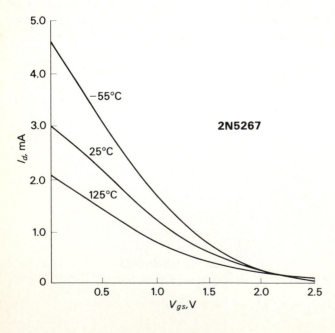

Figure 4-25 Transfer characteristics of type 2N5267 p-channel FET at three different temperatures.

IGFETs and MOSFETs

Quite analogous control over current carriers in a semiconductor is possible without a *pn* junction by using a suitable metal gate electrode insulated from the semiconductor. Such insulated-gate field-effect transistors are known as *IGFETs*. The most successful practical units are made of silicon and employ a thin silicon dioxide layer as the insulator. The descriptive terminology for a metal-oxide-semiconductor field-effect transistor is *MOSFET*.

The MOSFET illustrated in Fig. 4-26*a* has *n*-type source and drain regions imbedded in a *p*-type crystal. The metal gate electrode is deposited on top of a thin oxide layer on the surface. Usually the *p* substrate is connected to the source and a negative potential is applied to the gate.

The negative gate potential repels electrons in the *n*-type channel and a *depletion* layer devoid of carriers forms. This depletion layer is quite analogous to the region between the *n*-type and *p*-type sides of a *pn* junction. Creation of the depletion layer by the gate voltage decreases the conductivity of the *n*-channel and the drain current is reduced. Pinchoff occurs when the depletion layer extends entirely across the *n*-layer and the drain characteristics of the *depletion* MOSFET are quite similar to those of the junction FET.

If the *n*-channel between source and drain is eliminated, Fig. 4-26*b*, successful operation is still obtained when the gate voltage is made positive with respect to the source and the *p*-type substrate. The drain current is zero without gate bias because of the *pn* junctions at the source and drain ends. As the gate potential is increased positively from zero, holes in the *p* substrate immediately under the gate are repelled and the surface layer tends to become *n*-type. At the threshold gate voltage, V_T, an *n*-channel forms and current passes between source and drain. Greater gate voltage increases the drain current further. Since the gate voltage increases the *n*-channel conductivity, this design is called an *enhancement* MOSFET.

Actually, a depletion MOSFET can operate in the enhancement mode as well, since the channel can be widened as well as narrowed by the gate potential. A convenient comparison between the several possible FET types is provided by

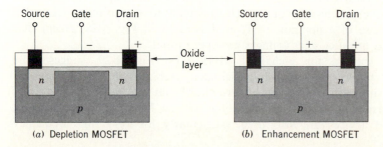

(*a*) Depletion MOSFET   (*b*) Enhancement MOSFET

Figure 4-26 *n*-channel MOSFETs (*a*) depletion and (*b*) enhancement versions.

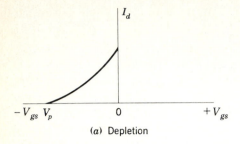

(a) Depletion

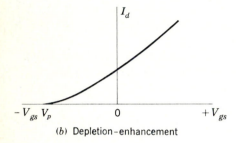

(b) Depletion-enhancement

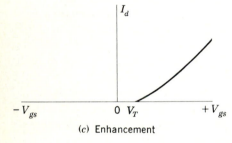

(c) Enhancement

Figure 4-27 Transfer characteristics of (a) depletion, (b) depletion-enhancement, and (c) enhancement FETs.

their transfer characteristics, Fig. 4-27. The junction FET employs the depletion mode, Fig. 4-27a, since the gate *pn* junction cannot have forward bias. The depletion MOSFET can be used with either gate polarity, Fig. 4-27b, while the enhancement MOSFET requires positive gate bias, Fig. 4-27c. Note, however, that except for lateral displacement, all three transfer characteristics are quite similar. The drain characteristics of a practical depletion-enhancement MOSFET are similar to those of a junction FET, as illustrated in Fig. 4-28, except that both positive and negative gate voltages are operational.

The circuit symbol for *n*-channel MOSFETs, Fig. 4-29, suggests the insulated-gate structure. As might be anticipated, *p*-channel MOSFETs are indicated by reversing the direction of the arrowhead that suggests the direction of conventional current in the channel. Although silicon MOSFETs predominate, gallium arsenide devices are also used, particularly for very high frequency applications.

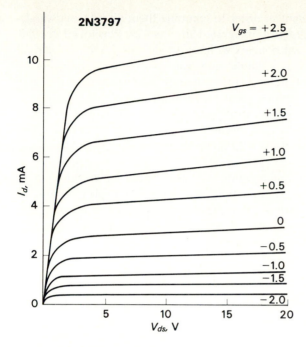

Figure 4-28 Drain characteristics of type 2N3797 n-channel MOSFET.

INTEGRATED CIRCUITS

Principles of Integrated Circuits

The properties of semiconductors, most particularly silicon, make it possible to produce an entire electronic circuit within one single crystal. Such an *integrated circuit* miniaturizes electronic networks and also reduces the number of individual components in complex electronic circuits. This is so because an entire integrated circuit represents, in effect, only one component. The basis for the integrated-circuit development is that many electronic components, such as transistors, diodes, and resistors, can be made of silicon having suitably disposed n-type and p-type regions. It remains only to position the various components in

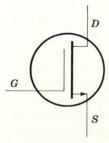

Figure 4-29 Circuit symbol for n-channel MOSFET.

a judicious geometric arrangement in order to combine them all into one single-crystal wafer or *chip*. The first practical integrated circuits were developed in 1958.

The close juxtaposition of components within a single chip leads to electrical coupling between them. It proves possible, however, to isolate components by suitable arrangement of the *p*-type and *n*-type regions. Consider the case of a resistor of *n*-type silicon in a *p*-type crystal, for example. Clearly the resistor is isolated from the crystal by the *pn* junction between the resistor and the chip. This isolation is most effective only at dc because of *pn*-junction capacitance. The stray coupling introduced by all the junction capacitances is taken into account in the design of the circuit.

Actually, the capacitance of a *pn* junction may also serve as a capacitor in an integrated circuit. Inductors of small inductance in the form of a spiral conductor are also possible. Most often, however, the integrated circuit per se is composed of transistors, diodes, and resistors, and the necessary capacitors and inductors are connected as separate, discrete components.

Fabrication Processes

Silicon integrated circuits are fabricated by processes similar to those used in the case of planar transistors. Appropriate patterns in the oxide layer prior to impurity diffusions are produced by first coating the oxide layer with a special photopolymeric material. A photopolymer becomes resistant to etching when exposed to strong light. Thus, when the photopolymer is exposed to light through a suitable mask, the unexposed portions can be subsequently removed. Because of the extremely tiny device dimensions in integrated circuits, dry etching processes based on an electrical discharge in a gas, so-called *plasma etching*, have replaced simple chemical etching used to fabricate planar transistors. The corresponding pattern in the oxide layer is etched away with a different plasma composition, in turn exposing the underlying silicon. After gaseous diffusion of *n*-type of *p*-type impurities, the wafer may be reoxidized and the process repeated to develop an additional array of doped regions.

The photographic process produces integrated circuits of extremely small dimensions. Accurate, large-scale masks are optically demagnified to microscopic dimensions. This also means that many integrated circuits may be fabricated simultaneously from the same silicon wafer simply by constructing a mask having the desired circuit repeated many times. Subsequent to the last impurity-diffusion step, a thin metal layer is deposited (also through a photopolymeric layer) to interconnect the various components and to provide terminals for external connections.

This process is best illustrated by following actual steps in the fabrication of a simple circuit, Fig. 4-30. To begin, a *p*-type single-crystal chip is exposed to silicon vapor containing *n*-type impurity atoms. The result, Fig. 4-30*a*, is an *n*-type *epitaxial* layer. Epitaxial means that the *n*-type layer has the same crystal structure as the *p*-type substrate. The surface is oxidized, masked, and etched, and a *p*-type diffusion isolates several *n* regions, Fig. 4-30*b*. These steps are repeated

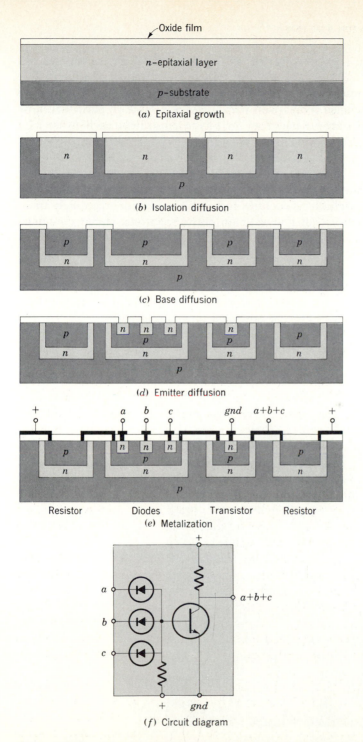

Oxide film

n–epitaxial layer

p–substrate

(*a*) Epitaxial growth

(*b*) Isolation diffusion

(*c*) Base diffusion

(*d*) Emitter diffusion

+   *a*   *b*   *c*   *gnd*   *a+b+c*   +

Resistor   Diodes   Transistor   Resistor

(*e*) Metalization

(*f*) Circuit diagram

Figure 4-30 Processing steps (*a*) through (*e*) result in complete integrated circuit (*f*).

127

in Fig. 4-30c and Fig. 4-30d. Finally, Fig. 4-30e, the metallic interconnections are added.

This procedure results in the circuit diagrammed in Fig. 4-30f, as can be appreciated from a detailed comparison with Fig. 4-30e. Note, in particular, how the resistors and diodes are isolated from the p-type substrate in each instance by a pn junction. The circuit, incidentally, is termed a DTL NAND gate, which is considered in greater detail in Chap. 9.

Considerably more elaborate and versatile fabrication techniques have been developed to complement the multiple diffusion process described above. These permit much more complicated structures to be produced with exceedingly small active regions. The final result is always an entire electronic circuit contained within a single chip.

Practical Circuits

It is possible to recognize the various components in the micrograph of a practical integrated circuit, Fig. 4-31. This so-called *difference-amplifier* circuit is diagrammed in Fig. 4-32a and is analyzed in the next chapter. Note that the metallic interconnections appear lighter in the micrograph and that it is often necessary to

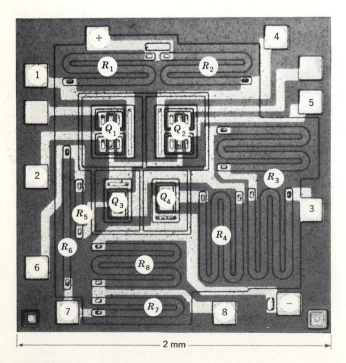

Figure 4-31 Photomicrograph of integrated circuit difference amplifier in Fig. 4-32. (*Motorola, Inc.*)

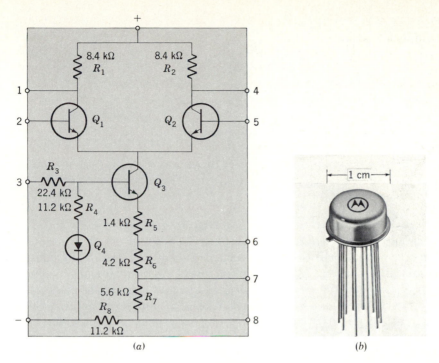

Figure 4-32 (a) Difference-amplifier circuit and (b) practical unit.

zigzag the resistors in order to achieve the desired resistance. Although this is a relatively simple circuit, it is noteworthy that all 12 components are contained in one small case about the size of a single transistor, Fig. 4-32b.

The most popular housing for integrated circuits is the so-called *dual in-line pin*, or *DIP*, structure, Fig. 4-33. Cases with 6, 14, 16, or as many as 40 pins in this configuration are standard, depending upon the complexity of the integrated circuit contained within. Often several independent circuits commonly used together are fabricated on a single chip, as, for example, the type 7404 hex inverter illustrated in Fig. 4-34. This device contains six transistor amplifiers quite independent of each other except for a common voltage-supply connection. The interpretation of the triangular symbols for the amplifiers and the application of such circuits are considered in following chapters.

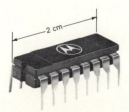

Figure 4-33 Dual in-line pin (DIP) case for integrated circuit. (*Motorola, Inc.*)

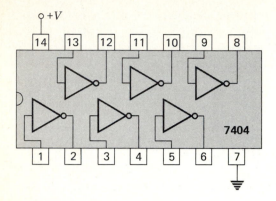

Figure 4-34 Top view of pin connections of type 7404 hex inverter integrated circuit in DIP case.

Much more complicated circuits are available as integrated circuits and these are also discussed in subsequent sections. Each such circuit usually can be employed in several ways and is often associated with other integrated circuits to perform an entire complex task. In certain instances, most notably in digital computers, the combination of several integrated circuits can itself be made as an integrated circuit. Such *large-scale integration,* or *LSI,* further miniaturizes electronic circuits and yields the advantage that signal propagation throughout the circuit is very rapid. LSI circuits containing 10,000 or more transistors are feasible. A complete digital LSI integrated circuit based on so-called I^2L logic

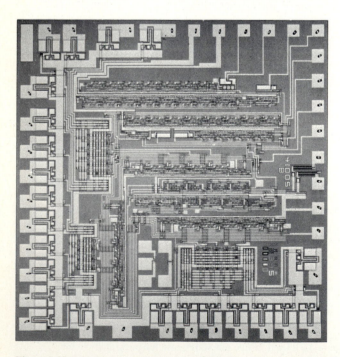

Figure 4-35 LSI integrated-circuit chip. (*University of Utah Research Institute.*)

(Chap. 9) illustrated in Fig. 4-35 is contained on a silicon chip of about the same size as the difference amplifier in Fig. 4-31.

Advances in fabrication technologies, most notably in MOSFET integrated circuits, have led to *very large scale integration, VLSI,* which incorporates 100,000 (and in special applications 10,000,000) devices per chip. In particular, so-called complementary symmetry MOSFET integrated circuits prove to be very advantageous in digital VLSI applications, as discussed in Chaps. 9 and 11.

SUGGESTIONS FOR FURTHER READING

Leonid V. Azaroff and James J. Brophy: "Electronic Processes in Materials," McGraw-Hill Book Company, New York, 1963.
L. Solymar and D. Walsh: "Lectures on the Electrical Properties of Materials, 2/e," Oxford Press, Inc., New York, 1980.
Rodney Bruce Sorkin: "Integrated Electronics," McGraw-Hill Book Company, New York, 1970.
Edward Yang: "Microelectronic Devices," McGraw-Hill Book Company, New York, 1988.

EXERCISES

4-1 Plot the forward characteristics of a germanium and a silicon *pn* junction up to a current of 10 mA. Use the rectifier equation and take $I_0 = 10^{-6}$ A for germanium and $I_0 = 10^{-12}$ A for silicon. How much more forward bias is required in the case of the silicon diode to make the current 10 mA? Repeat for 1 mA.
    Answer: 0.4 V; 0.4 V

4-2 By differentiating the rectifier equation for a junction diode, Eq. (4-7), determine an expression for the junction resistance $R = (dI/dV)^{-1}$. Given that $e/kT = 38$ V^{-1} at room temperature, calculate the rectification ratio at a potential of 1 V.
    Answer: 10^{33}

4-3 Sketch the energy-band model for an *npn* transistor in equilibrium and biased for transistor operation.

4-4* Calculate and plot the drain characteristics of a type 2N2499 FET using Eqs. (4-13) and (4-14). Do this by evaluating I_{dss} and V_p from the $V_{gs} = 0$ curve in Fig. 4-23. Compare with Fig. 4-23.
    Answer: 9 mA; 6.6 V

4-5 Calculate and plot the grounded-base collector characteristics of a silicon transistor at room temperature and for collector potentials up to a maximum of 0.5 V. Assume the current gain is unity and that $I_0 = 1$ μA.

4-6 Repeat Exercise 4-5 for a temperature of 150°C. Assume that I_0 increases according to Eq. (4-8) and take $V_0 = 0.7$ V.

4-7 Explain briefly why a transistor amplifies when connected in the common emitter configuration.

4-8 Determine numerical values for I_{dss} and V_p in Eq. (4-14) for a type 2N3797 MOSFET using drain characteristics in Fig. 4-28.
    Answer: 3 mA, -2.4 V.

4-9 Calculate the transfer characteristic of a type 2N2499 FET using Eq. (4-14). Do this by using values of I_{dss} and V_p from Exercise 4-4. Compare with an experimental curve derived from Fig. 4-23 at a drain potential of 15 V.

4-10* Plot experimental transfer characteristics for all transistors for which current-voltage characteristics are given in Figs. 4-13, 4-23, and 6-41.

4-11 For each bipolar transistor in Exercise 4-10 calculate the change in output current per change in input signal near the middle of the range over which the data is given. Note any trend in the magnitude of this ratio with power handling capability.

Answer: Smaller at high power

4-12 Compare the ratio of the change in output current divided by the change in input voltage for the bipolar transistor of Fig. 4-13 and the FET of Fig. 4-23. Do this by dividing the current-gain ratio of the junction transistor by the emitter junction resistance to convert the input-signal current to an input-signal voltage.

Answer: 41 mho, 2.6×10^{-3} mho

4-13 Explain why the depletion region between the gate and the channel of a FET is wider at the drain end than at the source end.

4-14 Mention two major advantages of integrated circuits.

Answer: Small size and fewer individual devices.

4-15 Describe how components are isolated in an integrated circuit. Is the isolation better at high or low frequencies?

Answer: Low frequencies

AC-CIRCUIT ANALYSIS

The principles of ac circuits treated in Chap. 2 can be used to find the currents in any network. However, solving the differential equation pertaining to each network of practical circuits is cumbersome. More powerful techniques of ac-circuit analysis permit solutions for the circuit currents with much less labor. These techniques are, of course, based on the same differential equation of the circuit. The procedures are only slightly more complicated than for dc-circuit analysis, and both Ohm's law and Kirchhoff's rules are used in modified form. In fact, all the techniques of dc-network analysis treated in Chap. 1 are applicable.

IMPEDANCE

Ohm's Law for Alternating Current

The steady-state solutions to ac-circuit differential equations developed in Chap. 2 invariably contain several terms in sin ωt and cos ωt. These expressions can usually be simplified by introducing appropriate trigonometric relations, but the procedure proves to be laborious for any but the simplest of circuits. It turns out that ac-circuit analysis is greatly facilitated if sinusoidal currents and voltages are represented by means of complex numbers.

The way in which complex numbers can represent sinusoidal signals is derived from the components of the rotating vector generating the waveform as discussed in Chap. 2. According to Fig. 5-1, the horizontal and vertical components are identified with the real and imaginary parts of a complex number such that

$$v = \mathbf{V} = V_p(\cos \omega t + j \sin \omega t) = V_p e^{j\omega t} \tag{5-1}$$

where $j = \sqrt{-1}$. The symbol j, rather than i, is commonly used to signify $\sqrt{-1}$ in electronic-circuit analysis so as to avoid confusion with the conventional symbol for current. In effect, the complex number representation brings in both sin ωt and cos ωt terms simultaneously, as Eq. (5-1) illustrates.

The circuit differential equations may be solved quite directly when sinusoidal signals are represented by complex numbers. Consider, for example, the circuit differential equation for a simple RL series circuit, Eq. (2-18),

$$v = Ri + L \frac{di}{dt} \tag{5-2}$$

Inserting Eq. (5-1) and its equivalent for the current in the circuit, and allowing

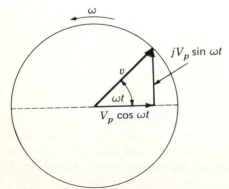

Figure 5-1 Complex number representation of sinusoidal voltage.

for a phase angle between the current and the voltage,

$$V_p e^{j(\omega t + \phi)} = RI_p e^{j\omega t} + j\omega L I_p e^{j\omega t} \tag{5-3}$$

$$= (R + j\omega L)I_p e^{j\omega t} \tag{5-4}$$

$$\mathbf{V} = (R + j\omega L)\mathbf{I} \tag{5-5}$$

The quantity in parentheses is called the complex impedance, **Z**, of the circuit. Thus the circuit differential equation reduces to

$$\mathbf{V} = \mathbf{ZI} \tag{5-6}$$

which is the ac form of Ohm's law. It relates the voltage to the current in terms of the complex impedance of the circuit. Thus the solution of ac networks is reduced simply to determining the complex impedance. The final answer for the voltage is obtained by choosing either the real or the imaginary part of the solution, according to whether the real (cos ωt) or the imaginary (sin ωt) component of the current is taken as the starting point.

Note that in the absence of reactance, Eq. (5-6) reduces to the standard dc form of Ohm's law. This suggests that series and parallel impedances combine in the same way as do series and parallel resistances. Thus the equivalent impedance of a series of individual impedances is

$$\mathbf{Z}_{eq} = \mathbf{Z}_1 + \mathbf{Z}_2 + \mathbf{Z}_3 + \cdots \tag{5-7}$$

Similarly, the equivalent of parallel impedances is

$$\frac{1}{\mathbf{Z}_{eq}} = \frac{1}{\mathbf{Z}_1} + \frac{1}{\mathbf{Z}_2} + \frac{1}{\mathbf{Z}_3} + \cdots \tag{5-8}$$

In applying Eqs. (5-7) and (5-8) due consideration must be given to the complex nature of impedances and the rules pertaining to how complex numbers combine. Several illustrative examples are considered in subsequent sections.

Complex Impedance

Equation (5-5) shows that the real part of the complex impedance is associated with resistance while the imaginary part stands for reactance. Therefore, the impedance angle is given by

$$\phi = \arctan \frac{X_L}{R} \tag{5-9}$$

where X_L is the inductive reactance. The term impedance comes from the fact that the effect of both reactance and resistance is to impede the current in a circuit. According to Eq. (5-6) the unit of impedance is the ohm.

It is useful to recall some of the properties of complex numbers in order to facilitate their use in circuit analysis. Two complex numbers are equal only if their real and imaginary parts are equal. Therefore, if it is true that

$$R_1 + jX_1 = R_2 + jX_2$$

then it must be that

$$R_1 = R_2 \quad \text{and} \quad X_1 = X_2 \tag{5-10}$$

From this it follows that the addition of two complex numbers is accomplished by separately adding the real and imaginary parts. That is,

$$(R_1 + jX_1) + (R_2 + jX_2) = (R_1 + R_2) + j(X_1 + X_2) \tag{5-11}$$

Clearly, the same rule applies to the subtraction of two complex numbers.
Complex numbers are multiplied in conventional fashion,

$$Z_1 Z_2 = Z_1 e^{j\phi_1} Z_2 e^{j\phi_2} = Z_1 Z_2 e^{j(\phi_1 + \phi_2)} \tag{5-12}$$

According to (5-12), the product of two complex numbers is a complex number of magnitude equal to the product of the individual magnitudes and with an angle that is the sum of the individual angles. Division of complex numbers also follows immediately.

$$\frac{Z_1}{Z_2} = \frac{Z_1 e^{j\phi_1}}{Z_2 e^{j\phi_2}} = \frac{Z_1}{Z_2} e^{j(\phi_1 - \phi_2)} \tag{5-13}$$

This relation states that the result of dividing two complex numbers is a complex number whose magnitude is the ratio of the individual magnitudes and whose angle is the difference between the individual angles.

It is often useful to *rationalize* the reciprocal of a complex number. Thus

$$\frac{1}{Z} = \frac{1}{Z(\cos \phi + j \sin \phi)} \tag{5-14}$$

This fraction is rationalized by multiplying numerator and denominator by the *complex conjugate*, obtained by replacing the imaginary part of the complex number by its negative. Therefore, Eq. (5-14) becomes

$$\frac{1}{Z} = \frac{1}{Z(\cos \phi + j \sin \phi)} \frac{\cos \phi - j \sin \phi}{\cos \phi - j \sin \phi} = \frac{1}{Z} \frac{\cos \phi - j \sin \phi}{\cos^2 \phi + \sin^2 \phi}$$

$$= \frac{1}{Z} (\cos \phi - j \sin \phi) \tag{5-15}$$

Evidently the reciprocal of a complex number is simply a complex number with a magnitude given by the reciprocal of the magnitude of the complex number and an angle which is its negative. Using the exponential form

$$\frac{1}{Z e^{j\phi}} = \frac{1}{Z} e^{-j\phi} \tag{5-16}$$

which could have been written immediately.

The equivalent forms of representing impedance,

$$\mathbf{Z} = Ze^{j\phi} = Z(\cos \phi + j \sin \phi) = R + jX \tag{5-17}$$

should be noted. Evidently addition and subtraction are easiest in the component form, $R + jX$, while multiplication and division are easiest in the *polar* form, $Ze^{j\phi}$. Equation (5-17) is used to pass directly from one representation to another as needed. These operations, together with that of rationalization, are useful in computing the equivalent complex impedance of circuits comprising a number of individual impedances, as is illustrated in the next section. Note also that a complex number may be written in terms of its real and imaginary parts as

$$\mathbf{Z} = Ze^{j\phi} = \sqrt{R^2 + X^2}\; e^{j\, \text{arctan}\,(X/R)} \tag{5-18}$$

In particular, the magnitude is equal to the square root of the sum of the squares of the real and imaginary components.

RLC CIRCUITS

Series Resonance

As the first example of ac-circuit analysis using the complex-impedance method, consider the series *RLC* circuit of Fig. 5-2. According to Kirchhoff's law

$$v = Ri + \frac{q}{C} + L\frac{di}{dt} \tag{5-19}$$

The differential equation of the circuit is, after differentiating with respect to t,

$$L\frac{d^2i}{dt^2} + R\frac{di}{dt} + \frac{1}{C}i = \frac{dv}{dt} \tag{5-20}$$

Note that this is a second-order differential equation. It can be solved for the steady-state current i by the procedure used in Chap. 2. The result is

$$i = \frac{V_p}{\sqrt{R^2 + (1/\omega C - \omega L)^2}} \sin (\omega t - \phi) \tag{5-21}$$

where

$$\phi = \text{arctan}\, \frac{\omega L - 1/\omega C}{R} \tag{5-22}$$

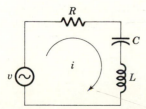

Figure 5-2 Series *RLC* circuit.

The complex-impedance technique for the solution of the same circuit proceeds as follows. The total impedance of the series combination is, from Eq. (5-7), taking the capacitive reactance to be $1/j\omega C$ (see Exercise 5-1),

$$\mathbf{Z} = R + j\omega L + \frac{1}{j\omega C} = R + j\left(\omega L - \frac{1}{\omega C}\right) \tag{5-23}$$

Using Ohm's law to calculate the current,

$$\mathbf{I} = \frac{\mathbf{V}}{\mathbf{Z}} = \frac{\mathbf{V}}{R + j(\omega L - 1/\omega C)} \tag{5-24}$$

Equation (5-24) is the final solution, obtained directly in only two simple steps. The voltage $\mathbf{V}$ usually is given in terms of its rms value, so that the rms value of $\mathbf{I}$ is obtained.

To illustrate more clearly the correspondence between this solution and that given by (5-21) and (5-22), (5-24) is rationalized,

$$\mathbf{I} = \frac{\mathbf{V}}{R + j(\omega L - 1/\omega C)} \frac{R - j(\omega L - 1/\omega C)}{R - j(\omega L - 1/\omega C)}$$

$$= \mathbf{V}' \frac{R - j(\omega L - 1/\omega C)}{R^2 + (\omega L - 1/\omega C)^2}$$

$$= \frac{\mathbf{V}}{\sqrt{R^2 + (\omega L - 1/\omega C)^2}}$$

$$\times \left[\frac{R}{\sqrt{R^2 + (\omega L - 1/\omega C)^2}} - j\frac{\omega L - 1/\omega C}{\sqrt{R^2 + (\omega L - 1/\omega C)^2}}\right] \tag{5-25}$$

Referring to the complex-impedance diagram of the circuit, Fig. 5-3, Eq. (5-25) may be written

$$\mathbf{I} = \frac{\mathbf{V}}{\sqrt{R^2 + (\omega L - 1/\omega C)^2}}(\cos \phi - j \sin \phi)$$

$$= \frac{V_p e^{j\omega t}}{\sqrt{R^2 + (\omega L - 1/\omega C)^2}} e^{-j\phi} = \frac{V_p}{\sqrt{R^2 + (\omega L - 1/\omega C)^2}} e^{j(\omega t - \phi)}$$

$$= \frac{V_p}{\sqrt{R^2 + (\omega L - 1/\omega C)^2}} \sin(\omega t - \phi) \tag{5-26}$$

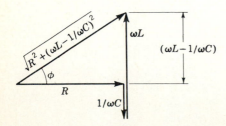

Figure 5-3 Complex impedance of *RLC* circuit.

where
$$\phi = \arctan \frac{\omega L - 1/\omega C}{R} \tag{5-27}$$

and the imaginary part is chosen to conform with $v = v_0 \sin \omega t$. The exact correspondence between the differential-equation solution, Eq. (5-21), and the result of the complex-impedance approach, Eq. (5-26), is evident. In subsequent sections the latter technique is used exclusively because of its simplicity.

Consider the amplitude of the current in the RLC circuit as a function of the frequency of the applied voltage. According to Eq. (5-26), the current is very small at low frequencies ($\omega \to 0$), because the capacitive reactance is great. Similarly, the current is small at high frequencies ($\omega \to \infty$), because the inductive reactance becomes large. Between these two extremes, the current is a maximum when the inductive reactance equals the capacitive reactance,

$$\omega L = \frac{1}{\omega C} \tag{5-28}$$

At this frequency the circuit is said to be in *resonance*, and the current is given by

$$I = \frac{V}{R} \tag{5-29}$$

The circuit appears to be a pure resistance, and the current is in phase with the applied voltage according to Eq. (5-27). The *resonant frequency* is, from Eq. (5-28),

$$\omega_0 = 2\pi f_0 = \frac{1}{\sqrt{LC}} \tag{5-30}$$

To illustrate the properties of a series resonant circuit, consider the current variation in the specific circuit of Fig. 5-4. After substituting the component values on the diagram into Eq. (5-26), it is found that the current in the circuit changes with frequency as illustrated in Fig. 5-5. The current maximum is $I = \frac{10}{100} = 0.1$ A at the resonant frequency,

$$f_0 = \frac{1}{2\pi\sqrt{LC}} = \frac{1}{6.28\sqrt{250 \times 10^{-3} \times 0.1 \times 10^{-6}}} = 1000 \text{ Hz} \tag{5-31}$$

and the decrease on either side of resonance is clearly evident.

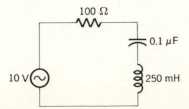

100 Ω

0.1 μF

10 V

250 mH

Figure 5-4

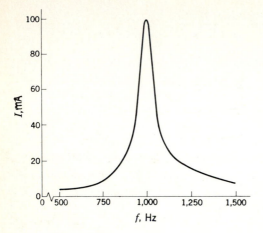

Figure 5-5 Resonance curve of circuit in Fig. 5-4.

The voltage drops around the circuit at resonance further illustrate an important feature of alternating currents. The drop across the resistor is

$$V_R = RI = 100 \times 0.1 = 10 \text{ V} \tag{5-32}$$

The corresponding voltage drop across the capacitor is the current times the capacitive reactance

$$V_C = I \frac{1}{\omega C} = \frac{0.1}{6.28 \times 10^3 \times 10^{-7}} = 158 \text{ V} \tag{5-33}$$

and, similarly, the voltage across the inductor is

$$V_L = I\omega L = 0.1 \times 6.28 \times 10^3 \times 0.25 = 158 \text{ V} \tag{5-34}$$

It is evident that the rms voltage drops around the circuit do not sum to zero. In fact, the voltages across both reactances rise to large values at resonance. Note, however, that the phase angle of the capacitor voltage is $+90°$ with respect to the current, while the phase angle of the voltage across the inductor is $-90°$. This means that the instantaneous voltages across the two reactances cancel each other and that the drop across the resistance equals the applied emf, as confirmed by Eq. (5-32). Kirchhoff's voltage rule is valid when the phases of the currents and voltages in ac circuits are taken into account.

Kirchhoff's voltage rule is, of course, equally valid at frequencies away from resonance. For example, consider the voltage drops around the circuit in Fig. 5-4 at a frequency of 1200 Hz. At this frequency the phase angle is, from Eq. (5-27),

$$\phi = \arctan (2\pi \times 1.2 \times 10^3 \times 250 \times 10^{-3} - 1/2\pi \times 1.2 \times 10^3 \times 0.1 \times 10^{-7})/100$$

$$= \arctan (559)/100 = 79.85° \text{ or } 1.39 \text{ radians}$$

The rms current in the circuit is given by Eq. (5-26),

$$\mathbf{I} = 10(100^2 + 559^2)^{-1/2}e^{-j1.39} = 1.76 \times 10^{-2}e^{-j1.39}$$

and the voltage drop across the resistor is

$$\mathbf{V}_R = IR = 100 \times 1.76 \times 10^{-2}e^{-j1.39}$$

$$= 1.76(\cos 1.39 - j \sin 1.39) = 0.31 - j1.73$$

Similarly, the voltage drops across the inductor and the capacitor are

$$\mathbf{V}_L = j\omega L\mathbf{I} = e^{j\pi/2}(2\pi \times 1.2 \times 10^3 \times 250 \times 10^{-3})1.76 \times 10^{-2}e^{-j1.39}$$

$$= 33.2e^{j0.177} = 33.2(\cos 0.177 + j \sin 0.177)$$

$$= 32.6 + j5.84$$

and

$$\mathbf{V}_C = \mathbf{I}/(1/j\omega C) = e^{-j\pi/2}(2\pi \times 1.2 \times 10^3 \times 0.1 \times 10^{-6})1.76 \times 10^{-2}e^{-j1.39}$$

$$= 23.3e^{-j2.97} = 23.3(\cos 2.97 - j \sin 2.97)$$

$$= -23 - j4.1$$

The sum of the voltage drops around the circuit is

$$\mathbf{V}_R + \mathbf{V}_L + \mathbf{V}_C = (0.31 + 32.6 - 23) + j(-1.73 + 5.84 - 4.1)$$

$$= 9.91 + j0.01 \cong 10$$

which is just the applied voltage, as expected.

Parallel Resonance

Resonance also occurs in a parallel circuit, such as the one shown in Fig. 5-6. The current in the circuit is obtained by first computing the total complex impedance. Since the inductance and capacitance are connected in parallel, their equivalent impedance is found with the aid of Eq. (5-8),

$$\frac{1}{\mathbf{Z}_1} = \frac{1}{1/j\omega C} + \frac{1}{j\omega L} = j\omega C + \frac{1}{j\omega L}$$

$$= \frac{-\omega^2 LC + 1}{j\omega L} \qquad (5\text{-}35)$$

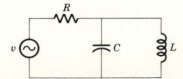

Figure 5-6 Parallel resonant circuit.

Therefore, the impedance of the LC combination is

$$\mathbf{Z}_1 = j\frac{\omega L}{1 - \omega^2 LC} \tag{5-36}$$

According to Eq. (5-36) the impedance is very large, actually innfinite, when

$$\omega_0^2 LC = 1$$

or

$$\omega_0 = \frac{1}{\sqrt{LC}} \tag{5-37}$$

This is the same relation found in the case of series resonance, Eq. (5-30). Note, however, that in series resonance the impedance is a minimum, according to Eq. (5-23), while in parallel resonance Eq. (5-36) shows that the impedance is a maximum at the resonant frequency.

The total impedance of the circuit includes the series combination of R with $\mathbf{Z}_1$

$$\mathbf{Z} = R + j\frac{\omega L}{1 - \omega^2 LC} \tag{5-38}$$

so that the current in the circuit is

$$\mathbf{I} = \frac{\mathbf{V}}{R + j\omega L/(1 - \omega^2 LC)} \tag{5-39}$$

According to Eq. (5-39) the current is zero at resonance, when the impedance becomes infinite. The voltage $\mathbf{V}$ then appears across the LC combination, independent of the value of R. This feature of the parallel resonant circuit is extensively used in practical circuits. The output voltage $\mathbf{V}_o$ is simply the current times the impedance $\mathbf{Z}_1$ or

$$\mathbf{V}_o = \mathbf{I}\mathbf{Z}_1 = \frac{\mathbf{V}}{R + j\omega L/(1 - \omega^2 LC)}\frac{j\omega L}{1 - \omega^2 LC}$$

$$= \mathbf{V}\frac{j\omega L}{R(1 - \omega^2 LC) + j\omega L} \tag{5-40}$$

Rationalizing

$$\mathbf{V}_o = \mathbf{V}\frac{j\omega L}{R(1 - \omega^2 LC) + j\omega L}\frac{R(1 - \omega^2 LC) - j\omega L}{R(1 - \omega^2 LC) - j\omega L}$$

$$= \mathbf{V}\frac{(\omega L)^2 + j\omega LR(1 - \omega^2 LC)}{(\omega L)^2 + R^2(1 - \omega^2 LC)^2}$$

$$= \mathbf{V}\frac{1 + j(R/\omega L)(1 - \omega^2 LC)}{1 + (R/\omega L)^2(1 - \omega^2 LC)^2} \tag{5-41}$$

The magnitude of the output voltage is found by taking the square root of the sum of the squares of the real and imaginary parts. Therefore, the ratio of the

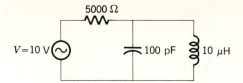

Figure 5-7

output voltage to the input voltage becomes

$$\frac{V_o}{V} = \left\{ \frac{1 + (R/\omega L)^2(1 - \omega^2 LC)^2}{[1 + (R/\omega L)^2(1 - \omega^2 LC)^2]^2} \right\}^{1/2}$$

$$= \frac{1}{[1 + (R/\omega L)^2(1 - \omega^2 LC)^2]^{1/2}} \tag{5-42}$$

The behavior of Eq. (5-42) is illustrated with the specific component values given on the parallel-circuit diagram of Fig. 5-7. The output voltage rises to equal the input voltage at the resonant frequency, as shown in Fig. 5-8. Parallel resonance is commonly used in electronic circuits to achieve a high impedance which develops an appreciable signal voltage at resonance. Resonance is also important in circuits designed to emphasize one single frequency over all others. By adjusting the value of, say, the capacitance, the circuit may be *tuned* to different frequencies. This is the principle used to select different channels in radio and TV receivers.

It is interesting to compare the currents in various components of a parallel resonant circuit at resonance. According to Eq. (5-42) the full input voltage is applied across both the capacitor and inductance at resonance, so that the currents are, respectively

$$I_C = \frac{V}{1/\omega C} \tag{5-43}$$

and

$$I_L = \frac{V}{\omega L} \tag{5-44}$$

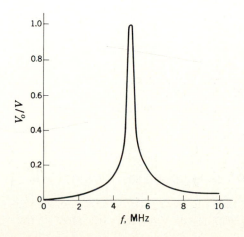

Figure 5-8 Resonance curve of circuit in Fig. 5-7.

Yet the total current in the circuit is zero, according to Eq. (5-39). This is another situation where the phase angle of the currents must be considered. Since the capacitive reactance equals the inductive reactance at resonance, the magnitudes of the currents in the inductance and capacitance are equal. Their phase angles are such that at, say, the upper branch point in Fig. 5-6 these two currents cancel at every instant and no current is present in the resistor. In this sense a circulating current exists in the parallel LC combination at resonance.

Here again, Kirchhoff's rules are valid at frequencies away from resonance. To illustrate, calculate the currents in the components of the circuit in Fig. 5-7 at a frequency of 4.5×10^3 Hz, or an angular frequency of $2\pi f = 2.83 \times 10^7$ radians. The current in R is given by Eq. (5-39),

$$\mathbf{I}_R = 10/(5000 + j2.83 \times 10^7 \times 10 \times 10^{-6})/[1 - (2.83 \times 10^7)^2 \times 10 \times 10^{-6} \times 100 \times 10^{-12}]$$

$$= 10/(5000 + j141)$$

Rationalizing,

$$\mathbf{I}_R = 10(5000 + j141)^{-1} \times (5000 - j141)(5000 - j1410)^{-1}$$

$$= (5 - j1.41) \times 10^4/2.7 \times 10^7 = 1.85 \times 10^{-3} - j5.22 \times 10^{-4}$$

The voltage across both the inductor and the capacitor is given by Eq. (5-41),

$$\mathbf{V}_0 =$$

$$10 \frac{1 + j(5000/2.83 \times 10^7 \times 10 \times 10^{-6})[1 - (2.83 \times 10^7)^2 \times 10 \times 10^{-6} \times 100 \times 10^{-12}]}{1 + (5000/2.83 \times 10^7 \times 10 \times 10^{-6})^2[1 - (2.83 \times 10^7)^2 \times 10 \times 10^{-6} \times 100 \times 10^{-12}]^2}$$

$$= 0.735 + j2.61 = (0.735^2 + 2.61^2)^{1/2}e^{j \arctan 2.61/0.735}$$

$$= 2.71e^{j1.3}$$

The current in the inductor is

$$\mathbf{I}_L = \mathbf{V}_0/j\omega L = 2.71e^{j1.3}e^{-\pi/2}/2\pi \times 4.5 \times 10^3 \times 10 \times 10^{-6}$$

$$= 9.59 \times 10^{-3}e^{-j0.275} = 9.59 \times 10^{-3}(\cos 0.275 - j \sin 0.275)$$

$$= 9.23 \times 10^{-3} - j2.60 \times 10^{-3}$$

and in the capacitor,

$$\mathbf{I}_C = \mathbf{V}_0/(1/j\omega C) = j\omega C\mathbf{V}_0$$

$$= e^{j\pi/2}(2\pi \times 4.5 \times 10^3 \times 100 \times 10^{-12})2.71e^{j1.3} = 7.66 \times 10^{-3}e^{j2.87}$$

$$= 7.66 \times 10^{-3}e^{j2.87} = 7.66 \times 10^{-3}(\cos 2.87 + j \sin 2.87)$$

$$= -7.38 \times 10^{-3} + j2.07 \times 10^{-3}$$

Summing the currents at the upper node,

$$\mathbf{I}_R - \mathbf{I}_L - \mathbf{I}_C = (1.85 - 9.23 + 7.38)10^{-3} + j(-0.522 + 2.60 - 2.07)10^{-3} = 0$$

which verifies Kirchhoff's current rule.

Q Factor

The resistance in a resonant circuit is significant in determining the current at frequencies removed from the resonant frequency, as well as at resonance. This is an important consideration since, for example, the resistance of the turns of wire is present in all inductors. The effect is similar in both series and parallel resonance cases, but the former is simpler mathematically and can be used to illustrate all the important features.

Consider the magnitude of the current in a series circuit, Fig. 5-2, given by Eq. (5-26)

$$I = \frac{V}{[R^2 + (\omega L - 1/\omega C)^2]^{1/2}} \tag{5-45}$$

This can be rearranged as

$$I = \frac{V}{R} \frac{1}{\sqrt{1 + (\omega L/R)^2(1 - 1/\omega^2 LC)^2}} \tag{5-46}$$

The ratio of inductive reactance to resistance is called the Q of the circuit. For present purposes, it is convenient to consider the Q at resonance,

$$Q_0 = \frac{\omega_0 L}{R} \tag{5-47}$$

Introducing (5-47) and the resonant frequency from (5-30) into (5-46),

$$\frac{I}{I_M} = \frac{1}{\{1 + Q_0^2(\omega/\omega_0)^2[1 - (\omega_0/\omega)^2]^2\}^{1/2}} \tag{5-48}$$

where I_M is the maximum current at resonance. Finally, an alternative form to Eq. (5-48) is, after multiplying the two quantities in ω/ω_0,

$$\frac{I}{I_M} = \left[1 + Q_0^2\left(\frac{\omega}{\omega_0} - \frac{\omega_0}{\omega}\right)^2\right]^{-1/2} \tag{5-49}$$

This expression, plotted in Fig. 5-9 for representative values of Q_0, shows that a large Q_0 (that is, a small value of R) leads to a very sharply resonant circuit. Under this condition the bandwidth of the resonant circuit, that is, the frequency interval between half-power points, is quite small.

Values of Q_0 in the range from 10 to 100 are common in electronic circuits and are useful in very selective tuned circuits, as illustrated by the two lower curves of Fig. 5-9. Special resonant components can have Q's as high as several thousand, and these result in very high frequency selectivity indeed. On the other hand, it is sometimes desirable to broaden the frequency response of a resonant circuit so that it is useful over a wide band of frequencies on either side of the resonant frequency. In this case a resistor is purposely included to yield a lower Q than that of the inductor alone.

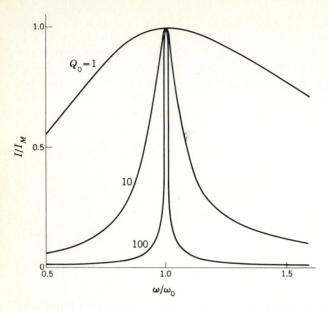

Figure 5-9 Sharpness of resonance curve is greatest for high-Q circuits.

BRIDGE CIRCUITS

Figure 5-10 shows the ac analog of the dc Wheatstone bridge. It has a complex impedance in each arm and uses a sine-wave generator and an ac detector, such as an oscilloscope. Analysis of this circuit proceeds as in the dc case considered in Chap. 1, except that complex impedances and currents are used. Only the balance

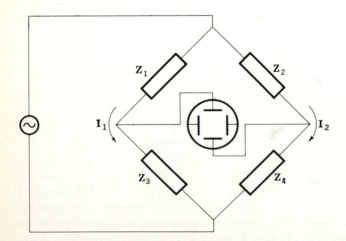

Figure 5-10 The ac Wheatstone bridge uses sine-wave generator and ac detector such as oscilloscope.

condition is of interest here, and this is obtained in the following way. At balance, the voltage across the detector is zero, which means that the current in this branch is zero and therefore that the current in $\mathbf{Z}_1$ is equal to that in $\mathbf{Z}_3$; also, the current in $\mathbf{Z}_2$ is equal to that in $\mathbf{Z}_4$. Furthermore, since the voltage across the detector is zero, the voltage drops across $\mathbf{Z}_1$ and $\mathbf{Z}_2$ are equal and so are the voltage drops across $\mathbf{Z}_3$ and $\mathbf{Z}_4$.

Equating the voltage drops across corresponding arms of the bridge

$$\mathbf{Z}_1\mathbf{I}_1 = \mathbf{Z}_2\mathbf{I}_2 \tag{5-50}$$

and

$$\mathbf{Z}_3\mathbf{I}_1 = \mathbf{Z}_4\mathbf{I}_2 \tag{5-51}$$

Dividing Eq. (5-50) by Eq. (5-51), the condition for balance is found to be

$$\frac{\mathbf{Z}_1}{\mathbf{Z}_3} = \frac{\mathbf{Z}_2}{\mathbf{Z}_4} \tag{5-52}$$

which should be compared with the condition for balance of the dc Wheatstone bridge, Eq. (1-46). According to Eq. (5-52) the balance condition involves the equality of two complex numbers. This means that both the real and imaginary parts must be equal and implies that two independent balance adjustments are necessary in ac bridge circuits. Specific illustrations of this situation are taken up in the following sections.

Inductance and Capacitance Bridge

Bridge circuits in Figs. 5-11 and 5-12 can be used to measure inductance and capacitance in the same way a Wheatstone bridge is used to measure resistance. Consider the inductance bridge, Fig. 5-11, and note the comparison with Fig. 5-10. The impedance $R_u + j\omega L_u$ represents the inductance and resistance of an unknown inductor. The balance condition, Eq. (5-52), means

$$\frac{R_1}{R_2} = \frac{R_3 + j\omega L_3}{R_u + j\omega L_u} \tag{5-53}$$

Cross-multiplying,

$$R_1 R_u + j\omega L_1 L_u = R_2 R_3 + j\omega R_2 L_3 \tag{5-54}$$

After equating real and imaginary parts, the two balance conditions are written as

$$R_1 R_u = R_2 R_3 \quad \text{and} \quad R_1 L_u = R_2 L_3 \tag{5-55}$$

The first condition is the same as the Wheatstone bridge condition and may be carried out with dc instruments. The second equation is obtained using ac excitation of the bridge subsequent to obtaining the dc balance.

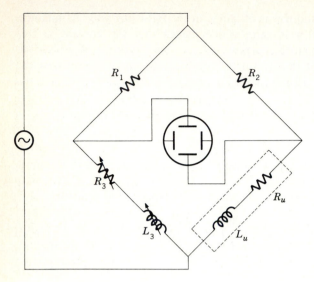

Figure 5-11 Inductance bridge measures resistance and inductance of unknown coil.

Rewriting Eqs. (5-55),

$$R_u = \frac{R_2}{R_1} R_3 \quad \text{and} \quad L_u = \frac{R_2}{R_1} L_3 \tag{5-56}$$

which illustrates how the ratio R_2/R_1 acts as a multiplying factor on the variable components R_3 and L_3 to yield the unknown resistance and inductance. This suggests that both R_3 and L_3 should be variable, but since variable inductors are difficult to make, the following procedure is more satisfactory. Either R_2 or R_1 is adjusted until balance is obtained with a given value of L_3. Then R_3 is adjusted to satisfy the dc balance condition. In this way the unknown inductance is

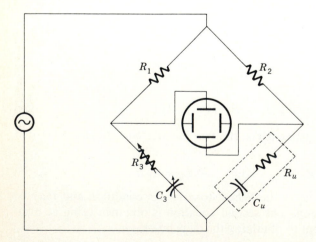

Figure 5-12 Capacitance bridge.

compared with the standard inductance L_3, and the only variable components needed are resistances.

An interesting feature of the balance conditions, Eqs. (5-56), is that they are independent of frequency, so that the generator frequency may have any convenient value. It is usually best to choose a frequency such that the inductive reactance is approximately equal to the resistances. A similar situation exists in the case of the capacitance bridge, Fig. 5-12, which is analyzed in Exercise 5-12. Since there is no dc path in the lower arms of the capacitance bridge, both balance adjustments are made with ac excitation (this can be done in the inductance bridge case as well). Although calibrated variable capacitors are usually used in this circuit, it is equally possible to employ a fixed standard capacitor together with variable resistors. This is particularly convenient when the same instrument is used as an inductance bridge as well as a capacitance bridge by replacing the standard capacitor C_3 with a standard inductance L_3. Since it is also possible to employ a standard resistor in this position, such an instrument is quite versatile in that it can measure resistance, capacitance, and inductance.

Wien Bridge

A bridge that has a parallel combination in one arm and a series combination in an adjacent arm is known as a *Wien* bridge. A useful example employing only resistors and capacitors, shown in Fig. 5-13, is analyzed by first calculating the impedances $\mathbf{Z}_3$ and $\mathbf{Z}_4$. Considering the parallel combination first,

$$\frac{1}{\mathbf{Z}_3} = j\omega C_3 + \frac{1}{R_3} = \frac{1 + j\omega R_3 C_3}{R_3} \qquad (5\text{-}57)$$

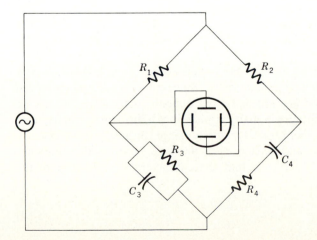

Figure 5-13 Wien bridge.

which is rationalized to give

$$\mathbf{Z}_3 = \frac{R_3}{1 + (\omega R_3 C_3)^2} (1 - j\omega R_3 C_3) \tag{5-58}$$

The series combination is

$$\mathbf{Z}_4 = R_4 - j\frac{1}{\omega C_4} \tag{5-59}$$

Inserting Eqs. (5-58) and (5-59) into the balance equation (5-52) results in

$$\frac{R_1}{R_2} = \frac{R_3/(1 + j\omega R_3 C_3)}{R_4 - j(1/\omega C_4)} \tag{5-60}$$

Cross-multiplying,

$$(1 + j\omega R_3 C_3)\left(R_4 - \frac{j}{\omega C_4}\right) = \frac{R_2}{R_1} R_3 \tag{5-61}$$

$$R_4 + \frac{R_3 C_3}{C_4} + j\left(\omega R_3 R_4 C_3 - \frac{1}{\omega C_4}\right) = \frac{R_2}{R_1} R_3 \tag{5-62}$$

Upon equating the real and imaginary parts, the balance conditions are found to be

$$\frac{C_3}{C_4} + \frac{R_4}{R_3} = \frac{R_2}{R_1}$$

and

$$\omega^2 R_3 C_3 R_4 C_4 = 1 \tag{5-63}$$

This result differs from those for the inductance and capacitance bridges in that the frequency ω appears in the balance equations. Therefore, it is possible to achieve balance by adjusting the frequency and only one component, say R_1, rather than using two variable impedances. Alternatively, by adjusting two of the components for balance, the bridge is capable of determining the frequency of a sine-wave source.

The Wien bridge is also useful as a frequency-selective network which has properties similar to those of a resonant circuit. Since inductors suitable for use over a wide frequency interval are expensive and difficult to construct, the Wien bridge has considerable advantages in many applications. This is particularly true for low-frequency circuits where inconveniently large inductance values are necessary. Often, the resistors and capacitors in the series and parallel branches are equal, as in Fig. 5-14. This means that the characteristic frequency of the network is, from Eq. (5-63),

$$\omega_0 = \frac{1}{RC} \tag{5-64}$$

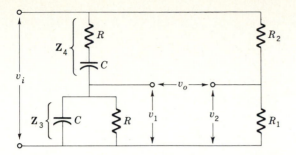

Figure 5-14 Wien bridge used as frequency-selective network.

The frequency-selective properties of the Wien bridge are illustrated by calculating the output voltage of the network as a function of the frequency of the input voltage. The output voltage is obtained by subtracting the voltages across Z_3 and R_1. The drop across Z_3 is simply the current in this arm times the impedance,

$$v_1 = \frac{v_i}{Z_3 + Z_4} Z_3 = \frac{v_i}{1 + Z_4/Z_3} \tag{5-65}$$

where it has been assumed that the current in the output circuit is negligible, even when the bridge is not balanced. This condition is satisfied in practice by connecting a high-impedance load to the output terminals. The impedance ratio in Eq. (5-65) is evaluated from Eqs. (5-58) and (5-59), after making all the R's and C's equal. Also, introducing ω_0,

$$\frac{Z_4}{Z_3} = \frac{R(1 - j\omega_0/\omega)}{R(1 - j\omega/\omega_0)/[1 + (\omega/\omega_0)^2]} = 2 + j\left(\frac{\omega}{\omega_0} - \frac{\omega_0}{\omega}\right) \tag{5-66}$$

Note that according to Eq. (5-66) the reactive term vanishes at the characteristic frequency when $\omega = \omega_0$. This means that the network is resistive at this frequency (and at this frequency only) and that the output voltage is in phase with the input voltage, just as for a resonant circuit.

The drop across R_1 is

$$v_2 = \frac{v_i}{R_1 + R_2} R_1 = \frac{v_i}{1 + R_2/R_1} \tag{5-67}$$

Finally, the output voltage is

$$v_o = v_i - v_2 = v_i\left[\frac{1}{3 + j(\omega/\omega_0 - \omega_0/\omega)} - \frac{1}{1 + R_2/R_1}\right] \tag{5-68}$$

The ratio of the output to the input v_o/v_i can be found after rationalizing. Two plots of Eq. (5-68) are presented in Fig. 5-15 for different values of the ratio R_2/R_1. The output voltage clearly goes through a minimum at $\omega = \omega_0$, very analogous to the situation in series resonance. The minimum is quite sharp and mathematically discontinuous in the case of true balance, $R_2 = 2R_1$. Therefore, the frequency selectivity of the circuit is quite good.

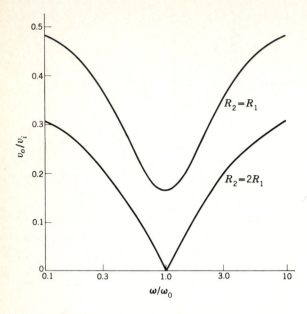

Figure 5-15 Frequency character-istic of Wien bridge network.

Bridged-T and Twin-T Networks

Although the so-called *bridged-T* filter is not a true bridge circuit, it has proper-ties similar to those of the Wien bridge. In contrast to the Wien bridge, the bridged-T, shown in Fig. 5-16, generally does not employ equal values of resist-ance and capacitance in the two positions. To solve for the output voltage, the current in the upper resistor is determined and the output is obtained by sub-tracting the voltage drop across R_2 from the input voltage.

Kirchhoff's law applied to the i_1 loop in Fig. 5-16 is

$$v_i = i_1 \frac{-j}{\omega C_1} - i_2 \frac{-j}{\omega C_1} + i_1 R_1 \tag{5-69}$$

Adding up the voltages around the second loop,

$$0 = i_1 \frac{-j}{\omega C_1} + i_2 \frac{-j}{\omega C_1} + i_2 R_2 + i_2 \frac{-j}{\omega C_2} \tag{5-70}$$

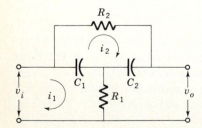

Figure 5-16 Bridged-T filter.

After simplification, i_1 is determined from (5-70)

$$i_1 = \left(1 + \frac{C_1}{C_2} + j\omega R_2 C_1\right)i_2 \tag{5-71}$$

and substituted into (5-69). The result is solved for i_2 and the output voltage is then given by

$$v_o = v_i - R_2 i_2 = v_i - \frac{v_i \omega R_2 C_2}{\omega C_2[R_1(1 + C_1/C_2) + R_2] + j(\omega^2 R_1 C_1 R_2 C_2 - 1)} \tag{5-72}$$

The imaginary part vanishes at the <mark>critical frequency</mark>

$$\omega_0 = \frac{1}{\sqrt{R_1 C_1 R_2 C_2}} \tag{5-73}$$

which should be compared with the corresponding expression for the Wien bridge, Eq. (5-63). After inserting (5-73) into (5-72), the complex ratio of output to input voltage may be written

$$\frac{v_0}{v_i} = 1 - \left\{\left[1 + \frac{R_1}{R_2}\left(1 + \frac{C_1}{C_2}\right)\right] + j\sqrt{\frac{R_1 C_1}{R_2 C_2}}\left(\frac{\omega}{\omega_0} - \frac{\omega_0}{\omega}\right)\right\}^{-1} \tag{5-74}$$

The frequency-response characteristic of the bridged-T filter, Eq. (5-74), is similar to that of the Wien bridge, except that the minimum is not as sharp at the characteristic frequency. Nevertheless, its relative simplicity, and the fact that both input and output terminals have a common connection, make the bridged-T filter a useful frequency-selective network.

Somewhat more complicated than the bridged-T filter is the *twin-T* filter shown in Fig. 5-17, which may be analyzed by the technique used in the previous section. The twin-T filter is equivalent in response characteristic to the Wien bridge, but the minimum is much sharper (see Exercise 5-14). It also has the advantage of a common input-output terminal. The choice between the Wien bridge, bridged-T network, or twin-T network for any given application depends on factors such as frequency range, desired performance, and complexity; all three circuits are commonly used.

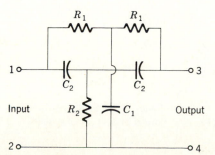

Figure 5-17 Twin-T filter.

TRANSFORMERS

Mutual Inductance

Suppose a changing magnetic flux resulting from current in one circuit is intercepted by another circuit. Then an emf is induced in the second circuit, and the
mutual inductance between circuit 1 and circuit 2 is defined by analogy with Eq.
(2-15) as

$$v_2 = M \frac{di_1}{dt} \qquad (5\text{-}75)$$

where v_2 is the induced emf in circuit 2 produced by the current i_1 in circuit 1. A
very important application of mutual inductance is the *transformer*, which has
two multiturn coils wound on the same iron core. This makes the mutual inductance between the two coils as large as possible. Schematically, a transformer
appears as in Fig. 5-18a, with a *primary* winding which is part of one circuit and
a *secondary* winding which is part of the second circuit. The symbol for a transformer is shown in Fig. 5-18b.

Consider an ideal transformer, in which all of the magnetic flux from the
primary winding is intercepted by the secondary winding. Suppose the secondary
winding is open-circuited and the primary is connected to a sinusoidal voltage
source. Current in the primary winding is determined by the inductance of
the primary. The voltage induced in the primary winding v_1 is proportional
to the primary inductance according to Eq. (2-15), and the inductance depends
upon the number of turns on the primary winding. Since all of the flux is also
intercepted by the secondary winding, the voltage v_2 induced in the secondary
depends upon the number of turns in the same way. That is,

$$\frac{v_1}{v_2} = \frac{n_1}{n_2} \qquad (5\text{-}76)$$

Note that it is possible to achieve either a *step-up* transformer or a *step-down*
transformer, depending upon whether $n_1 < n_2$ or $n_1 > n_2$, so that the secondary
voltage is respectively greater or less than the primary voltage.

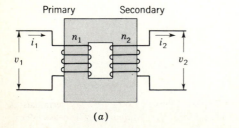

(a)

(b)

Figure 5-18 (a) Sketch of transformer and (b) circuit symbol.

Suppose now the secondary is connected to a load such as a resistor. Current in the secondary circuit results in I^2R losses in the resistor, and this power must come from the primary circuit. The way this comes about is as follows. Both the primary current and the secondary current set up magnetic flux in the core. The magnetic flux caused by the secondary current acts to oppose the magnetic flux set up by the primary current. The weaker magnetic flux means that the induced primary voltage is correspondingly smaller. Therefore the voltage source connected to the primary winding increases the primary current until the voltages around the primary circuit are zero again as required by Kirchhoff's rules. Thus current in the secondary winding requires current in the primary winding and the peak magnetic flux remains constant at its no-load value. The near cancellation of fluxes also means that the induced voltages are essentially independent of frequency. Since the rate of change of current increases with frequency, Eq. (5-75) suggests that the secondary voltage should increase with frequency. The induced voltage in the primary winding increases with frequency correspondingly, however, so that the magnitude of the primary current is reduced, and the secondary voltage remains independent of frequency.

Note that the total magnetic flux in the core does not change with current, because the magnetic flux from the primary and secondary currents are equal and opposite. Since both inductances depend upon the number of turns, it follows that

$$n_1 i_1 = n_2 i_2 \tag{5-77}$$

The primary current in Eq. (5-77) really refers only to the additional current accompanying a load on the secondary. The current under no-load conditions is usually so small, however, that it may be neglected.

Transformer Ratio

The ratio of the number of turns on the secondary to that on the primary, called the *transformer ratio*, is particularly significant when the impedance connected to the secondary is compared with the apparent impedance in the primary circuit. Consider the situation depicted in Fig. 5-19, where the secondary load is a simple resistance R.

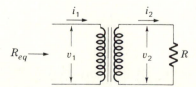

Figure 5-19 Apparent resistance at primary terminals of transformer differs from resistance connected to secondary.

The magnitude of R determines the secondary current

$$i_2 = \frac{v_2}{R} \tag{5-78}$$

Substituting for i_2 from Eq. (5-77) and for v_2 from Eq. (5-76),

$$\frac{n_1}{n_2} i_1 = \frac{n_2 v_1}{n_1 R}$$

or

$$i_1 = \left(\frac{n_2}{n_1}\right)^2 \frac{v_1}{R} \tag{5-79}$$

Viewed from the primary side, the current i_1 is in an equivalent resistance R_{eq} such that

$$i_1 = \frac{v_1}{R_{eq}} \tag{5-80}$$

Comparing (5-79) and (5-80), the secondary load resistance R appears to be a resistance on the primary side given by

$$R_{eq} = \left(\frac{n_1}{n_2}\right)^2 R \tag{5-81}$$

This means that a transformer may be used to match the resistances in a circuit to obtain maximum power transfer between a given power source and a fixed-load resistance. This very useful property is applied, for example, in coupling the large internal resistance of an amplifier to the low resistance of a loudspeaker load. Complex impedances are transformed by the turns ratio in exactly the same way as the resistance in Eq. (5-81).

According to Eq. (5-81) the transformer appears from the primary side to be a pure resistance. In particular, the self-inductance of the primary winding is not evident. The reason for this is that the fluxes caused by current in the primary and secondary windings cancel each other. This means that a transformer acts as a device which changes the effective resistance from one circuit to another but which has no inductance of itself. Note also that the primary and secondary circuits in a transformer are not connected electrically and that the two circuits are isolated for direct currents.

Practical Transformers

Most transformers are constructed in a fashion similar to iron-core chokes described in Chap. 2, except, of course, that more than one winding is present on the core. For large transformer ratios it is common practice to make the low-

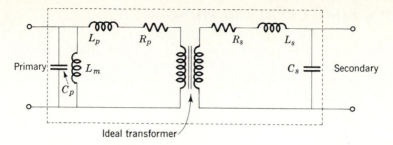

Figure 5-20 Equivalent circuit of practical transformer.

voltage, high-current winding of heavy-gauge wire in order to reduce I^2R losses in this winding. The other winding is a large number of turns of fine wire since only a small current is present. Laminated cores are ordinarily used to reduce eddy-current losses, but transformers for frequencies in excess of 100,000 Hz generally employ high-resistivity ferrite cores.

Transformers with one primary winding and several separate secondary windings are used to supply different voltages to transistor and integrated circuits. Such a *power* transformer may, for example, have a primary winding suitable for connection to a 115-V 60-Hz source with secondaries providing 100 and 5.0 V. Applications of such power transformers are studied in a later chapter.

Although for many purposes the inherent inductive effects of the transformer windings may be neglected, in careful work it is necessary to account for the properties of the transformer more exactly. An equivalent circuit of a practical transformer is illustrated in Fig. 5-20. The inductances in the primary and secondary circuits are caused by leakage magnetic flux which does not link both windings, so that the opposing fluxes do not quite cancel. The resistances are included to account for the resistance of the wire in the windings. The inductance L_m accounts for the small magnetizing current corresponding to the no-load primary current. The capacitors on the primary and secondary sides result from the layer-to-layer capacitances between windings.

According to this equivalent circuit, a transformer is ineffective at low frequencies where the reactance of L_m becomes so small that current is shunted from the primary winding of the ideal transformer. High-frequency performance is impaired by the leakage flux inductances and winding capacitances. In spite of these limitations, transformers can be designed for effective performance over useful frequency intervals. It is common practice to indicate the approximate impedance level at which the primary and secondary windings are designed to be used, even though Eq. (5-81) suggests that only the transformer ratio is significant. The specification indicates the conditions under which the transformer inductances and resistances in Fig. 5-20 are negligible in comparison with primary and secondary load impedances, and, correspondingly, the transformer operates very nearly like an ideal transformer.

SUGGESTIONS FOR FURTHER READING

P. Chirlian: "Basic Network Theory," McGraw-Hill Book Company, New York, 1969.

G. Lancaster: "DC and AC Circuits," Oxford University Press, New York, 1980.

Clayton R. Paul, Syed A. Nasar, and L. E. Unnewehr: "Introduction to Electrical Engineering," McGraw-Hill Book Company, New York, 1986.

Basil H. Vassos and Galen W. Ewing: "Analog and Digital Electronics for Scientists," John Wiley & Sons, Inc., New York, 1985.

EXERCISES

5-1 Starting with Eq. (2-31) for the RC filter and using the method leading to the ac form of Ohm's law, Eq. (5-6), determine the complex reactance of a simple capacitor. [*Hint:* Take $R = 0$ and use Eq. (5-1).]

Answer: $1/j\omega C$

5-2 Calculate the equivalent impedance of the network shown in Fig. 5-21. Is the reactive term capacitive or inductive?

Answer: Capacitive

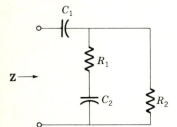

Figure 5-21

5-3 Determine the rms current in the 1000-Ω resistor of the circuit in Fig. 5-22. Is the current inductive or capacitive?

Answer: 6.5 mA; capacitive

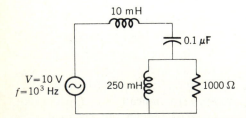

Figure 5-22

5-4 Calculate the equivalent impedance of the circuit in Fig. 5-23 at a frequency of 100 Hz. Repeat for 1000 Hz.

Answer: $0.198 + j2.63 \times 10^6 \ \Omega$; $384 + j2.47 \times 10^9 \ \Omega$

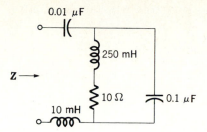

Figure 5-23

5-5 Solve Eq. (5-20) for the series *RLC* circuit by the differential equation technique to arrive at Eq. (5-21).

5-6* Determine the current and the rms voltages across each component in the circuit of Fig. 5-4 at a frequency of 900 Hz. Show that, considering the phase angles of the voltages, Kirchhoff's rule concerning the sum of the voltages around a loop is valid.

Answer: $2.75 \times 10^{-2} e^{j1.29}$ A; $2.75 e^{j1.29}$ V; $48.6^{-j0.27}$ V; $38.8 e^{j2.87}$ V

5-7* Repeat Exercise 5-6 at a frequency of 1100 Hz.

Answer: $3.36 \times 10^{-2} e^{-j1.23}$ A; $3.36 e^{-j1.23}$ V; $48.6 e^{-j2.8}$ V; $58 e^{j0.034}$ V.

5-8 The impedance at resonance of a parallel resonant circuit is limited by the resistance of the windings of the inductor. Derive an expression for the impedance of the circuit of Fig. 5-24 and calculate the value at resonance appropriate to the components given on the circuit diagram.

Answer: 4×10^5 Ω

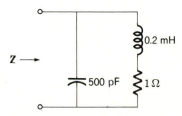

Figure 5-24

5-9 Compute the currents in the components of the circuit of Fig. 5-7 at resonance. Repeat for a frequency $\omega = \omega_0/2$. In both cases show that, considering the phase angles of the currents, Kirchhoff's current rule is valid.

Answer: 0 A; $3.16 \times 10^{-2} e^{-j1.57}$ A; $3.16 \times 10^{-2} e^{j1.57}$ A; $2 \times 10^{-3} e^{-j0.042}$ A;

$2.67 \times 10^{-3} e^{-j0.042}$ A; $6.67 \times 10^{-4} e^{j3.10}$ A

5-10 Is the equivalent impedance of a parallel resonant circuit inductive or reactive below resonance? Above resonance? *Hint:* Use Eq. (5-36). Compare with a series resonant circuit.

Answer: Inductive; capacitive

5-11 Using Eq. (5-49) determine the upper and lower half-power frequencies for $Q_0 = 1$, 10, and 100. Express the bandwidth in terms of a percentage of the center (resonant) frequency.

Answer: $1.62\omega_0$, $0.62\omega_0$, 100%; $1.05\omega_0$, $0.95\omega_0$, 10%; $1.005\omega_0$, $0.995\omega_0$, 1%

5-12 Determine the balance conditions for the capacitance bridge, Fig. 5-12.

Answer: $R_2 R_3 = R_1 R_u$; $R_1/R_2 = C_u/C_3$

5-13* Plot the frequency and phase characteristics of a bridged-T filter using Eq. (5-74). Select component values such that the characteristic frequency is 1 kHz while $R_2 = 1000 R_1$, $C_1 = C_2$, and $R_1 = 10^4$ Ω.

5-14* Plot the frequency and phase characteristics of the twin-T filter, Fig. 5-17, for which the response characteristic is

$$\frac{v_o}{v_i} = \left(1 + j\omega_0 R_1 \frac{C_1 + 2C_2}{\omega/\omega_0 - \omega_0/\omega}\right)^{-1} \tag{5-82}$$

where the resonance frequency is given by

$$\omega_0^2 = \frac{2}{R_1^2 C_1 C_2} \tag{5-83}$$

when $R_1 C_1 = 4R_2 C_2$. Take $C_1 = 2C_2$ and choose appropriate component values to make the characteristic frequency equal to 1 kHz. Derive Eq. (5-82).

5-15 Sketch the approximate gain-versus-frequency curve for the circuit in Fig. 5-25. You may treat the circuit as being composed of two independent RC filters. Why?

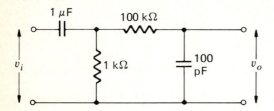

Figure 5-25

SIX

TRANSISTOR AMPLIFIERS

The advent of electronics is reckoned from the discovery that the current in a vacuum diode can be controlled by introducing a third electrode to make the triode an effective amplifier (see Appendix 1). The possibilities inherent in amplifier circuits were greatly expanded by the invention of the transistor in 1948. Since then a wide variety of semiconductor devices has supplanted vacuum tubes in most applications.

Transistors are nonlinear devices, and their operation in any circuit is determined by graphical analysis using the description of their electrical properties given by current-voltage characteristics. The analysis differs in detail for voltage-controlled devices such as the field-effect transistor compared to current-controlled devices such as the bipolar transistor, but is not different in principle. Furthermore, in either case it proves possible to develop useful equivalent-circuit representations which are very helpful for circuit analysis.

FET AMPLIFIERS

The Operating Point

The operation of a FET as an amplifier is examined most easily with the aid of the elementary amplifier circuit in Fig. 6-1 which employs a *p*-channel FET. In this circuit the drain is maintained at a negative potential with respect to the source by the battery V_{dd} while the gate is biased positively by the gate bias battery V_{gg}. Variations in gate voltage resulting from an input signal ΔV_{gs} produce changes in the drain current which are observed as a voltage signal ΔV_{ds} across the load resistor R_L.

The circuit is analyzed using the graphical load-line technique discussed in Chap. 3. The voltage equation around the output circuit is

$$V_{dd} - I_d R_L + V_{ds} = 0 \tag{6-1}$$

or

$$I_d = \frac{V_{dd}}{R_L} + \frac{1}{R_L} V_{ds} \tag{6-2}$$

which is the load line. Equation (6-2), together with the drain characteristics of the FET, represent two relations in the two unknowns V_{ds} and I_d. By plotting the load line on the drain characteristics as in Fig. 6-2, the drain current corresponding to any gate voltage is given by the intersection of the load line with the characteristic curve for that gate voltage.

The *operating point* is the intersection of the load line with the characteristic curve corresponding to the gate bias voltage. As the gate voltage varies in accordance with an applied input signal, the drain-current excursions move back and forth along the load line so that Eq. (6-2) is satisfied at every instant. In Fig. 6-2, for example, an input signal ΔV_{gs} of 0.4 V causes the drain current to change by approximately 0.6 mA and results in an output signal ΔV_{ds} of about 6 V. That is, the circuit amplifies the input signal by a factor of $6/0.4 = 15$.

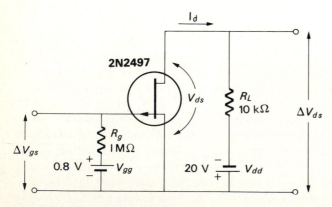

Figure 6-1 Simple *p*-channel FET amplifier.

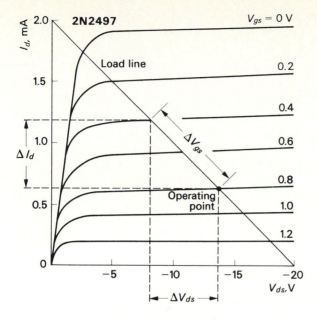

Figure 6-2 Operating point is intersection of load line with gate bias curve.

Note that the input power is very small because the gate current is negligible. In contrast, the output power, which is equal to the product of the change in drain current times the change in drain voltage, may be quite appreciable. This power is derived from the drain supply battery, V_{dd}, and is controlled by the valve-like action of the gate.

A separate gate bias battery is economically impractical in most circuits. Instead, gate bias can be obtained by inserting a resistor in series with the source, Fig. 6-3. The gate potential is set by a voltage divider connected to the drain voltage source. Then the current through the source resistor R_s results in a gate bias of the proper polarity and magnitude to maintain the gate junction under reverse bias. A useful feature of this circuit is that if, for example, a change in ambient temperature increases the drain current, the voltage drop across the source resistor increases the gate bias, which tends to return the drain current to the original value. That is, the drain current and the operating point are stabilized against external influences.

Note that the source resistor is shunted by a large capacitor to prevent ac signals caused by ac drain currents in the source resistor from appearing in the gate circuit. Also, capacitors are included at the input and output to isolate the amplifier from external dc voltages that could change the operating point.

The operating point is determined by writing the equation for the load line from the dc voltage drops in the output circuit,

$$V_{dd} - I_d R_L - V_{ds} - I_d R_s = 0 \tag{6-3}$$

Since, however, the gate bias voltage is not known, in contrast to the simpler circuit previously analyzed, another relation is required. This is found from the dc

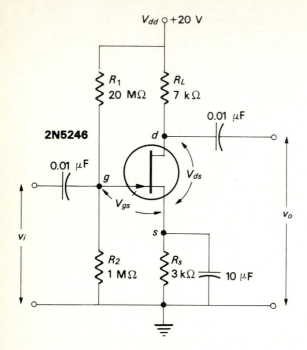

Figure 6-3 Practical *n*-channel FET amplifier using source bias.

voltage drops around the input circuit and the voltage-divider action of R_1 and R_2,

$$\frac{V_{dd} R_2}{R_1 + R_2} - V_{gs} - I_d R_s = 0 \qquad (6\text{-}4)$$

Both equations may be put in the same form. The load line is, from Eq. (6-3),

$$I_d = \frac{V_{dd}}{R_s + R_L} - \frac{1}{R_s + R_L} V_{ds} \qquad (6\text{-}5)$$

and the *bias line*, from Eq. (6-4), is

$$I_d = \frac{V_{dd} R_2}{(R_1 + R_2) R_s} - \frac{1}{R_s} V_{gs} \qquad (6\text{-}6)$$

Note that Eq. (6-5) is similar to Eq. (6-2), and that Eq. (6-6) has the same form. These expressions, together with the current-voltage characteristics of the FET, represent three relations in the three unknowns, V_{ds}, I_d, and V_{gs}, and are solved graphically.

The bias line, Eq. (6-6), is a straight line on the transfer characteristic, Fig. 6-4, and the intersection gives the drain current and gate voltage immediately. The drain voltage is then found from the intersection of the gate bias curve with the load line plotted on the drain characteristics. Thus the operating point is completely determined.

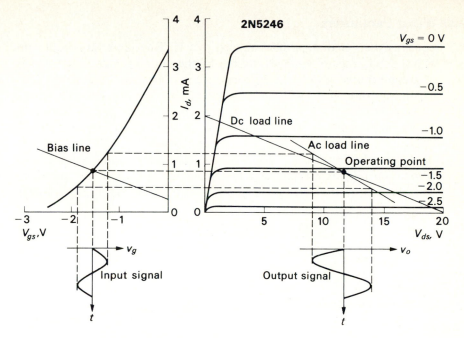

Figure 6-4 Intersection of bias line with transfer characteristic yields operating point of source-bias amplifier.

An input signal applied to the gate results in corresponding changes in the drain current, as shown by the transfer characteristic in Fig. 6-4. As in the previous circuit, drain signal currents through the load resistor produce an output signal. This is illustrated graphically in Fig. 6-4 by plotting the *ac load line* on the drain characteristics. The ac load line is simply the dc load line, Eq. (6-5), with the source resistor set equal to zero to account for the shunting action of the source bypass capacitor at signal frequencies.

Note that a sinusoidal input signal results in a sinusoidal output signal of larger amplitude. The output waveform is not an exact amplified replica of the input signal, however, because of curvature in the transfer characteristic. This *distortion* is minimized by proper circuit design and suitable choice of the operating point. Note also in Fig. 6-4 that an increase in gate voltage produces a decrease in the output voltage. This means that the circuit introduces a 180° phase shift between input and output signals.

It is inconvenient to measure electrode voltages with respect to the source in circuits employing source bias. This is particularly true when more than one transistor is used in a circuit. Rather, usual practice is to refer all potentials to a common point called *ground*. The ground point is considered to be electrically neutral so that, for example, the ground points of two separate circuits may be connected with no influence upon the operation of either circuit. A typical use of the circuit symbol for the ground point in Fig. 6-3 is at the junction of the lower input and output terminals and the bottom end of the source resistor.

Small-Signal Parameters

Very frequently the signal amplitudes applied to a FET amplifier are small compared with the full range of voltages covered by the drain characteristics. In this situation graphical analysis of FET performance is inaccurate because the drain characteristics are not given with sufficient precision. A more satisfactory procedure is to replace the FET with an equivalent circuit which can be studied by conventional circuit analysis. After the operating point is determined graphically, small departures about the operating point caused by small signals are treated by assuming that the FET is a linear device. According to the drain characteristics, drain current depends upon both the gate voltage and the drain voltage. Therefore, a small change in drain current, ΔI_d, away from the operating point may be written

$$\Delta I_d = k_1 \, \Delta V_{gs} + k_2 \, \Delta V_{ds} \tag{6-7}$$

where ΔV_{gs} and ΔV_{ds} are small changes in the electrode potentials and k_1 and k_2 are constants.

The ratio $\Delta I_d/\Delta V_{ds}$ can be identified as the reciprocal of an equivalent resistance called the *drain resistance*,

$$\frac{1}{r_d} = \frac{\Delta I_d}{\Delta V_{ds}} \tag{6-8}$$

Similarly, the ratio $\Delta I_d/\Delta V_{gs}$ gives the change in drain current resulting from a change in voltage in the gate circuit. It has the dimensions of conductance and is called the *mutual transconductance*,

$$g_m = \frac{\Delta I_d}{\Delta V_{gs}} \tag{6-9}$$

The total change in drain current is, from Eqs. (6-7) to (6-9),

$$\Delta I_d = g_m \, \Delta V_{gs} + \frac{1}{r_d} \, \Delta V_{ds} \tag{6-10}$$

Consider now that the small departures about the operating point are ac signals, so that Eq. (6-10) becomes

$$i_d = g_m v_{gs} + \frac{1}{r_d} v_{ds} \tag{6-11}$$

This expression may be interpreted as the circuit equation for a current generator in parallel with a resistor, as in Fig. 6-5; summing the currents at the upper node yields Eq. (6-11) directly. Figure 6-5 is the *ac equivalent circuit* of the FET. Note that it has the configuration of the Norton equivalent circuit discussed in Chap. 1, which is consistent with the constant-current properties of the FET drain characteristics.

The mutual transconductance and the drain resistance are called the *small-signal parameters* of the FET. The mutual transconductance is just the slope of

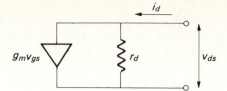

Figure 6-5 FET ac equivalent circuit.

the transfer characteristic at the operating point, and the drain resistance is the reciprocal of the drain characteristic slope at the operating point. Typical values of small-signal parameters given in Table 6-1 show that the several different types of FETs are comparable. A rather wide range of parameters may be achieved by suitable choice of the FET design, however. Actually, since the FET is a nonlinear device, the magnitudes of the small-signal parameters depend upon the dc drain current as well, Fig. 6-6. For this reason, it is usually necessary to evaluate g_m and r_d graphically at the operating point. According to Fig. 6-6 quite appreciable variations in the small-signal parameters can be obtained by selecting the operating point. This is important in the design of FET circuits having specific performance requirements.

It is illustrative to analyze the simple FET amplifier in Fig. 6-3 by means of the ac equivalent circuit technique. According to Fig. 6-5, the equivalent circuit of the amplifier is obtained by replacing the FET with a constant-current generator in parallel with a resistor, Fig. 6-7a. Note that the gate terminal is not connected to the remainder of the equivalent circuit. This is a consequence of the very high input resistance of the reverse-biased gate junction. This circuit ignores the capacitance of the gate junction, however, and is only accurate at frequencies low enough that its capacitive reactance can be neglected.

Similarly, it is usually assumed that the capacitive reactances of the coupling capacitors and the source bypass capacitor are negligible at signal frequencies of interest. This leads to the simpler equivalent circuit shown in Fig. 6-7b. In spite of these approximations, the circuit in Fig. 6-7b proves to be a useful representation of the actual amplifier for most purposes.

Table 6-1 FET small-signal parameters

Type		g_m, 10^{-6} mho	r_d, Ω
2N5484	n-channel FET	2000	50,000
2N5268	p-channel FET	1700	15,000
2N3797	n-channel MOSFET (depletion)	2300	40,000
3N157	p-channel MOSFET (enhancement)	2000	17,000

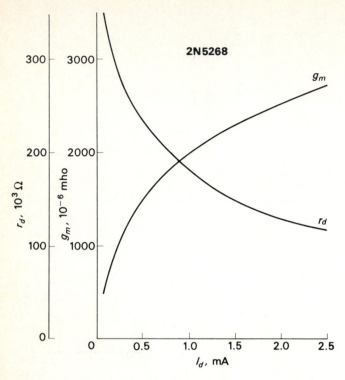

Figure 6-6 Variation of small-signal parameters of type 2N5268 p-channel FET with drain current.

The output signal may be written immediately as the voltage drop across the parallel combination of the load resistance and the drain resistance,

$$v_o = -g_m v_{gs} \frac{r_d R_L}{r_d + R_L} \tag{6-12}$$

The ratio of the output signal to the input signal is called the *gain* of the amplifier. Since the gate voltage is equal to the input signal, the gain is

$$a = \frac{v_o}{v_i} = -g_m \frac{r_d R_L}{r_d + R_L} \tag{6-13}$$

Most often the drain resistance is much greater than the load resistor so that Eq. (6-13) reduces to

$$a = -g_m R_L \tag{6-14}$$

According to Table 6-1, a gain of about $7000 \times 2000 \times 10^{-6} = 14$ can be anticipated. A more precise value is calculated after determining the actual mutual transconductance at the operating point. The minus sign in Eq. (6-14) represents the 180° phase shift between input and output signals, as discussed

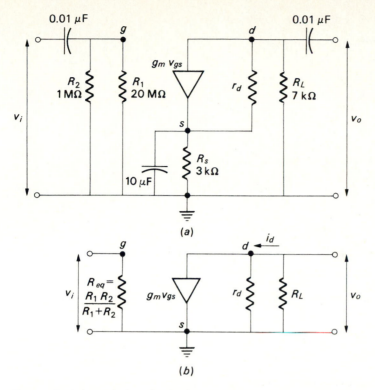

Figure 6-7 (a) Ac equivalent circuit of FET amplifier in Fig. 6-3 and (b) simplified version assuming capacitive reactances are negligible.

previously. An important feature of the FET amplifier is the large input impedance which is essentially equal to R_{eq} in Fig. 6-7b. Also of interest is the output impedance, which is equal to the load resistor so long as the drain resistance is large, as assumed in arriving at Eq. (6-14).

Source Follower

The source is common to both input and output terminals in this basic FET amplifier circuit insofar as ac signals are concerned. This is the most common configuration and is sometimes referred to as the *grounded-source* connection. Another possible configuration is the *grounded-drain* amplifier, Fig. 6-8. The drain load resistor is omitted and the output signal is developed across the source resistor. Note that in this arrangement the drain is at ground potential insofar as ac signals are concerned and is common to both input and output circuits. The drain remains at a dc potential with respect to the source and to ground, however. Gate bias is developed across R_1, and the operating point is found in the same fashion as for the grounded-source amplifier. The circuit is commonly

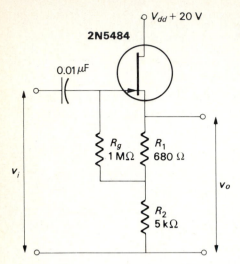

Figure 6-8 FET source-follower amplifier.

called a *source follower* because the output signal at the source follows the signal at the gate very closely.

The equivalent circuit of the source follower is shown in Fig. 6-9a for comparison with Fig. 6-8, and rearranged for greater clarity in Fig. 6-9b. The output

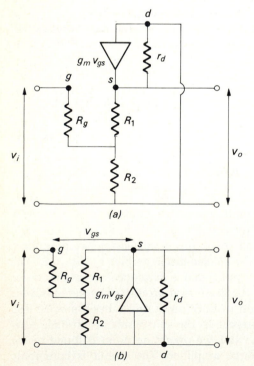

Figure 6-9 (a) Equivalent circuit of source-follower amplifier and (b) same circuit rearranged.

signal is given by the voltage drop across the combination of r_d in parallel with R_1 and R_2 in series.

$$v_o = g_m v_{gs} \frac{r_d(R_1 + R_2)}{r_d + R_1 + R_2} \cong g_m v_{gs} R_2 \tag{6-15}$$

where the approximation is valid since in practice the drain resistance is greater than $R_1 + R_2$ and R_2 is much greater than R_1. The gate signal is equal to the input signal less the voltage across the source resistor.

$$v_{gs} = v_i - g_m v_{gs} R_2 \tag{6-16}$$

where r_d is again taken to be large in comparison to $R_1 + R_2$. Equation (6-16) is solved for v_{gs} and inserted into Eq. (6-15). The result is

$$v_o = g_m R_2 \frac{v_i}{1 + g_m R_2} = \frac{g_m R_2}{1 + g_m R_2} v_i \cong v_i \tag{6-17}$$

The approximation applies for $g_m R_2 \gg 1$, as is usually the case.

According to Eq. (6-17), the output signal at the source is equal to the input signal at the gate, so that the name *source follower* is appropriate indeed. Although it is true that the voltage gain of the circuit is only equal to unity, the power gain is large. This is so because the source voltage replicates the input signal in a much lower resistance circuit. The ac signal power is proportional to the square of the signal voltage divided by resistance, so that the power amplification is great and the circuit can deliver appreciable power to a load.

Since the output voltage is included in the input circuit and the output signal opposes the input signal, the effective input impedance of a source follower is very large, much greater than just R_g. It turns out also that the effective output impedance of the circuit is very low, much less than $R_1 + R_2$. Thus, the source follower is an impedance-matching amplifier with a high input impedance and a low output impedance and for this reason proves useful at both the input and output of multiple-stage amplifier circuits. This aspect is explored more quantitatively in Chap. 7. In addition, Eq. (6-17) shows that circuit performance is relatively unaffected by changes in circuit parameters or by changes in the FET small-signal parameters, so long as the drain resistance and the mutual transconductance are large enough for the approximations to remain valid.

BIPOLAR-TRANSISTOR AMPLIFIERS

Bias Circuits

The emitter junction of a bipolar transistor requires forward bias and the collector junction reverse bias. Consider, for example, the simple grounded-emitter amplifier, Fig. 6-10, in which base bias current is supplied by the resistor R_B. Because the forward resistance of the emitter junction is very small, the base current is given by V_{cc}/R_B. An approximate value of collector current is then βI_b,

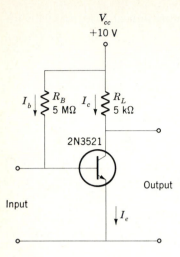

Figure 6-10 Simple grounded-emitter amplifier.

and the operating point is determined. Since the current gain depends somewhat upon the operating point, it is more accurate to plot the load line corresponding to R_L on the appropriate collector characteristics, such as Fig. 4-13. The intersection of the load line with the base current curve for $I_b = V_{cc}/R_B$ is the operating point.

This simple bias circuit is not generally satisfactory because the operating point shifts drastically with temperature. Comparison of collector characteristics at an elevated temperature, Fig. 4-14, with those for room temperature, Fig. 4-13, reveals that a much greater collector current exists at the higher temperature. Since the base bias current is fixed by the circuit, it is possible for the operating point to move into an unusable region of the transistor characteristics.

The most satisfactory transistor bias circuit is obtained by including a resistor in the emitter circuit, Fig. 6-11a. The voltage drop across R_E tends to bias the emitter junction in the reverse direction and the voltage divider comprising R_1 and R_2 sets the base voltage so that the base-emitter potential is in the forward direction. This circuit has the advantage that an increase in transistor current, increases the voltage drop across R_E so that the base bias current is reduced, which reduces the collector current, quite analogous to the case of source bias of the FET.

The quiescent operating point is found using the dc equivalent circuit in Fig. 6-11b. First, the appropriate load line is calculated from the voltage equation around the output circuit,

$$I_c = \frac{V_{cc}}{R_E + R_L} - \frac{1}{R_E + R_L} V_{ce} \tag{6-18}$$

Next, the bias line is derived from voltage drops around the input circuit,

$$I_c = \frac{V_{eq} - V_{be}}{R_E} - \frac{R_{eq}}{R_E} I_b \tag{6-19}$$

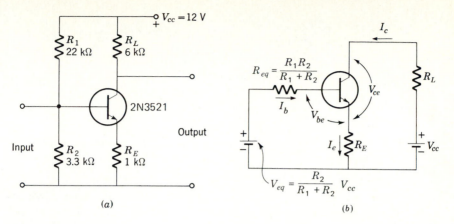

Figure 6-11 (*a*) Practical transistor bias circuit and (*b*) rearrangement to determine operating point.

In arriving at Eqs. (6-18) and (6-19) it is assumed that $I_e = I_c$, for simplicity. The bias line may be plotted on the transfer characteristic and the operating point determined from the intersection, as in the case of the FET. Actually, however, the transfer characteristics of bipolar transistors are so linear that it is easier to use a simple analytical expression for the transfer characteristic,

$$I_c = \beta I_b \tag{6-20}$$

In Eq. (6-20), β is often specified by the manufacturer and in any event can be estimated directly from the collector characteristics. After inserting Eq. (6-20) into Eq. (6-19), the expression may be solved for the base current,

$$I_b = \frac{V_{eq} - V_{be}}{R_{eq} + \beta R_E} \tag{6-21}$$

Knowing I_b and I_c, a value for V_{ce} is then calculated from Eq. (6-18) and the operating point is completely determined. Note that V_{be} may be considered to be a constant equal to 0.2 V for germanium transistors and to 0.6 V in the case of silicon transistors, as can be judged from the forward characteristics of the *pn* junctions in Fig. 3-2.

Hybrid Parameters

The electrical characteristics of a bipolar transistor can be represented by a *T-equivalent circuit*, Fig. 6-12, which corresponds to the grounded-base configuration already illustrated in Fig. 4-11. In the T-equivalent circuit the constant-current generator αi_e is in parallel with the collector-junction resistance r_c. The collector resistance is the reciprocal of the slope of the collector characteristic curves and in practice is of the order of 1 to 10 MΩ. The emitter resistance r_e is the forward resistance of the emitter junction, which turns out to be about 26 Ω at a quiescent bias current of 1 mA. The base resistance in the T-equivalent

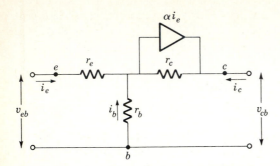

Figure 6-12 T-equivalent circuit of bipolar transistor.

circuit arises from two sources; the ohmic resistance of the base region (which may be appreciable since the base is so thin) and a feedback effect between the collector and the base. The origin of this effect is the decrease in base width as the collector junction widens with collector voltage. A typical value for the base resistance arising from these combined effects is 500 Ω.

The T-equivalent circuit for transistors is used in circuit analysis analogously to the FET ac equivalent circuit. The T equivalent is particularly appropriate because its parameters are directly related to the basic physical structure of the transistor. Note that the T-equivalent circuit involves a direct connection between the input and output terminals, which differs from the FET case. This may be illustrated explicitly by solving the T-equivalent circuit, Fig. 6-12, for the various currents and voltages. Using Kirchhoff's rules,

$$i_e + i_b + i_c = 0$$

$$v_{eb} = i_e r_e - i_b r_b \tag{6-22}$$

$$v_{cb} = (\alpha i_e + i_c)r_c - i_b r_b$$

Solving for i_c and v_{eb},

$$i_c = -\alpha i_e + \frac{1}{r_c} v_{cb} \tag{6-23}$$

$$v_{eb} = [r_e + (1 - \alpha)r_b]i_e + \frac{r_b}{r_c} v_{cb} \tag{6-24}$$

where the approximation $r_c \gg r_b$ has been introduced. Thus, two equations are necessary to describe the operation of a bipolar transistor, whereas a single expression is sufficient in the case of the FET. Equation (6-23) is quite analogous to the corresponding expression for a FET and should be compared with Eq. (6-11). The other relation, Eq. (6-24), indicates that the input voltage depends upon the output voltage, v_{cb}, as well as upon the input current i_e. This is a direct result of the connection between input and output in a bipolar transistor.

Although the T-equivalent circuit is a satisfactory representation of transistor operation, it is fairly difficult to determine the various parameters of the circuit by direct measurements on actual transistors. For this reason an equivalent circuit which involves so-called *hybrid parameters* is more commonly employed in circuit analysis. This representation may be illustrated by writing Eqs. (6-23) and (6-24) as

$$i_c = h_{fb} i_e + h_{ob} v_{cb} \tag{6-25}$$

$$v_{eb} = h_{ib} i_e + h_{rb} v_{cb} \tag{6-26}$$

where the h's are the hybrid parameters. The significance of the h-parameter subscripts is as follows: the subscripts i, r, f, and o refer to input, reverse, forward, and output, respectively, which may be understood by examining the meaning of each coefficient in Eqs. (6-25) and (6-26). The subscript b signifies the common-base configuration for which these equations have been developed. This designation distinguishes these parameters from those appropriate for the common-emitter and common-collector connections discussed below. A summary of the subscript notations is presented in Table 6-2.

According to Eq. (6-26) it is possible to determine h_{ib} from the ratio of the emitter-base voltage to the emitter current v_{eb}/i_e with the collector shorted to ground so that $v_{cb} = 0$. Similarly, the ratio of the collector current to emitter current i_c/i_e with the collector shorted yields h_{fb}, using Eq. (6-25). When the emitter is open-circuited, $i_e = 0$, so that $h_{rb} = v_{eb}/v_{cb}$ and $h_{ob} = i_c/v_{cb}$. Note that the conditions $v_{cb} = 0$ and $i_e = 0$ refer to ac signals; the dc potentials are maintained at the proper operating point for the transistor. Because of the small emitter resistance and large collector resistance of a bipolar transistor it is particularly easy to achieve an open-circuited emitter or a short-circuited collector for ac signals while maintaining the desired dc potentials. In this way the small-signal h parameters can be measured using ac bridge techniques. The great virtue of the hybrid parameters is the fact that they can be measured directly with relative ease.

The *hybrid equivalent circuit* representing the grounded-base configuration is given in Fig. 6-13, as can be determined by inspecting Eqs. (6-25) and (6-26).

Table 6-2 Subscript notation for h parameters

Subscript	Meaning
i	Input parameter
r	Reverse parameter
f	Forward parameter
o	Output parameter
e	Common emitter
b	Common base
c	Common collector

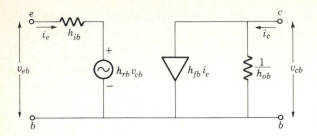

Figure 6-13 Grounded-base hybrid equivalent circuit.

Kirchhoff's rule applied to the input loop yields Eq. (6-26) directly. The voltage equation around the output loop gives

$$v_{cb} = \frac{1}{h_{ob}} (i_c - h_{fb} i_e) \tag{6-27}$$

Solving for i_c yields Eq. (6-25). Therefore, this hybrid equivalent circuit does represent the operation of the transistor as given by Eqs. (6-25) and (6-26). These equations, in turn, are based on the fundamental processes associated with pn junctions in the transistor structure.

The direct connection between input and output is not so immediately apparent in the hybrid equivalent circuit as compared with the T equivalent. Note, however, that the voltage generator $h_{rb} v_{cb}$ in the input circuit involves the collector-base voltage and that the current generator in the output circuit $h_{fb} i_e$ includes the emitter current. Thus, the input and output circuits are indeed coupled. According to Fig. 6-13, h_{ib} represents a resistance, h_{ob} is a conductance, while h_{rb} and h_{fb} are simple numerics. These quantities represent different physical entities and are therefore termed hybrid parameters. The relation between hybrid parameters and T-equivalent parameters may be determined directly by comparing Eq. (6-25) with (6-23) and Eq. (6-26) with (6-24).

The common-emitter configuration is more widely used than is the common-base circuit. While it is possible to rearrange the equivalent circuit appropriate for the common-base configuration to apply to the common-emitter case, it is much more convenient to employ the same form of equivalent circuit and adjust parameters of the circuit for the common-emitter configuration. The relations between grounded-emitter h parameters and those corresponding to the grounded-base circuit are developed by straightforward circuit analysis analogous to the transition from the T-equivalent circuit to the grounded-base case.

The h parameters depend upon the operating point of the transistor as well as upon the particular transistor type. Illustrative values of the common-emitter parameters and their variation with emitter current are given in Fig. 6-14. Note that the forward current gain h_{fe} and the reverse voltage amplification factor h_{re} are sensibly constant, while both the input resistance h_{ie} and the output conductance h_{oe} vary considerably.

For most practical purposes h_{re} and h_{oe} are small enough to be ignored, particularly if R_L is also small enough so that $h_{oe} R_L \leq 0.1$. In this case the

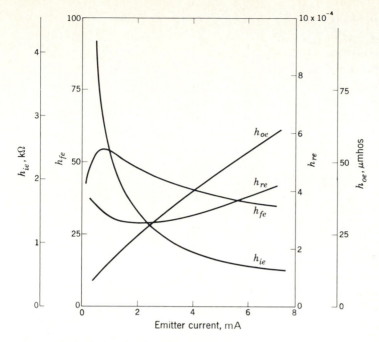

Figure 6-14 Variation of hybrid parameters with emitter current.

simplified hybrid equivalent circuit in Fig. 6-15a is appropriate for the grounded-emitter configuration. Note that this circuit is just Fig. 6-13 with $h_{re} = h_{oe} = 0$. The analogous approximate hybrid equivalents for the common-base and common-collector connections are obtained directly from Fig. 6-15a by grounding the appropriate terminal, as illustrated in Fig. 6-15b and Fig. 6-15c, respectively. In this way, only the two parameters, h_{fe} and h_{ie}, need be known to analyze any of the three circuit configurations.

Bipolar-Transistor Circuits

The complete circuit of a practical common-emitter amplifier using an *npn* transistor is shown in Fig. 6-16. The performance of this circuit as an amplifier is derived from analysis of the hybrid equivalent circuit in Fig. 6-15a. Note that the effect of R_1 and R_2 is neglected in comparison with the transistor input resistance, h_{ie}. The voltage gain is just the output voltage divided by the input signal,

$$a = \frac{v_o}{v_i} = \frac{-(h_{fe} i_b)R_L}{i_b h_{ie}} = -h_{fe} \frac{R_L}{h_{ie}} \tag{6-28}$$

According to Eq. (6-28), the voltage gain is equal to the forward current gain of the transistor times the ratio of the load resistance to the input resistance. The gain of the common-emitter amplifier can be appreciable since both factors are

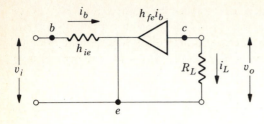

(a) Common emitter

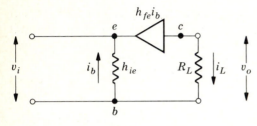

(b) Common base

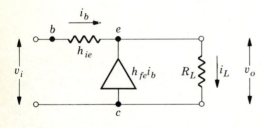

(c) Common collector

Figure 6-15 Approximate hybrid equivalent circuit for (a) common-emitter, (b) common-base, and (c) common-collector bipolar transistor amplifiers.

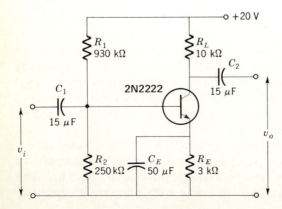

Figure 6-16 Practical common-emitter amplifier using npn type 2N2222 transistor.

large. The minus sign signifies that the input and output signals are 180° out of phase.

The bipolar transistor is basically a current-controlled device, and the current gain, g, which is the ratio of the output current to input current, is also important. According to Fig. 6-15a,

$$g = \frac{i_o}{i_i} = \frac{-h_{fe} i_b}{i_b} = -h_{fe} \tag{6-29}$$

This is just the forward current gain of the transistor, and again the minus sign represents phase shift between input and output.

The input resistance of bipolar transistor amplifiers is much smaller than that corresponding to FET circuits. It is determined from the ratio of the input voltage to the input current, or

$$R_i = \frac{v_i}{i_i} = \frac{i_b h_{ie}}{i_b} = h_{ie} \tag{6-30}$$

Analysis of the full hybrid equivalent circuit yields a more accurate expression for the input resistance,

$$R_i = h_{ie} - h_{re} \frac{h_{fe} R_L}{1 + h_{oe} R_L} \tag{6-31}$$

According to Eq. (6-31), the input resistance depends somewhat upon the output load resistance, an indication of the coupling between input and output inherent in bipolar transistors. For most practical purposes, the simpler version, Eq. (6-30), is sufficiently accurate.

The output resistance of the amplifier is determined from the ratio of the open-circuit ($R_L = \infty$) output voltage to the short-circuit ($R_L = 0$) output current, according to Eq. (1-53). Thus,

$$R_o = \frac{-h_{fe} i_b R_L}{-h_{fe} i_b} = R_L \tag{6-32}$$

A slightly more accurate expression includes the output resistance of the transistor, $1/h_{oe}$, in parallel with R_L.

In summary, the common-emitter amplifier yields both voltage and current gain. It has a modestly high input resistance and large output resistance. Because of the favorable values of these four quantities, it is the most commonly used bipolar-transistor amplifier circuit. Columns 1 and 4 of Table 6-3 summarize these expressions for the properties of the common-emitter amplifier together with typical values derived from Fig. 6-14.

A typical common-base, or grounded-base, amplifier circuit using an *npn* transistor is shown in Fig. 6-17. Comparing this circuit with the common-emitter configuration, Fig. 6-16, reveals that the dc bias arrangements are identical. This means that bias considerations and the techniques for determining the operating

Table 6-3 Properties of bipolar transistor amplifiers

	Common emitter	Grounded base	Emitter follower	Numerical values*		
				CE	GB	EF
Voltage gain	$-h_{fe}\dfrac{R_L}{h_{ie}}$	$h_{fe}\dfrac{R_L}{h_{ie}}$	$1 - \dfrac{h_{ie}}{R_i}$	-330	330	0.997
Current gain	$-h_{fe}$	$\dfrac{h_{fe}}{1 + h_{fe}}$	$1 + h_{fe}$	-50	0.98	51
Input resistance, ohms	h_{ie}	$\dfrac{h_{ie}}{1 + h_{fe}}$	$h_{ie} + (1 + h_{fe})R_L$	1500	29	5.1×10^5
Output resistance, ohms	R_L	R_L	$\dfrac{h_{ie} + R_s}{1 + h_{fe}}$	10^4	10^4	31

* Using Fig. 6-14 for $I_E = 2\text{ mA}$; $R_L = 10^4\ \Omega$, $R_s = 100\ \Omega$.

point previously described for the common-emitter amplifier apply to the grounded-base circuit as well.

Capacitor C_B bypasses the base resistor R_2, and R_1 is shorted out for ac signals by the low impedance of the battery. Consequently in the hybrid equiva- lent circuit, Fig. 6-15b, these resistors are absent. Analysis of this circuit proceeds exactly as for the grounded-emitter amplifier, Fig. 6-15a.

According to cols. 2 and 5 of Table 6-3, the voltage gain of the grounded-base amplifier is the same as that of the grounded-emitter circuit. The current gain, however, is about unity and there is no phase shift between input and output signals. Also, the input resistance is very small, which is usually undesirable, while the output resistance is, again, equal to the load resistance.

The common-collector amplifier, Fig. 6-18, is more often termed an *emitter follower* by analogy to the source-follower amplifier. Here again, bias considera- tions are identical with those previously discussed, and the appropriate hybrid equivalent circuit is shown in Fig. 6-15c. As might be anticipated from the FET case, circuit analysis results summarized in cols. 3 and 6 of Table 6-3 show a unity voltage gain and an appreciable current gain without phase shift.

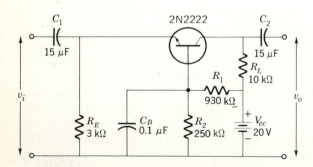

Figure 6-17 Common-base tran- sistor amplifier. Bias circuit is identical to common-emitter case.

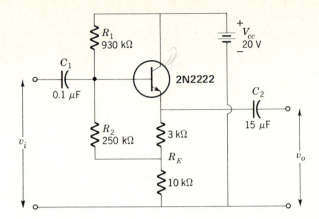

Figure 6-18 Common-collector transistor amplifier. This circuit is also called emitter follower.

The input and output resistances of the emitter-follower amplifier are especially noteworthy. The input resistance is essentially the load resistance multiplied by the forward current gain, and is correspondingly large. To derive the output resistance it is necessary to include the resistance of the input voltage source, R_s, in order to determine the open-circuit output voltage. Referring to Fig. 6-15c, the output resistance is

$$R_o = \frac{i_b(R_s + h_{ie})}{i_b + h_{fe} i_b} = \frac{h_{ie} + R_s}{1 + h_{fe}} \tag{6-33}$$

According to Eq. (6-33) the output resistance is approximately the resistance of the source divided by the forward current gain and is quite small. Clearly, the emitter follower acts as an impedance transformer between input and output. It makes the output load resistance appear much larger at the input terminals, and, correspondingly, it makes the source resistance appear to be much smaller at the output terminals.

SPECIAL AMPLIFIERS

Difference Amplifier

A very popular and versatile circuit often used as the input stage in laboratory instruments is the *difference amplifier*, Fig. 6-19, so named because it develops an output signal which is proportional to the difference between two input voltages. The two signals are introduced to the base terminals of a pair of similar transistors coupled together through a common-emitter resistor. The output signal is taken between the collector terminals.

The useful properties of this amplifier are illustrated by analyzing the equivalent circuit, Fig. 6-20. Application of Kirchhoff's voltage rule to the input loop

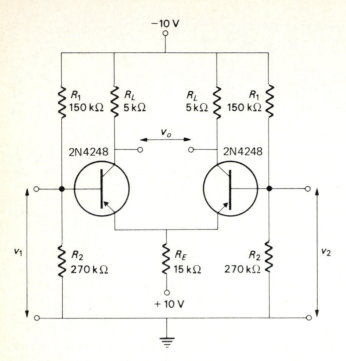

Figure 6-19 Bipolar-transistor difference amplifier.

$b_1 e_1 e_2 b_2$ yields

$$v_1 - i_{b_1} h_{ie} + i_{b_2} h_{ie} - v_2 = 0$$
$$v_1 - v_2 = h_{ie}(i_{b_1} - i_{b_2})$$

(6-34)

Similarly, around the output loop,

$$v_o + h_{f_e} i_{b_1} R_L - h_{f_e} i_{b_2} R_L = 0$$
$$v_o = -h_{f_e}(i_{b_1} - i_{b_2})R_L$$

(6-35)

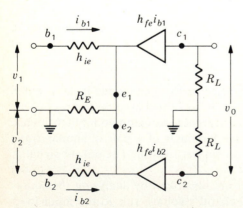

Figure 6-20 Equivalent circuit of difference amplifier.

Substituting for the quantity $(i_{b_1} - i_{b_2})$ in Eq. (6-35) from (6-34) and solving for the output voltage,

$$v_o = -h_{fe} \frac{R_L}{h_{ie}} (v_1 - v_2) \tag{6-36}$$

This expression is similar to Eq. (6-28) for the voltage gain of a single grounded-emitter amplifier.

According to Eq. (6-36), the difference amplifier rejects any voltage signals that are common to both terminals, such that $v_1 = v_2$. At the same time a signal applied between the input terminals is amplified normally since in this case $v_1 = v_i/2$ and $v_2 = -v_i/2$,

$$v_o = -h_{fe} \frac{R_L}{h_{ie}} v_i \tag{6-37}$$

The same result holds if the signal is applied between either input terminal and ground so that, for example, $v_1 = v_i$ and $v_2 = 0$. Since the circuit rejects *common-mode* signals resulting from, say, stray electric fields caused by the 60-Hz power mains, it is useful as the input stage of sensitive amplifiers. In addition, the input may be connected to signal sources that have neither terminal grounded, which often proves to be convenient.

The ability of the difference amplifier to reject common-mode signals rests on exact symmetry in the two halves of the circuit. Since this is rarely achieved in practical amplifiers, it is useful to examine the effect of small asymmetries. Suppose, for example, that the forward current gain parameters differ slightly, $h_{fe_1} \neq h_{fe_2}$. In this case, Eq. (6-35) becomes

$$v_o' = -(h_{fe_1} i_{b_1} - h_{fe_2} i_{b_2}) R_L \tag{6-38}$$

An additional expression is needed to solve for all the unknowns, and this is obtained from the voltage drops around the input loop $b_1 e_1 R_E$,

$$v_1 - i_{b_1} h_{ie} - R_E (h_{fe_1} i_{b_1} + h_{fe_2} i_{b_2}) = 0 \tag{6-39}$$

Here the effect of the base currents in R_E have been neglected in comparison to the collector currents.

Equation (6-35) yields an expression for i_{b_2}, which, when substituted into Eq. (6-39) can be solved for i_{b_1}. Both expressions are substituted into Eq. (6-38), and the output voltage becomes

$$v_o' = -h_{fe_1} \frac{R_L}{h_{ie}} \frac{h_{ie}(v_1 - v_2 h_{fe_2}/h_{fe_1}) + 2h_{fe_2} R_E(v_1 - v_2)}{h_{ie} + (h_{fe_1} + h_{fe_2}) R_E} \tag{6-40}$$

Note that Eq. (6-40) reduces to the exactly balanced case, Eq. (6-36), when $h_{fe_1} = h_{fe_2}$. Suppose now that $v_1 = v_2 = v_i$; then

$$v_o' = -h_{fe_1} \frac{R_L}{h_{ie}} \frac{h_{ie}(1 - h_{fe_1}/h_{fe_2}) v_i}{h_{ie} + (h_{fe_1} + h_{fe_2}) R_E} \tag{6-41}$$

According to Eq. (6-41), an output signal results from the circuit imbalance.

The ratio of the difference signal output, Eq. (6-36), to the common-mode signal, Eq. (6-41),

$$\frac{v_o}{v_o'} = \frac{h_{ie} + (h_{fe_1} + h_{fe_2})R_E}{h_{ie}(1 - h_{fe_1}/h_{fe_2})}$$

$$\cong \frac{2h_{fe}/h_{ie}}{1 - h_{fe_1}/h_{fe_2}} R_E \qquad (6\text{-}42)$$

is called the *common-mode rejection ratio*. The value of the ratio is very large when the circuit is balanced, that is, when $h_{fe_1} \cong h_{fe_2}$. Conversely, for a given asymmetry, the magnitude of the ratio and the ability of the amplifier to reject common-mode signals is enhanced by large values of the common-emitter resistor, R_E.

Complementary Symmetry

One of the more intriguing circuit applications of transistors is based on the combination of *npn* and *pnp* transistors, with their symmetrically inverted bias and signal-voltage polarities. Consider, for example, the *complementary-symmetry* circuit, Fig. 6-21, employing *npn* and *pnp* grounded-emitter amplifiers having common input and output connections. The base bias on both transistors is zero so in the absence of signal the transistors are cut off. Therefore, current is present in each transistor only when the input signal voltage biases its emitter junction in the forward direction. This happens on alternate half-cycles of the input voltage waveform because of the opposite polarities of the two transistors. Thus, the *npn* transistor delivers current to the load resistor when the *pnp* unit is cut off, and vice versa. The output signal is a replica of the input waveform, even though each transistor operates only half the time.

This particularly simple circuit is an efficient power amplifier since the quiescent current is zero and each transistor operates over the entire range of its characteristics. Furthermore, I^2R losses are small because the dc current in the load resistor is zero. Complementary-symmetry FET amplifiers are equally useful, particularly in connection with digital circuits discussed in Chap. 9.

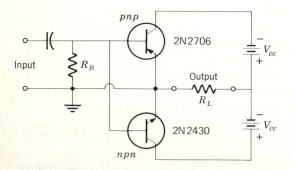

Figure 6-21 Complementary-symmetry amplifier employing *pnp* and *npn* transistors.

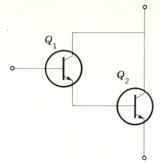

Figure 6-22 Darlington-connected amplifier.

Darlington Connection

The combination of two similar transistors in the so-called *Darlington connection*, Fig. 6-22, has many favorable circuit properties. Note that Q_2 is directly connected to Q_1 and the base-to-collector potential of Q_2 provides the emitter-collector voltage for Q_1. Furthermore, the output emitter current of Q_1 is the base input current of Q_2.

The circuit may be viewed as an emitter follower, Q_1, followed by a grounded-emitter amplifier, Q_2. The combination produces a very large current gain, $h_{fc} \times h_{fe} \cong \beta^2$. The circuit also has the voltage gain of the common emitter with the large input impedance of the emitter-follower amplifier, both of which are desirable features.

For most purposes the combination may be considered to be a single device. Indeed, units having the two transistors in one case are commercially available. Proper bias potentials are applied by a conventional transistor bias circuit such as Fig. 6-11 in which the combination of Q_1 and Q_2 replaces the single grounded-emitter transistor.

Cascading

The transistor circuits discussed in previous sections are ideally suited to amplify voltage signals with minimum waveform distortion. Gain factors greater than those possible with a single-stage amplifier are obtained by *cascading* several amplifier stages. The output of one amplifier stage is amplified by another stage or stages until the desired signal voltage level is achieved.

Consider, for example, the two-stage cascaded bipolar-transistor amplifier, Fig. 6-23. Two individual circuits are connected with the coupling capacitor C_{c2}. This capacitor passes the amplified ac signal from Q_1 to the base of Q_2. At the same time it blocks the collector voltage of Q_1 from the base of the second transistor. Similarly, capacitors C_{c1} and C_{c3} isolate the input and output circuits insofar as dc potentials are concerned.

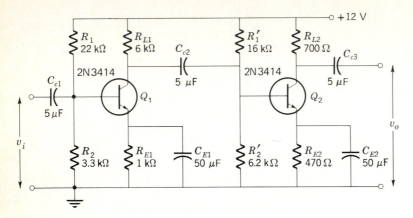

Figure 6-23 Two-stage cascaded transistor amplifier.

Cascaded bipolar-transistor voltage amplifiers most often employ the grounded-emitter configuration because of the combined voltage and current gain of this circuit. Neither the common-base nor the emitter-follower configuration achieves as great overall voltage amplification when cascaded. Note that the bias resistors of the second stage in Fig. 6-23 are different from those of the first stage, even though both transistors are identical. The operating points are set at different places in order to obtain most favorable values of the *h* parameters in each stage.

The entire ac equivalent circuit of this amplifier may be drawn and the performance of the system determined from a complete ac-circuit analysis. Actually, this procedure is unwieldy because of the number of loops in the network and is rarely attempted. Rather, the circuit is analyzed in several separate steps, each of which has a minimum of mathematical complexity. This has the further advantage that the important effects can be isolated and more clearly examined.

For example, the reactances of both emitter bypass capacitors are assumed small enough to be negligible. Accordingly, these components are absent in the ac equivalent circuit of the amplifier, Fig. 6-24. Similarly, the reactances of the coupling capacitors are taken to be negligible. The output voltage may then be written immediately as

$$v_o = a_2 v_o' = a_1 a_2 v_i \tag{6-43}$$

where a_1 and a_2 are the gain factors of the first stage and the second stage, respectively. Evidently, the overall gain of a cascaded amplifier is just equal to the product of the gains of the individual stages.

Note that in calculating the gain of each stage the output load of the first stage involves the input impedance of the second. This means that the input impedance of this stage must be determined first. Since the input impedance of a bipolar-transistor amplifier depends upon the output load impedance, it is necessary to work backward through the circuit starting from the output terminals.

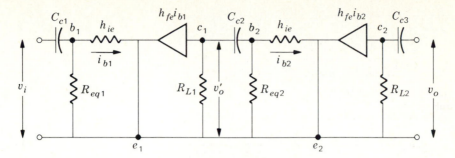

Figure 6-24 Equivalent circuit of two-stage transistor amplifier of Fig. 6-23.

With the output load specified, the gain and input impedance of the second stage are calculated using results developed earlier. The input impedance is part of the load for the preceding stage, and its gain and input impedance are calculated accordingly. This procedure is not required in the case of cascaded FET stages because the very large input impedance inherent in the FET does not load the preceding stage. Thus the overall gain of a FET amplifier is just the product of the individual gains of isolated stages. In either case, the result is termed the *midband gain* of the amplifier because the reactances of the capacitors are taken to be negligible. This approximation applies to signal frequencies which are neither so low that the reactances cannot be ignored nor so high that other effects reduce the gain.

Low- and High-Frequency Gain

At sufficiently low frequencies the capacitive reactances may no longer be neglected. The effect of the emitter bypass capacitors is usually of greater significance than that of the coupling capacitors, although both tend to reduce the gain at low frequencies. Because of the large input resistance of FET amplifiers, the coupling capacitors are dominant in this case. The low input impedance of bipolar transistors requires coupling capacitors of a few tens of microfarads, and practical coupling capacitors are limited to values below about 0.5 μF in the case of FET circuits by leakage currents. Emitter and source bypass capacitors may be much larger, 100 μF or more, because leakage currents are less critical in this portion of the circuit.

Because the overall gain of cascaded stages is the product of individual stage gains, it is only necessary to examine the effect of the coupling capacitor reactance for an isolated amplifier stage. According to the ac equivalent circuit, Fig. 6-24, this effect can be treated by considering the simple RC circuit comprising the coupling capacitor and the input impedance of the amplifier. This portion is shown separately in Fig. 6-25 for clarity. The input impedance R_i is simply the gate resistor in the case of the FET amplifier, but in bipolar-transistor stages it includes the input impedance of the transistor as well. In either case, Fig. 6-25 is a RC high-pass filter already discussed in Chap. 2. The gain of this amplifier stage

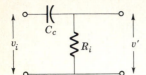

Figure 6-25

including the effect of the high-pass filter is given by

$$a(f) = \frac{v_o}{v_i} = \frac{av'}{v_i} = \frac{a}{\sqrt{1 + (f_0/f)^2}} \tag{6-44}$$

where the frequency-response characteristic of a high-pass filter, Eq. (2-45), is used and f_0 is the half-power frequency. Note that the gain is reduced from the mid-band value, a, when the signal frequency is smaller than the half-power frequency.

At the same time, a phase shift is introduced between the input and output signal, which is also significant in determining waveform distortion of the ampli-fier. Recall that in the Fourier analysis of a complex signal waveform, both the amplitudes and relative phases of all frequency components must be preserved if the output wave is to be an amplified replica of the input signal. It is also important to recognize that Eq. (6-44) applies to an individual stage. The low-frequency response of the entire amplifier is always poorer than that of any individual stage because the gain of cascaded stages is the product of individual stage gains.

The high-frequency gain of any amplifier is reduced by stray capacitive effects that are not purposely made part of the circuit. Referring to the simple FET amplifier stage, Fig. 6-26, these are the gate-source capacitance, C_1, the gate-drain capacitance, C_2, and the drain-source capacitance, C_3 of the FET. Also included in C_1 and C_3 are stray capacitances between the wires and components attached to the gate and drain terminals. All three of these capacitors shunt the signal at frequencies high enough for the capacitive reactances to be small.

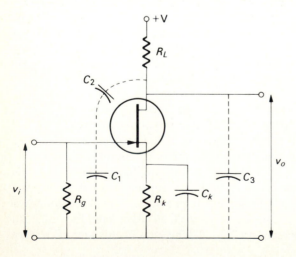

Figure 6-26 Stray capacitances in FET amplifier.

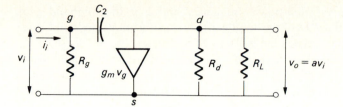

Figure 6-27

The effect of the gate-drain capacitance is particularly important. Consider the pertinent equivalent circuit, Fig. 6-27, in which C_2 is connected between gate and drain terminals and the other stray capacitances are ignored. The input impedance of the amplifier stage is calculated assuming that the reactance of C_2 is the dominant factor,

$$\mathbf{Z}_i = \frac{v_i}{i_i} = \frac{v_i}{(v_i + av_i)/(1/j\omega C_2)}$$

$$= \frac{1}{j\omega(1 + a)C_2} \tag{6-45}$$

This result indicates that the effect of C_2 is as though a larger capacitance, $(1 + a)C_2$, is connected from gate to ground. The increase in effective shunt capacitance caused by the gain of the stage is called the *Miller effect* and is the dominating factor in determining the high-frequency response of both FET and bipolar-transistor amplifiers.

The appropriate high-frequency equivalent circuit includes a shunt capacitance

$$C_s = C_1 + C_3 + (1 + a)C_2 \tag{6-46}$$

from the gate or base terminal to ground. The effect of C_s is ascertained after isolating the input circuit, Fig. 6-28. In this circuit R_{eq} represents the entire shunt resistance to ground comprising the output resistance of the preceding stage and the input resistance of the transistor, as well as the equivalent resistance of the bias resistors. Figure 6-28 is a RC low-pass filter, so that the gain is, from Eq. (2-27),

$$a(f) = \frac{v_o}{v_i} = \frac{av'}{v_i} = \frac{a}{\sqrt{1 + (f/f_0)^2}} \tag{6-47}$$

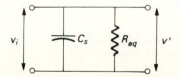

Figure 6-28

where, here again, f_0 is the half-power frequency and a is the midband gain. According to Eq. (6-47) the gain of the amplifier is reduced from its midband value when the signal frequency exceeds the half-power frequency.

Using Eqs. (6-44) and (6-47), the *frequency response* of any voltage amplifier is similar to that illustrated in Fig. 6-29a. The *low-frequency cutoff* for each stage is determined from Eq. (6-44) while the *high-frequency cutoff* for each stage is found using Eq. (6-47). The *bandpass* of the complete amplifier is the frequency interval between the high- and low-frequency points where the gain falls to $1/\sqrt{2}$ of the midband gain. It is conventional to employ logarithmic scales on both axes of bandpass characteristics such as Fig. 6-29a because the range of gains and frequencies is so great. The vertical scale is often put in terms of a unit called the *bel*, named after Alexander Graham Bell, the inventor of the telephone. Actually, a unit one-tenth as large, the *decibel*, abbreviated dB, proves more convenient in

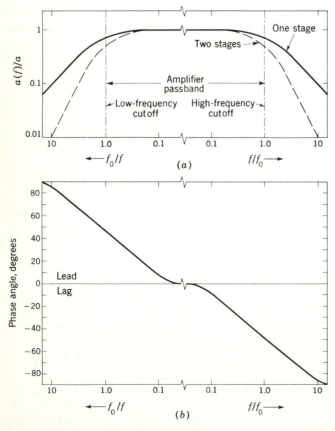

Figure 6-29 (*a*) Frequency-response characteristic of single amplifier stage. Characteristic of two-stage amplifier shown dashed. (*b*) Phase-shift characteristic of single amplifier stage.

practice. The decibel is defined as

$$dB = 20 \log a(f) \tag{6-48}$$

Correspondingly, the midband gain is often quoted in terms of decibels using the definition

$$dB = 20 \log \frac{v_o}{v_i} \tag{6-49}$$

The advantage of this unit is that the total gain in decibels of several amplifier stages is simply the sum of the individual gains in terms of decibels. Note that, according to Eq. (6-49), the amplifier gain is down 3 dB at the upper and lower half-power points.

The phase-shift characteristics of a single-stage amplifier are illustrated in Fig. 6-29b. The output signal leads the input at frequencies below the low-frequency cutoff and lags at frequencies above the high-frequency cutoff. The phase-shift characteristics of an entire amplifier are determined by adding the contributions from each stage.

Decoupling

When three or more stages of amplification are cascaded, it is necessary to *decouple* the power supply of the input stage from the remainder of the amplifier. The reason for this is that the supply voltage changes with current because of the effective internal impedance of the power supply.

Any small change in the power-supply voltage alters the bias on the first stage, and this change is amplified in the same fashion as an input signal. If the change in bias increases the current in the first stage, current in the second stage is reduced because of the 180° phase shift in the input amplifier. The current in the third stage is increased, however, because of another 180° phase shift in the second stage, and the change in the third stage is much larger than the original disturbance because of the gain of the amplifier. This additional load causes a decrease in the power-supply voltage and the process is cumulative. The changes continue until one transistor is driven to cutoff or into saturation, which reduces the overall gain to zero. The power-supply voltage then returns to normal and the process repeats itself. The result of this coupling from output to input is that the amplifier rapidly oscillates from cutoff to saturation at a rate that is a function of the circuit components.

A low-pass *RC* filter inserted in the power-supply lead to the first stage, Fig. 6-30, circumvents this difficulty. The time constant of this decoupling filter is selected so that power-supply variations are greatly attenuated. Actually, the characteristic filter frequency is put below the low-frequency cutoff of the amplifier where the gain is insufficient to support oscillations.

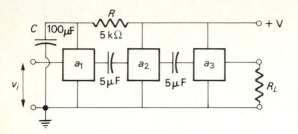

Figure 6-30 Simple *RC* filter eliminates instability in amplifier by reducing coupling from common power supply.

Transformer Coupling

When transistor amplifiers deliver appreciable amounts of power, it is no longer feasible to use resistors in the collector or drain circuit. The dc I^2R losses become significant at the high currents associated with large powers. Instead, a transformer couples the amplifier to the load. The dc collector current in the winding resistance introduces only a small power loss, and the output impedance of the amplifier is matched to the load by the transformer so that the actual load resistance can be any convenient value.

Transformer coupling is also useful in the two-transistor *push-pull* amplifier, Fig. 6-31, which has increased power output, efficiency, and less distortion than a single-transistor circuit. The center-tapped input transformer drives each of the transistors with signals 180° out of phase, which accounts for the name of the circuit. The amplified collector currents combine in the center-tapped output transformer to produce a load current waveform that is a replica of the input signal. This action has already been noted in the complementary-symmetry amplifier. The input transformer also matches the driver stage to the input impedance of the amplifier.

Increased efficiency results when the push-pull amplifier is biased nearly to cutoff. This is known as *class B* operation to distinguish the circuit performance from that previously described in which current is present in the transistor during the entire input-signal cycle. Because each transistor is biased near cutoff, the

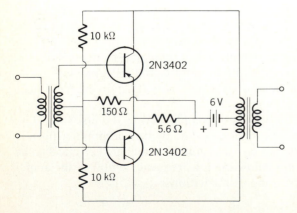

Figure 6-31 Push-pull power amplifier.

quiescent current is very small and the signal-voltage and current excursions can be equal to the maximum permissible collector voltage and current.

Coupled resonant circuits, which are in effect tuned transformers, are useful when it is only necessary to amplify a single frequency or a narrow band of frequencies. The impedance of a parallel resonant circuit is very great at resonance, as discussed in Chap. 5, so that appreciable gain is achieved at the resonant frequency. Stray circuit capacitances are incorporated into the resonant circuits so that good performance at high frequencies is possible. The operation of a typical tuned amplifier is analyzed in Exercise 6-13.

PULSE AMPLIFIERS

Signals which consist of a series of rapid transitions from one voltage level to another as, for example, square waves, are equally important as sinusoidal waveforms. Although such *pulse* waveforms may be resolved into their sinusoidal harmonic components using the method of Fourier analysis, it usually proves more convenient to consider the signal pulses themselves. An ideal pulse amplifier introduces no change in the pulse shape other than increasing its amplitude, and practical circuits can come close to this performance.

Rise Time

The important characteristics of pulse amplifiers, gain and frequency response, are similar to those already discussed in connection with conventional amplifiers. To illustrate, in particular, the significance of frequency response, consider the ideal square pulse shown in Fig. 6-32 and the nonideal pulse shape resulting from passing this waveform through a practical amplifier. Note that the amplified pulse does not rise instantaneously to the maximum value. This is a result of the high-frequency cutoff of the amplifier caused by shunt capacitance. Consider the case

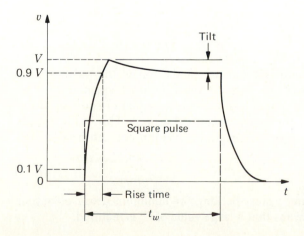

Figure 6-32 Comparison of ideal square voltage pulse with amplified, nonideal pulse.

of a single-stage amplifier, which is represented by the input circuit of Fig. 6-28. The waveform of the output pulse v' due to a square voltage pulse v_i starting at $t = 0$ is found by the transient circuit analysis technique described in Chap. 2. According to Eq. (2-59), the output signal increases exponentially at a rate given by the circuit time constant

$$\tau = R_{eq} C_s \tag{6-50}$$

It is desirable to make the time constant as small as possible to minimize distortion of the leading edge of the pulse.

At the end of the input pulse, the output signal decays exponentially to zero at a rate again determined by the circuit time constant. Note, however, that the preceding stage may be completely cut off at the end of the pulse so that R_{eq} is larger since it then does not include the output resistance of the preceding stage. This means that the leading and trailing edges of the pulse waveform are likely to have different time constants; the rise time is usually more rapid. This disparity is minimized in bipolar-transistor pulse amplifiers because the low input impedance dominates.

It is convenient to define the *rise time* t_r of a pulse as the time required for the voltage to go from 10 to 90 percent of its final value, as indicated in Fig. 6-32. A useful approximate expression for t_r may be developed from Eq. (2-59) by setting

$$0.9 = 1 - e^{-t_r/\tau} \tag{6-51}$$

Solving for t_r and introducing the high-frequency cutoff, $2\pi f_0 = 1/\tau$, of the amplifier from Eq. (2-46), the result is

$$t_r = \frac{\ln 10}{2\pi f_0} \cong \frac{1}{3 f_0} \tag{6-52}$$

According to Eq. (6-52) an amplifier with a good high-frequency response is necessary if the pulse rise time is short. For example, if the pulse rise time is 1 μs, the amplifier bandwidth must extend to 370 kHz. In this connection it should be remembered that the bandwidth of a multiple-stage amplifier is always poorer than that of any individual stage.

Tilt

A second form of pulse distortion is departure of the top of the pulse from a constant value, called *tilt*, in Fig. 6-32. This is a result of low-frequency cutoff of the amplifier. In effect, the pulse length represents a dc signal of short duration and the appropriate single-stage equivalent circuit is Fig. 6-25. This is a high-pass filter and the voltage transient is an exponential waveform characterized by the time constant

$$\tau = R_i C_c \tag{6-53}$$

Clearly, the circuit time constant τ must be long for minimum decay of output voltage during the pulse. This means that a large value of C_c is required.

The tilt is small in most practical situations, so the exponential .can be approximated by the first two terms of a series expansion,

$$v' = v_i\left(1 - \frac{t}{\tau}\right)$$

(6-54)

The percentage tilt $P = \Delta v'/v_i$ at the end of the pulse duration t_w is then

$$P = \frac{\Delta v'}{v_i} = \frac{t_w}{\tau}$$

(6-55)

Introducing the low-frequency cutoff $2\pi f_0 = 1/\tau$,

$$t_w = \frac{P}{2\pi f_0}$$

(6-56)

Notice that a long pulse requires an amplifier with an excellent low-frequency response. For example, suppose that tilt distortion must be less than 1 percent. Then the low-frequency cutoff of an amplifier required to handle 1-ms pulses must extend down to 1.6 Hz.

The results of these two sections provide a convenient technique for determining response characteristics of amplifiers. A square wave is applied to the input, and the output waveform is displayed on an oscilloscope. The frequency of the square wave is reduced until tilt is measurable in the output waveform and the low-frequency cutoff determined using Eq. (6-56). Similarly, the square-wave frequency is increased until the rise time of the amplified pulse is observed on the oscilloscope. Then Eq. (6-52) gives the high-frequency cutoff of the amplifier. Of course, it is assumed that the square-wave signal source and the response characteristics of the oscilloscope are sufficiently good so that they introduce negligible pulse distortion.

Such *square-wave testing* is particularly advantageous in adjusting amplifiers for optimum response because the results of any changes are immediately apparent in the output pulse waveform. This makes repeated laborious determinations of the sine-wave frequency-response characteristic unnecessary. Waveforms similar to those of Figs. 2-28 and 2-30 are often seen, depending upon the relative values of the square-wave frequency and amplifier cutoff frequency.

DC AMPLIFIERS

Direct Coupling

All the amplifier circuits discussed to this point have zero gain for dc signals because of the infinite reactance of the coupling networks at zero frequency. Amplification of dc or very slowly varying signals is achieved by eliminating the coupling networks entirely. In addition to the dc response of such a *direct-coupled* amplifier, the high-frequency performance is also enhanced. This comes

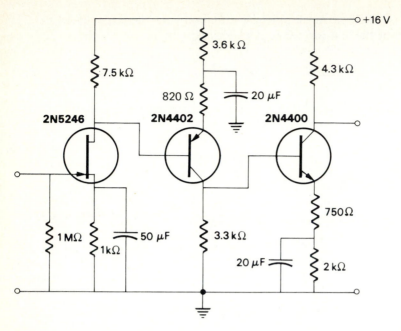

Figure 6-33 Three-stage dc amplifier.

about because stray capacitances associated with the coupling networks are eliminated. For this reason, direct coupling is often employed in ac amplifiers as well.

The Darlington connection already discussed is an elementary two-stage dc
amplifier. More elaborate circuits are necessary to achieve larger values of voltage gain. Consider, for example, the three-stage direct-coupled amplifier in Fig.
6-33, which consists of two grounded-emitter transistors preceded by a FET
input stage. Note that the two grounded-emitter stages use complementary symmetry to simplify interconnections and keep the dc voltage levels in each stage to
reasonable values. A major difficulty with such cascaded dc amplifiers is *drift*
caused by small variations in transistor characteristics with time or temperature
and changes in the supply potentials. The input transistor is particularly suscep-
tible because any small changes in the input are amplified by succeeding stages in
the amplifier.

It is common practice to stabilize the quiescent point of direct-coupled ac
amplifiers by returning a portion of the output voltage to the input stage in such
a fashion that drifts tend to be compensated. This is accomplished in Fig. 6-34 by
the resistor R_1 which sets the base bias on the input transistor. Note that if the
emitter voltage at the second stage, V_E, increases for some reason, the base
current of the first stage increases correspondingly. This results in a decrease in
base current of the second transistor which tends to return V_E to its original
value. The stabilizing action of R_1 can be examined best with the equivalent
circuit of the input-stage base circuit, which is similar to Fig. 6-11b. The voltage

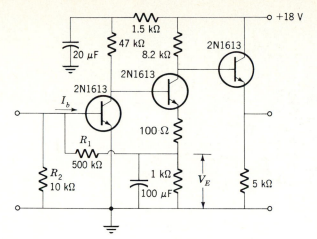

Figure 6-34 Direct-coupled ac amplifier.

drops around the equivalent circuit are

$$\frac{R_2}{R_1 + R_2} V_E - I_b \frac{R_1 R_2}{R_1 + R_2} - V_{be} = 0$$

where V_{be} is the emitter-base voltage. Solving for V_E

$$V_E = \left(1 + \frac{R_1}{R_2}\right) V_{be} + R_1 I_b \qquad (6\text{-}57)$$

Now V_{be} is a relatively constant quantity. If the circuit is arranged so that the first term in Eq. (6-57) is much larger than the second, V_E is constant and the amplifier is stabilized. After calculating V_E from Eq. (6-57), the currents in each transistor may be determined from the transistor characteristics and the quiescent point of the amplifier is completely determined.

Note, however, that in achieving stability against drift the amplifier does not respond to dc signals. If the second term in Eq. (6-57) is negligible, an input signal I_b does not cause any change in V_E; hence the output signal is zero. This is not so for ac signals where the 100-μF emitter bypass capacitor shorts out V_E, and no signal is returned to the input, so that the circuit is an effective ac amplifier. The low-frequency cutoff is, from Eq. (2-46), $1/2\pi \times 100 \times 10^{-6} \times 10^3 = 1.6$ Hz.

Drifts in dc amplifiers caused by variations in device characteristics and supply voltages are minimized by using a balanced difference amplifier, Fig. 6-35. Changes in one side of the circuit tend to be compensated by similar changes on the other side. Furthermore, the output terminals are at the same potential when the input signal is zero. Note the similarity of Fig. 6-35 to Fig. 4-32a and Fig. 6-19.

The purpose of transistor Q_3 is to provide a constant-current source for the difference amplifier to further reduce drift tendencies. In addition, the ac collector impedance of Q_3 is very large, which greatly reduces the effect of circuit asymmetry, according to Eq. (6-42). This is more convenient than a large emitter

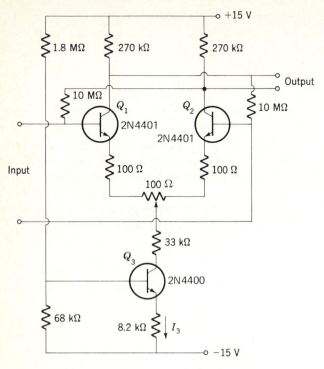

Figure 6-35 Difference amplifier is balanced dc amplifier.

resistor since the dc voltage drop across Q_3 is much smaller than that associated with a resistance equal to $1/h_{oe}$. Note that the base bias resistors for Q_1 and Q_2 are returned to the opposite collectors. This tends to stabilize the operating point without reducing gain since the collector potentials change in opposite directions in response to an input signal.

The output signal of the difference amplifier is, using Eq. (6-36),

$$v_0 = \frac{h_{fe}}{h_{ie}} R_L(v_2 - v_1) = g_m R_L(v_2 - v_1) \tag{6-58}$$

It is instructive to evaluate the transconductance of the transistor using the rectifier equation, Eq. (4-7),

$$g_m = \frac{dI_c}{dV_e} = \alpha \frac{dI_e}{dV_e} = \frac{\alpha e}{kT} I_o e^{eV_e/kT} = \frac{\alpha e}{kT} I_e \tag{6-59}$$

The circuit is balanced so that $2I_e = I_3$, and the output signal is then

$$v_o = \frac{\alpha e}{2kT} I_3 R_L(v_2 - v_1) \tag{6-60}$$

According to Eq. (6-60) the gain of the amplifier is set by the current in Q_3. This current, in turn, is determined by the base bias on Q_3 and may be adjusted to suit gain requirements.

Chopper Amplifiers

To circumvent the drift and instabilities inherent in direct-coupled amplifiers, it is useful to convert a dc input signal to an ac voltage which can be amplified by a standard ac-coupled circuit. Subsequently, the amplifier output is rectified to recover the amplified dc input signal. The superior stability features of ac-coupled amplifiers provide much greater gains than are practical with dc amplifiers.

A convenient technique for converting a dc voltage to an ac signal is with a mechanical *chopper*, which is a rapidly vibrating switch driven by an electromagnet, Fig. 6-36. As the switch alternately opens and closes, a square wave with an amplitude equal to the dc signal is produced. The frequency of the square wave corresponds to the chopping frequency, which is most often 60 Hz, since it is convenient to drive the chopper from the power line. The amplified ac output signal is converted back to a dc signal using a peak-reading voltmeter circuit similar to Fig. 3-28.

The sensitivity of chopper amplifiers to very low-level dc signals is limited by switching noise effects at the contacts. The mechanical chopper has a very low resistance when the contact is closed and a very high resistance when the contact is open and is, therefore, quite an effective chopper. In effect the chopper resistance in combination with the series resistance R_s in Fig. 6-36 represents an alternating voltage divider. A large change in the dividing ratio caused by the opening and closing of the switch means that even feeble input signals can override contact noise effects.

In many applications the chopper amplifier must accurately reflect changes in the input signal. As a rule of thumb, the upper frequency limit of a chopper amplifier is about one-fourth of the chopping frequency. Mechanical choppers are limited to 60 Hz, although 400-Hz units are occasionally used. To achieve higher

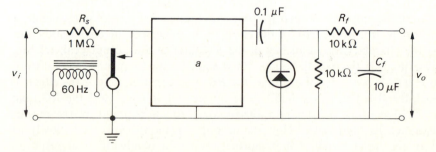

Figure 6-36 Chopper amplifier uses electromagnetically driven switch to convert dc input voltage to ac signal.

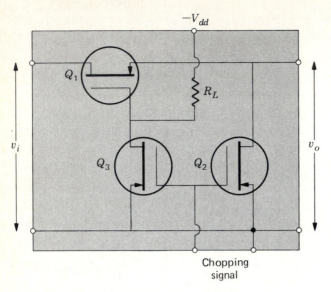

Figure 6-37 Integrated-circuit MOSFET chopper.

chopping frequencies nonmechanical choppers employing diodes, transistors, photoconductors, etc., are used in various circuits.

It is possible, for example, to replace the chopper in Fig. 6-36 with a MOSFET which is driven alternately into saturation or cutoff by a square-wave signal. A more satisfactory approach is the integrated-circuit MOSFET chopper in Fig. 6-37 wherein Q_1 and Q_2 are the two elements of the input voltage divider. The purpose of Q_3 is to introduce a 180° phase shift of the chopping signal applied to Q_1 so that when Q_2 is conducting Q_1 is cut off and vice versa. Since, therefore, both arms of the input voltage divider are changed simultaneously, the chopping ratio is enhanced.

A major advantage of chopper amplifiers is their very low internal-noise levels which means that even minute input signals can be amplified. As discussed in Chap. 10, interference caused by random-noise effects is proportional to the bandwidth of the amplifier, so that it proves advantageous to employ the smallest possible bandwidth. In the case of the chopper amplifier the effective bandwidth is set by the characteristic frequency of the RC filter of the output peak-reading voltmeter. In effect, the overall bandwidth is equal to the frequency interval from dc (zero Hz) to $1/2\pi RC$. If, for example, the RC time constant is 10 s, the effective bandwidth is 0.016 Hz, a very small value indeed. Of course, when the RC time constant is 10 s, a time interval of approximately 30 s is required for the output signal to reach its final value. Thus the amplifier responds only very slowly to changes in the input signal. This reciprocity between bandwidth and response time is a general property of all measuring systems.

A difficulty with the simple chopper amplifier is that the output signal is independent of the polarity of the input signal. This is so because the ac amplifier

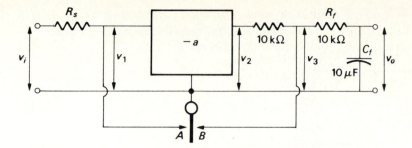

Figure 6-38 Synchronous chopper amplifier preserves polarity of input signal.

yields an ac output signal for either polarity of input voltage. The defect is corrected in chopper amplifiers which employ a second contact to rectify the output, Fig. 6-38. As the mechanical chopper alternately closes contacts A and B, the input signal is converted to a square wave, and the amplified ac signal is converted back to a dc signal. An electronic version of this *synchronous rectifier* is analyzed in the next section.

Lock-In Amplifier

The principles of the synchronous chopper amplifier are employed in an electronic circuit which has found wide use in instrumentation systems. The version illustrated in Fig. 6-39 uses a diode modulator to convert the input signal to a form suitable for the amplifier and a synchronous diode demodulator to recover the amplified version of the input signal. For generality, we consider the input signal to be a slowly varying signal; if dc signals are important, the input transformer can be replaced by a center-tapped resistor.

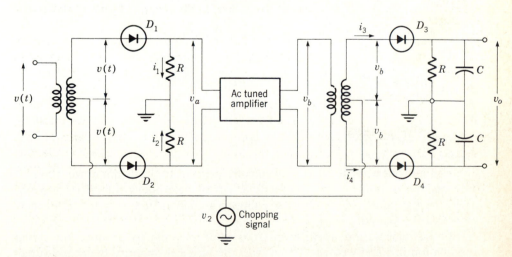

Figure 6-39 Electronic synchronous chopper amplifier.

The input circuit is analyzed using techniques discussed in Chap. 3. It is assumed that the diode characteristic can be represented by a quadratic expression. Then the current in the diode D_1 is, from Eq. (3-16),

$$i_1 = a_1 v(t) + a_1 V_2 \sin \omega_2 t + a_2 v^2(t) + a_2 V_2^2 \sin^2 \omega_2 t + 2a_2 v(t)V_2 \sin \omega_2 t$$

$$(6\text{-}61)$$

where a_1 and a_2 are constants related to the diode characteristic, $v_2 = V_2 \sin \omega_2 t$ corresponds to the chopping signal, and $v(t)$ is the input signal. An identical expression applies to the current in D_2, except that the polarity of $v(t)$ is reversed with respect to v_2. The voltage signal applied to the amplifier is

$$v_a = R(i_1 - i_2) \tag{6-62}$$

Substituting for i_1 and i_2, many terms cancel because of the reversed polarity of $v(t)$ in D_2. The result is

$$v_a = 2Ra_1 v(t) + 4Ra_2 v(t)V_2 \sin \omega_2 t \tag{6-63}$$

The first term in Eq. (6-63) is not transmitted by the amplifier since it is assumed that ω_2 is much larger than any of the frequency components associated with $v(t)$. Note that the second term is simply a modulated sine wave of frequency ω_2. The amplitude variations correspond to the input signal.

Ignoring the capacitors C for a moment, the current in D_3 is, again using Eq. (3-16),

$$i_3 = a_1 v_b + a_1 V_2 \sin \omega_2 t + a_2 v_b^2 + a_2 V_2^2 \sin^2 \omega_2 t + 2a_2 v_b V_2 \sin \omega_2 t \tag{6-64}$$

The current in D_4 is similar except that the polarity of v_b with respect to v_2 is reversed. Consequently, the output voltage becomes, where $v_b = kv_a$,

$$v_o = R(i_3 - i_4) = 2a_1 Rv_b + 16a_2^2 kR^2 v(t)V_2^2 \sin^2 \omega_2 t \tag{6-65}$$

Inserting the standard trigonometric identity $2 \sin^2 \omega t = 1 - \cos 2\omega t$,

$$v_o = 2a_1 Rv_b + 8a_2^2 kR^2 V_2^2 v(t)(1 - \cos 2\omega_2 t) \tag{6-66}$$

The filter capacitors eliminate the high-frequency terms in ω_2 and $2\omega_2$. Therefore, the filtered output signal is simply

$$v_0 = 8k(a_2 RV_2)^2 v(t) \tag{6-67}$$

According to Eq. (6-67) the output voltage is an amplified replica of the input signal.

The amplifier passband is conveniently peaked at ω_2 to minimize noise effects. Actually, however, the output filter circuit determines the effective bandwidth in the same way as for the chopper amplifier, and very-low-noise performance is achieved.

In instrumentation applications it is often possible to produce the initial modulation in some way associated with the physical quantity being measured. For example, the infrared beam of an infrared spectrometer is chopped by means of a rotating shutter before it strikes the infrared detector. The amplified signal from the detector is demodulated synchronously by a circuit similar to Fig. 6-39 using a voltage v_2 derived from the shaft of the rotating shutter. The low-noise performance of the circuit permits extremely weak infrared signals to be detected. In this form the circuit is usually called a *lock-in amplifier*, since the detector is locked in step with the input signal.

SUGGESTIONS FOR FURTHER READING

E. J. Angeleo, Jr.: "Electronics: BJT's, FET's and Microcircuits," McGraw-Hill Book Company, New York, 1969.
Jacob Millman and Arvin Grabel: "Microelectronics," McGraw-Hill Book Company, New York, 1987.
Jack J. Studer: "Electronic Circuits and Instrumentation Systems," John Wiley & Sons, Inc., New York, 1963.
Joseph A. Walston and John R. Miller (eds.): "Transistor Circuit Design," McGraw-Hill Book Company, New York, 1963.

EXERCISES

6-1* Determine the operating point of the FET grounded-source amplifier in Fig. 6-40. Use drain characteristics given in Fig. 4-23.
   Answer: -6 V, 4.5 mA, 1.8 V

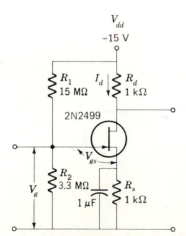

Figure 6-40

6-2* Determine the small-signal parameters at the operating point of the amplifier of Exercise 6-1. Calculate the gain of the amplifier.
   Answer: 2.5×10^{-3} mho, 1.5×10^4 ohms; 2.5.

6-3 Plot the transfer characteristics of the bipolar transistors whose collector characteristics are given in Figs. 4-13 and 6-41. In each case calculate a value for β.
   Answer: 500, 300

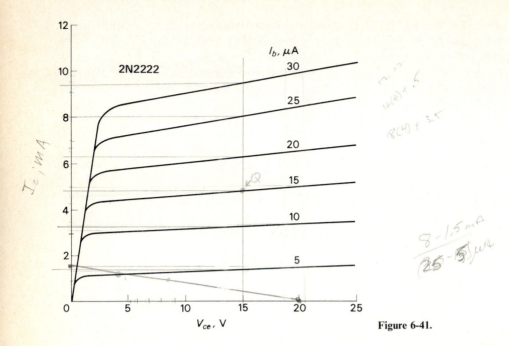

Figure 6-41.

6-4 Determine the operating point of the 2N2222 common-emitter amplifier, Fig. 6-16, graphically. Characteristic curves of the 2N2222 are in Fig. 6-41.

   Answer: 4 μA, 4 mA

6-5 Calculate the voltage and current gain and the input and output impedances of the amplifier of Exercise 6-4. The parameters are $h_{ie} = 3600\ \Omega$, $h_{fe} = 150$. If $h_{oe} = 1.4 \times 10^{-5}$ mho, calculate a more accurate value for the output resistance.

   Answer: 416; 150; 3600 Ω; $10^4\ \Omega$, $8.7 \times 10^3\ \Omega$

6-6 Show by a graphical argument exactly how much raising the supply voltage will increase the voltage gain of a transistor amplifier. Assume that the transistor has a $\beta = 100$ and that the operating point remains fixed at $I_c = 2$ mA, $V_{ce} = 5$ V for the two different supply voltages, 10 V and 20 V.

   Answer: 3

6-7 Derive expressions for the voltage and current gain, and input and output resistance of the grounded-base amplifier. Verify the numerical values listed in Table 6-3. Repeat for the emitter-follower amplifier.

6-8* Determine the operating point of a grounded-emitter amplifier using a type 2N3521 transistor if $R_1 = 90$ kΩ, $R_2 = 45$ kΩ, $R_L = 10$ kΩ, $R_E = 6$ kΩ, and $V_{cc} = 36$ V. Use collector characteristics for the 2N3521 given in Fig. 4-13. Repeat if the ambient temperature is increased to 125°C using collector characteristics given in Fig. 4-14. Calculate the voltage gain of the amplifier in both cases. Is this circuit stabilized against changes in temperature?

   Answer: 5.9 V, 1.9 mA; 5.9 V, 1.9 mA; 750, 750; yes

6-9* Rearrange the circuit of Exercise 6-8 into an emitter-follower amplifier, and calculate the input and output impedances. Suppose a type 2N2222 transistor is inadvertently used in this circuit. Determine the operating point using the collector characteristics in Fig. 6-41 and calculate the input and output impedances. Is circuit performance sensitive to the change in transistors?

   Answer: $7.2 \times 10^6\ \Omega$, 13 Ω; 5.5 V, 1.9 mA; $5.1 \times 10^6\ \Omega$, 18 Ω; no

6-10 Analyze the Darlington amplifier using the approximate hybrid equivalent circuit and find an expression for the output voltage as a function of input current. The collector load resistance is R_L.

   Answer: $-(1 + h_{fe})^2 R_L i_b$

6-11 Plot the low-frequency response of the transistor amplifier of Fig. 6-42. Compare with the response of the amplifier with R_3 and C_3 removed. Use values of h parameters given in Exercise 6-5.

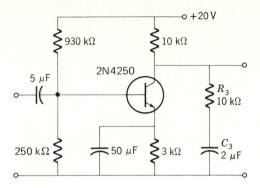

Figure 6-42 Combination $R_3 C_3$ improves low-frequency response of amplifier because reactance of C_3 becomes large.

6-12 Over what frequency range is the gain of an amplifier within 0.5 percent of the midband gain?
Answer: $10f_0'$, $0.1f_0'$

6-13 Sketch the equivalent circuit of the transistor tuned amplifier, Fig. 6-43. Calculate the peak gain assuming that the reactances of the tuned circuit are infinite at resonance. Use h parameters given in Exercise 6-5.
Answer: 3.7×10^5

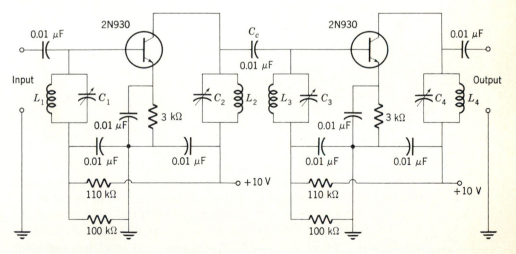

Figure 6-43 Two-stage tuned amplifier.

6-14 Select appropriate square-wave test frequencies to confirm that an amplifier has a passband extending from 20 to 20,000 Hz. Sketch the expected output waveforms at both frequencies.
Answer: 314 Hz; 3 kHz

6-15 Show that the lock-in amplifier, Fig. 6-39, acts as a phase detector by calculating the output signal if $v_a = V_i \sin(\omega_2 t + \phi)$. In this application, the input modulator is not used.
Answer: $V_o = (2Ra_2 k V_2 V_i) \cos \phi$

SEVEN

OPERATIONAL AMPLIFIERS

The performance of transistor amplifiers is enhanced in many respects by returning a fraction of the output signal to the input terminals. This process is called feedback. The feedback signal may either augment the input signal or tend to cancel it, and the latter, called negative feedback, is the primary concern of this chapter. Improved frequency-response characteristics and reduced waveform distortion are attained with negative feedback. In addition, amplifier performance is much less dependent upon changes in transistor parameters caused by aging or temperature effects.

A particular form of negative feedback, known as operational feedback, is used in amplifiers which perform mathematical operations such as addition or integration on an input signal. Commercially available integrated-circuit units have come to be called operational amplifiers. Operational amplifiers are widely used in measurement and control applications, as well as in electronic analog computers.

NEGATIVE FEEDBACK

Voltage Feedback

The circuit alterations of a standard amplifier which return a portion of the output signal to the input may be analyzed by the techniques developed in previous chapters. It is more illustrative, however, to isolate the feedback portion of the circuit and treat it separately. Consider the feedback amplifier, Fig. 7-1, comprising a standard amplifier with a gain a and a feedback network indicated by the box marked β. According to this circuit, a voltage βv_o is added to the input signal v_i so that the total input signal to the amplifier is

$$v_1 = v_i + \beta v_o \tag{7-1}$$

Introducing the fact that $v_o = av_1$,

$$v_o = av_i + a\beta v_o \tag{7-2}$$

so that

$$v_o = \frac{a}{1 - \beta a}\, v_i \tag{7-3}$$

According to Eq. (7-3) the overall gain of the amplifier with feedback,

$$a' = \frac{a}{1 - \beta a} \tag{7-4}$$

may be greater or smaller than that of the amplifier alone, depending upon the algebraic sign of βa.

The condition of greatest interest in this chapter is *negative feedback*, when βa is a negative quantity. In this case, Eq. (7-4) shows that the overall gain is reduced because, in effect, the feedback voltage cancels a portion of the input signal. If the amplifier gain is very large, so that $-\beta a \gg 1$, the overall gain reduces to

$$a' = \frac{1}{\beta} \tag{7-5}$$

This means that the gain depends only upon the properties of the feedback circuit. Most often, the feedback network is a simple combination of resistors

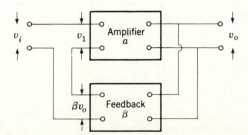

Figure 7-1 Block diagram of feedback amplifier.

and/or capacitors. Therefore, the gain is independent of variations in transistor parameters in the amplifier. In addition to this desirable improvement in stability, the gain may be calculated from circuit values of the feedback network alone. Thus it is not necessary to know, for example, the *h* parameters of all transistors in the circuit.

Negative feedback is also effective in reducing waveform distortion in amplifiers. Waveform distortion results from a nonlinear transfer characteristic, which may be interpreted as a smaller gain where the slope of the transfer characteristic is less, and as a larger gain where the slope of the transfer characteristic is greater. According to Eq. (7-5), however, the gain of an amplifier with feedback is essentially independent of variations caused by nonlinearities in transistor characteristics. Therefore, the transfer characteristic is more linear and distortion is reduced.

The benefits of negative feedback are obtained at the expense of reduced gain. This is not a serious loss, however, because large amplifications are easily obtained in transistor amplifiers. As discussed in Chap. 10, the maximum usable gain is limited by random-noise effects anyway and it is not difficult to achieve the largest amplification that can be used effectively, even with feedback included.

In the foregoing, a portion of the output voltage is returned to the input terminals, a condition referred to as *voltage feedback*. The two-stage transistor amplifier, Fig. 7-2, uses voltage feedback introduced by the resistor R_F connecting the output terminal with the input circuit of the first stage. The feedback voltage is introduced into the emitter circuit of the first stage because the output voltage of a two-stage amplifier is in phase with the input signal. The feedback factor is

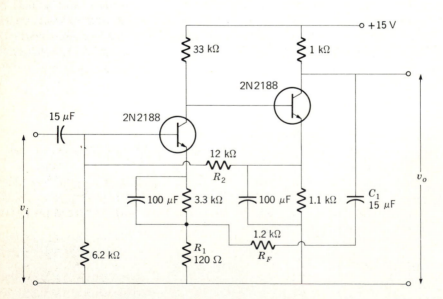

Figure 7-2 Two-stage feedback amplifier. Feedback is determined by ratio R_1/R_F.

the result of the resistor divider made up of R_F and R_1, so that

$$\beta = \frac{R_1}{R_F + R_1} \cong \frac{R_1}{R_F} \qquad (7\text{-}6)$$

Capacitor C_1 isolates the dc components of the two stages. Also, note that dc interstage coupling is used in this amplifier, which is quite independent of negative-feedback considerations. Resistor R_2 sets the bias on the first stage and provides dc feedback which helps stabilize the direct-coupled amplifier against slow drifts, as discussed in Chap. 6. It is not part of the ac feedback circuit.

The effect of the feedback ratio on the frequency-response characteristic is illustrated in Fig. 7-3. Without feedback (resistor R_F removed) the midband gain of the amplifier is 1000 and the high-frequency cutoff is 100 kHz. As feedback is increased by using smaller values of R_F the midband gain is reduced and the frequency response is extended to lower and higher frequencies. With a 1200-Ω feedback resistor the gain is 10 and the upper-frequency cutoff is extended to 15 MHz. Under this condition $\beta = 120/1200 = 0.1$, so $a\beta = 100$. According to Eq. (7-5) the midband gain is $1/\beta$, in agreement with the experimental response curves. Thus, a stable very wideband amplifier is possible through the use of feedback.

Note that in Fig. 7-3 the gain increases slightly at the extreme of the response curve when a large feedback ratio is used. This is a result of phase shift in the amplifier at these frequencies. In effect, the negative feedback is reduced since the feedback voltage is no longer exactly 180° out of phase with the input signal. The phase characteristic of feedback amplifiers at frequencies outside of the passband

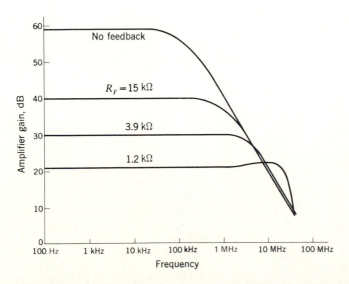

Figure 7-3 Effect of feedback on response characteristic of amplifier in Fig. 7-2.

is very important. Such irregularities in the response curve should be small, particularly when large feedback ratios are employed. Otherwise it becomes possible for the phase shift to become great enough to cause positive feedback at the frequency extremes, a condition which must be avoided if the amplifier is to remain stable. This matter is considered further in a later section.

If the feedback circuit is frequency-selective, it is possible to develop a specific frequency-response characteristic for the amplifier. Consider, for example, an amplifier having a bridged-T feedback network in one stage, Fig. 7-4. In this configuration a FET integrated-circuit difference amplifier (symbolized by the triangle) and a source-follower output stage are used to achieve isolation for the frequency-selective feedback network. The difference amplifier also provides a convenient terminal for the input signal.

The feedback voltage is 180° out of phase with the input signal, as required for negative feedback. According to the response curve of the bridged-T filter, the feedback voltage is a minimum at the characteristic frequency of the filter. Accordingly, the amplifier gain is a maximum at this frequency and the result is a tuned amplifier. The feedback ratio β is a function of frequency because of the characteristics of the bridged-T filter, and the feedback effect considerably enhances the selectivity of the filter, as illustrated in Fig. 7-5.

Feedback tuned amplifiers are commonly used at audio frequencies where high-Q inductances are difficult to construct because large values of inductance are required. Furthermore, it is a simple matter to tune the amplifier by making the resistors variable. This approach is also useful in integrated circuits where

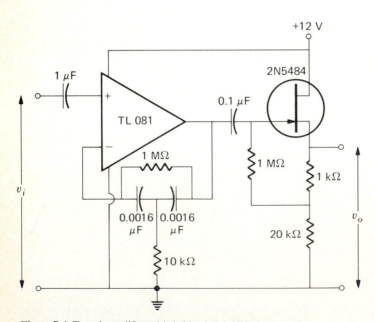

Figure 7-4 Tuned amplifier with bridged-T feedback network.

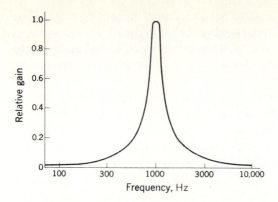

Figure 7-5 Frequency-response characteristic of bridged-T amplifier in Fig. 7-4.

inductances are difficult to produce. Other frequency-selective feedback networks, such as the twin-T or Wien bridge, are also used in tuned-feedback amplifiers.

Negative feedback alters the input and output impedance of an amplifier. To see how this comes about, replace the amplifier by its Thévenin equivalent, Fig. 7-6. The output voltage and the input signal are then

$$v_o = av_1 - i_o r_o \tag{7-7}$$

and
$$v_1 = v_i + \beta v_o \tag{7-8}$$

Equation (7-8) is introduced into Eq. (7-7),

$$v_o = av_i + a\beta v_o - i_o r_o$$

Solving for v_o,

$$v_o = \frac{a}{1 - a\beta} v_i - \frac{r_o}{1 - a\beta} i_o \tag{7-9}$$

Equation (7-9) is the Thévenin expression for the amplifier with negative feedback. Note that the coefficient of v_i is the gain which is in agreement with Eq.

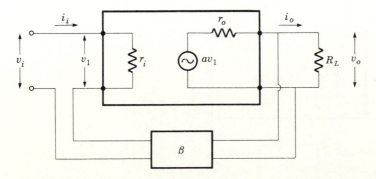

Figure 7-6 Circuit to determine effective input and output impedance of amplifier with voltage feedback.

(7-3). The coefficient of i_o is the effective output impedance. According to Eq. (7-9) the output impedance is reduced, and may become very small if $-a\beta$ is large. The effective input impedance is found by replacing v_o by $R_L i_o$ and v_1 by $r_i i_i$ in Eqs. (7-7) and (7-8). The result is solved for the ratio v_i/i_i, or the input impedance,

$$R_i = r_i\left(1 - \frac{a\beta}{1 + r_o/R_L}\right) \tag{7-10}$$

According to Eq. (7-10) the input impedance is increased by negative feedback. Both reduced output impedance and increased input impedance are favorable features in amplifiers, as discussed in previous chapters.

Current Feedback

It is possible to develop a feedback signal proportional to the output current rather than to the output voltage and this is called *current feedback*. Both voltage and current feedback can be employed in the same amplifier, Fig. 7-7. Voltage relations in the output circuit of Fig. 7-7 are

$$v_o = av_1 - i_o(r_o + R_F) \tag{7-11}$$

Similarly, Kirchhoff's voltage equation for the input circuit yields

$$v_1 = v_i + \beta_v v_o + \beta_i i_o R_F \tag{7-12}$$

Substituting Eq. (7-12) into Eq. (7-11) and solving for the output voltage,

$$v_o = \frac{a}{1 - a\beta_v} v_i - i_o \frac{r_o + (1 - a\beta_i)R_F}{1 - a\beta_v} \tag{7-13}$$

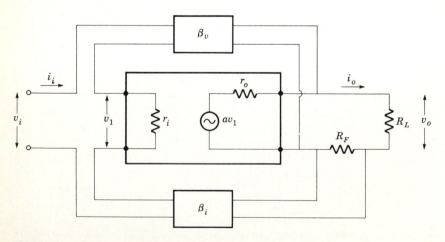

Figure 7-7 Combined voltage and current feedback.

Note that this expression incorporates Eq. (7-9) for voltage feedback alone, $\beta_i = 0$.

Combined current and voltage feedback permits a unique adjustment of the output impedance of the amplifier. The output impedance is given by the coefficient of i_o in Eq. (7-13) and can be made to vanish if

$$r_o + (1 - a\beta_i)R_F = 0 \tag{7-14}$$

or
$$a\beta_i = 1 + \frac{r_o}{R_F} \tag{7-15}$$

If Eq. (7-15) is satisfied, the amplifier has zero internal impedance and therefore can deliver maximum power to any load impedance. Note that this requires positive current feedback, according to Eq. (7-15). Positive current feedback is permissible in this case because of the stabilizing effect of negative voltage feedback which is also present.

Stability

The amplifier gain a and the feedback factor β are inherently complex numbers. That is, phase shifts associated with coupling capacitors and stray capacitance effects are present, particularly at frequencies outside of the passband of the amplifier. These phase shifts cause a departure from the 180° phase shift necessary for the feedback voltage to interfere destructively with the input signal. It can happen that the overall phase shift becomes zero (that is, 360°) so that $a\beta$ is positive and the feedback voltage augments the input signal. This is called *positive feedback* and leads to serious instability effects in feedback amplifiers.

Note, particularly, that if $a\beta = +1$, the output voltage, Eq. (7-3), becomes very large, even in the absence of an input signal. This means that positive feedback may cause an amplifier to oscillate. Oscillation is deleterious in amplifiers since the output voltage is not a replica of the input signal. Positive feedback is, however, a useful condition in oscillator circuits, as discussed in the next chapter.

Great care is taken in the design of feedback amplifiers to make them stable in the face of phase shifts leading to positive feedback at the frequency extremes of the amplifier bandpass. Fortunately there exists a rather straightforward criterion which can be applied to establish the stability of feedback amplifiers. According to Eq. (7-4), instability occurs if $a\beta$ is positive and equal to unity. It follows that if the absolute value of $a\beta$ drops below unity before the phase shift reaches $-360°$, the amplifier is stable. It is convenient to plot the magnitude and phase of $a\beta$ as a function of frequency, Fig. 7-8, in order to establish the relative positions at which $a\beta = 1$ and the phase shift equals zero. By this standard, the amplifier with characteristics represented by Fig. 7-8 is stable.

In order to apply this criterion for stability, the amplifier must be stable in the absence of feedback. Also, there exist certain situations in which a feedback

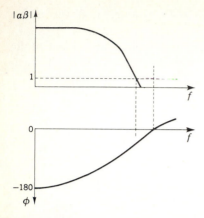

Figure 7-8 Feedback amplitude and phase characteristics of stable amplifier.

amplifier can be conditionally stable, that is, stable under some conditions of, say, loading, and not under others. For most practical circuit analysis, however, the stated condition is sufficient.

It is useful to consider the $a\beta$ characteristics of several amplifier types in the light of this stability requirement. Since in most cases the feedback ratio is independent of frequency, it is sufficient to examine the amplitude and phase characteristics of the amplifier gain separately. In the case of a single RC-coupled amplifier stage, Fig. 7-9a, the phase shift never exceeds 90°. This means the phase of $a\beta$ is always less than $90 + 180 = 270°$. Since this is smaller than 360°, a single-stage negative-feedback amplifier cannot be unstable.

Two identical RC-coupled stages in cascade have the ideal characteristic sketched in Fig. 7-9b. The phase shift reaches 180° at very low and very high frequencies, but the gain is very small at these extremes. Therefore, it is unlikely that $a\beta$ can be equal to unity, where the total phase shift is $180 + 180 = 360°$. Practical amplifiers always have stray capacitive effects that can introduce additional phase shift at high frequencies, however. It is possible that the additional phase shift caused by stray capacitances may result in unstable conditions at high frequencies if the gain is large.

The characteristics of three identical RC-coupled stages, Fig. 7-9c, are such that the extreme phase shift approaches 270°. Therefore, 180° phase shift is encountered at both low and high frequencies, and a three-stage feedback amplifier is certain to be unstable if the midband gain is large enough. It can be shown that the maximum value of $a\beta$ permitted at midband in this case is equal to 8, although even this value puts the amplifier on the verge of oscillation.

Many possibilities exist to remove this unfavorable difficulty other than minimizing the gain. For example, the bandwidth of two stages in a three-stage amplifier can be made very much greater than that of the remaining stage. This means that the phase characteristics are determined primarily by the single stage which is unconditionally stable. Alternatively, dc coupling can be used between two stages to eliminate the phase shift associated with one interstage coupling network. It is common practice to alter the feedback and gain characteristics,

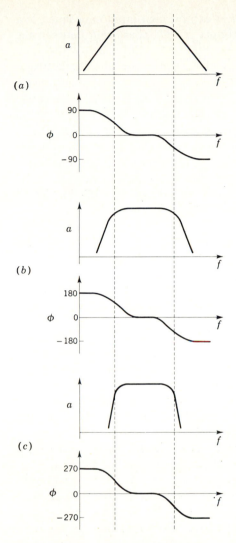

(a)

(b)

(c)

Figure 7-9 Gain and phase characteristics of (a) one-stage amplifier, (b) two-stage amplifier, and (c) three-stage amplifier.

particularly at high frequencies, by including small capacitances and resistances to change the phase and gain characteristics in a way that improves stability.

OPERATIONAL FEEDBACK

The Virtual Ground

A particularly versatile negative-feedback connection is called *operational feedback* because the circuit is capable of performing a number of mathematical operations on input signals. High-gain dc-coupled amplifiers are universally used in this application. Such *operational amplifiers* exhibit high input impedance, low output impedance, and conventionally introduce a 180° phase shift between the input signal and the output voltage.

To illustrate the features of an operational amplifier, consider the feedback circuit, Fig. 7-10, in which negative voltage feedback is produced by the resistor R_f connected between the output and input. Note that the feedback is negative because of the phase inversion in the amplifier. The feedback ratio in operational feedback can vary from unity for a high-impedance source to $R/(R + R_f)$ for a low-impedance source since the feedback voltage is effectively connected in parallel with the input-signal source. It is convenient to analyze the operational feedback circuit by applying Kirchhoff's current rule to the branch point S. Since the amplifier input impedance is large, the current in this branch is negligible, which means the current in R equals the current in R_f, or

$$\frac{v_i - v_1}{R} = \frac{v_1 - v_o}{R_f} \tag{7-16}$$

Introducing $v_1 = -v_o/a$ and rearranging,

$$v_o\left(1 + \frac{1}{a} + \frac{R_f}{aR}\right) = -\frac{R_f}{R} v_i \tag{7-17}$$

Since the gain is very large

$$v_o = -\frac{R_f}{R} v_i \tag{7-18}$$

which means that the output voltage is just the input signal multiplied by the constant factor $-R_f/R$. If precision resistors are used for R_f and R, the accuracy of this multiplication operation is quite good.

The branch point S has a special significance in operational amplifiers. This may be illustrated by determining the effective impedance between S and ground, which is given by the ratio of v_1 to the input current,

$$Z_s = \frac{v_1}{i_i} = \frac{v_1 R_f}{v_1 - v_o} = \frac{R_f}{1 - v_o/v_1} = \frac{R_f}{1 + a} \tag{7-19}$$

where the right side of Eq. (7-16) has been inserted for the input current. According to Eq. (7-19) the impedance of S to ground is very low if the gain is large.

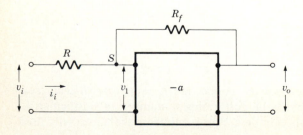

Figure 7-10 Block diagram of operational amplifier.

Typical values are $R_f = 10^4$ Ω and $a = 10^5$, so that the impedance is 0.1 Ω. The low impedance results from the negative feedback voltage, which cancels the input signal at S and tends to keep the branch point at ground potential. For this reason the point S is called a *virtual ground*. Although S is kept at ground potential by feedback action, no current to ground exists at this point. The virtual ground at S shows immediately that the impedance viewed from the input terminals is equal to R.

Mathematical Operations

The operational feedback circuit, Fig. 7-10, multiplies the input signal by the constant $-R_f/R$. It is conventional to simplify the circuit diagram, as in Fig. 7-11, by not showing the ground terminals specifically. The operational amplifier is indicated by a triangle pointing toward the output terminal. It is understood that the simplified circuit diagram, Fig. 7-11, actually implies the corresponding complete circuit, Fig. 7-10. If $R_f = R$ in the multiplier circuit, the signal is simply multiplied by -1. This is often useful in obtaining a signal multiplied by a positive constant wherein this operational amplifier precedes one which determines the multiplier. Multiplicative factors ranging from -0.1 to -10 are common, and the precision is determined principally by the accuracy of the two resistors.

The operational amplifier can also be used to sum several signals by connecting individual resistors to the branch point S, Fig. 7-12. In this circuit, the sum of the currents in R_1, R_2, and R_3 equals the current in R_f, since no current to ground exists at S. Furthermore, S is at ground potential, so that [compare Eq. (7-16)]

$$\frac{v_1}{R_1} + \frac{v_2}{R_2} + \frac{v_3}{R_3} = -\frac{v_o}{R_f} \tag{7-20}$$

$$v_o = -R_f\left(\frac{v_1}{R_1} + \frac{v_2}{R_2} + \frac{v_3}{R_3}\right) \tag{7-21}$$

Many signals can be added in this way. There is no interaction between individual signal sources since S is a virtual ground. Because the addition appears to take place at this point, S is commonly called the *summing point*. Note that each signal may be multiplied by the same factor (or -1), if $R_1 = R_2 = R_3$, etc; alternatively, individual factors may be selected, as convenient.

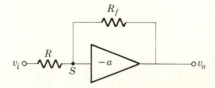

Figure 7-11 Conventional circuit symbol for operational amplifier.

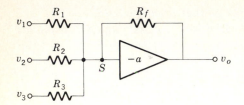

Figure 7-12 Summing circuit using operational amplifier.

If the feedback resistor is replaced by a capacitor, Fig. 7-13, the circuit performs the operation of integration. Using the fact that S is at ground potential,

$$v_o = \frac{q}{C} = \frac{1}{C} \int_0^t i \, dt = -\frac{1}{RC} \int_0^t v_i \, dt \qquad (7\text{-}22)$$

According to Eq. (7-22) the output voltage is the integral of the input signal. Note that there is no restriction placed on the frequency components of the input signal, as in the case of the simple RC integrator discussed in Chap. 2. It is only necessary that the amplifier bandwidth be large enough to handle all signal frequencies. The integral of the sum of several signals is obtained by introducing several input resistors, as in Fig. 7-12.

Interchanging R and C, Fig. 7-14, results in a differentiating circuit which gives the time derivative of the input signal. Again equating currents at the summing point,

$$-\frac{v_o}{R} = \frac{dq}{dt} = \frac{d}{dt} C v_i = C \frac{dv_i}{dt} \qquad (7\text{-}23)$$

$$v_o = -RC \frac{dv_i}{dt} \qquad (7\text{-}24)$$

The differentiator circuit is equally independent of frequency restrictions, assuming only that the gain of the amplifier is sufficiently large at all signal frequencies.

Simple Amplifiers

An operational amplifier makes an effective high-gain amplifier for use with small signals. If, however, the desired input impedance R, Fig. 7-11, is $10^5 \, \Omega$ and a gain

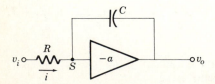

Figure 7-13 Integrating circuit.

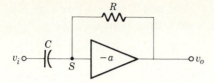

Figure 7-14 Differentiating circuit.

of 10^3 is needed, the feedback resistor must be $10^8\ \Omega$. Stable resistors of this magnitude are not readily available, so an alternate connection is used. As discussed in the next section, operational amplifiers conventionally employ a difference-amplifier input stage, and this permits the operational feedback connection to be separate from the signal input, Fig. 7-15a. The output signal is, according to Eq. (6-36),

$$v_o = a\left(v_i - \frac{v_o R}{R + R_f}\right)$$

or

$$v_o\left(\frac{1}{a} + \frac{R}{R + R_f}\right) = v_i \tag{7-25}$$

Since a is very large, the gain with feedback is just

$$\frac{v_o}{v_i} = 1 + \frac{R_f}{R} \tag{7-26}$$

If $R_f \gg R$, the amplification can be appreciable. At the same time, the input impedance is one-half that of the operational amplifier (see Exercise 7-9). Note also that the output is not inverted with respect to the input as is the case for the conventional operational feedback connection.

The advantages of a difference amplifier can also be realized in connection with operational feedback, Fig. 7-15b. In this circuit the output is determined by

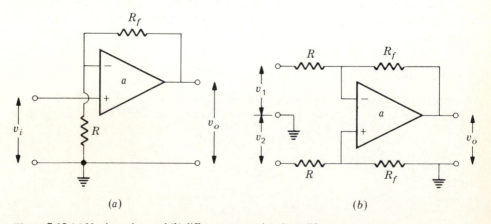

(a)   (b)

Figure 7-15 (a) Noninverting, and (b) difference operational amplifier.

adding the individual outputs due to v_1 and v_2. The effect of v_1 is given by Eq. (7-18) and that of v_2 by Eq. (7-26), so that the total output is just,

$$v_o = -\frac{R_f}{R} v_1 + \left(1 + \frac{R_f}{R}\right)\left(\frac{R_f}{R + R_f} v_2\right) \tag{7-27}$$

$$v_o = \frac{R_f}{R}(v_2 - v_1) \tag{7-28}$$

According to Eq. (7-28), the output is the difference of the input signals multiplied by the gain as established by the operational feedback connection.

OPERATIONAL AMPLIFIER CIRCUITS

Practical Amplifiers

Integrated-circuit operational amplifiers have largely supplanted discrete-device circuits, but it is useful to examine a typical discrete-device amplifier to display the main features of most practical units. The circuit, Fig. 7-16, includes a difference-amplifier input stage employing a constant-current source transistor to improve the common-mode rejection ratio. The 1-kΩ potentiometer in the drain

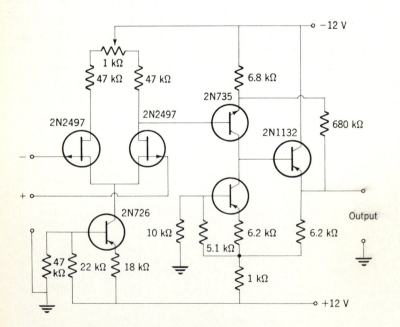

Figure 7-16 Operational amplifier with FET input stage.

circuit is used to balance the input stage so that the output is zero when no input signal is present. The second stage is an *npn* grounded-emitter amplifier using a *pnp* transistor as the collector load resistor. The amplifier is directly coupled to an emitter-follower output stage. Negative voltage feedback is provided by the 680-kΩ resistor connected from the output to the emitter of the second stage. In addition, positive current feedback is obtained by connecting the output emitter resistor to a portion of the emitter resistor of the collector load transistor.

This circuit has a large gain, 10^5, and a large input impedance, approximately 5 MΩ, by virtue of the FET input stage. Additionally, the output impedance is low because of the emitter-follower output and the effect of negative feedback. Positive current feedback is employed to achieve some gain in the third stage and to increase the maximum permissible output-voltage swing without distortion. Note that the input terminals include polarity markings. The negative terminal, usually referred to as the *inverting* input, causes an output signal of opposite polarity to the input signal. The positive terminal, called the *noninverting* input, produces an output signal in phase with the input. The inverting input is the normal input terminal since 180° phase shift is necessary in operational feedback.

Integrated-circuit operational amplifiers can be much more complex; the circuit of a typical unit, Fig. 7-17, employs a total of 20 transistors, for example. In this device, the input stage is a composite difference amplifier made up of Q_1Q_3 and Q_2Q_4 emitter-follower, grounded-base pairs which have Q_5 and Q_6 as collector loads. This arrangement yields high input impedance, large gain, and good balance. The second stage is a Darlington-connected $Q_{16}Q_{17}$ pair with Q_{13} as the load, which drives a complementary-symmetry, grounded-emitter output stage consisting of Q_{14} and Q_{20}. Bias for the output stage is provided by Q_{15}. Transistors Q_8 through Q_{12} provide stabilized, temperature-compensated bias for the first two stages. Transistors Q_{15} and Q_{19} act to limit the output current on positive and negative signals, respectively, as protection against output short circuits. Finally, Q_7 improves input-stage balance by providing a load for Q_5 similar to that on Q_6 produced by Q_{16}.

An *RC* phase compensation network can be connected between terminals 1 and 8 to prevent positive feedback at the extreme of the amplifier passband. This is particularly important when large feedback ratios are used, as previously noted in connection with Fig. 7-3. Also, an external balance potentiometer can be connected to terminals 1 and 5, although this is usually unnecessary except for very critical applications.

The inherent flexibility of integrated circuits makes such complex circuitry feasible at both low cost and small size. A typical silicon amplifier chip is illustrated in Fig. 7-18.

Even with proper design and well-stabilized supply voltages, it is difficult to prevent drifts in dc amplifiers. Drift voltages at the amplifier output mean that operational feedback no longer keeps the branch point *S* at virtual ground, which may prove serious in critical applications. Improved performance is achieved by using a chopper amplifier to stabilize the virtual ground voltage, Fig. 7-19. Here, the chopper amplifier, which is drift-free because it is ac-coupled, measures the

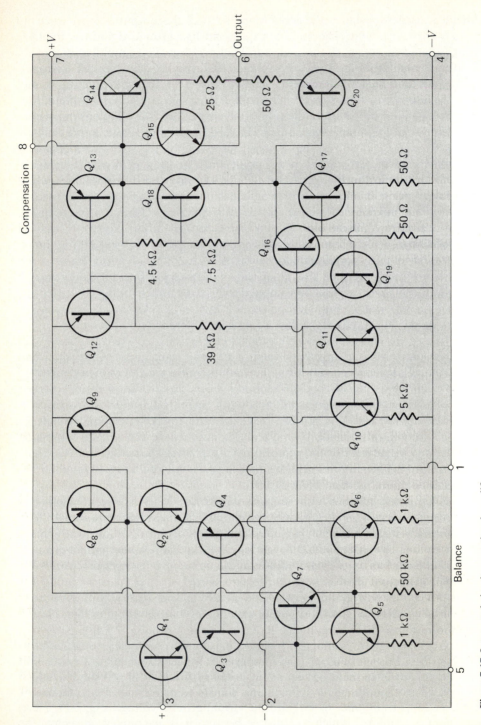

Figure 7-17 Integrated-circuit operational amplifier.

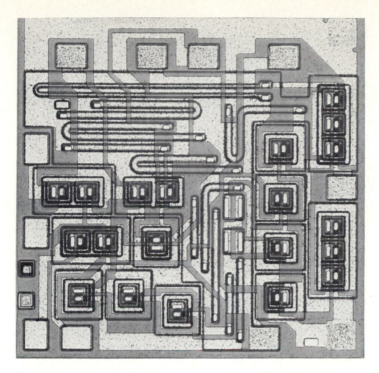

Figure 7-18 Integrated circuit operational amplifier. (*Motorola, Inc.*)

voltage at S and provides an amplified error signal to the noninverting input of the operational amplifier.

Note that a capacitor is interposed between S and the operational amplifier. This prevents the operational amplifier itself from responding to dc signals, but the chopper amplifier does so and, in feeding its output signal to the operational amplifier, becomes part of the operational feedback circuit. In effect, the operational amplifier is not dc-coupled and handles the high-frequency signals, while the chopper amplifier handles the dc and very-low-frequency signals. The dc gain is very large since it is the product of the gains of both amplifiers. This means that the overall response curve is nonuniform in that the low-frequency gain is larger than the gain at high frequencies. According to Eq. (7-18) this is unimportant since by feedback action the gain does not appear in the expression for the output signal, so long as the gain remains large at all frequencies of interest.

The combination of an operational amplifier stabilized by a chopper amplifier is often itself referred to as an operational amplifier. Thus the dashed line in Fig. 7-19 encloses a chopper-stabilized operational amplifier. Integrated-circuit chopper-stabilized operational amplifiers employing MOSFET choppers similar to Fig. 6-37 are commercially available.

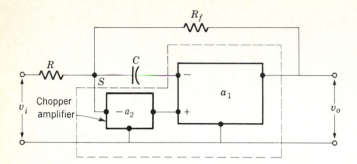

Figure 7-19 Chopper-stabilized operational amplifier.

It is not necessary to have a detailed understanding of the internal circuitry of operational amplifiers because, according to Eq. (7-18), only the parameters of the feedback connection determine the performance of the amplifier circuit. This is true so long as the gain is very large and the input resistance is large enough so that the input current can be neglected. In precision applications, it is useful to employ an equivalent circuit of the operational amplifier, Fig. 7-20, to determine the exact influence of finite gain, input resistance, and output resistance.

The input resistance, R_i, is of the order of 5 MΩ for practical bipolar devices and reaches 10^{12} Ω in the case of FET input stages. These values are large enough to be neglected in most applications, but clearly FET units approach the ideal more closely. The common-mode input resistance, R_{ic}, is much greater than R_i and most often is neglected entirely.

The output resistance, R_o, is 50 to 100 Ω for all integrated-circuit operational amplifiers, both bipolar and FET types. This is further reduced by feedback, according to Eq. (7-9), and in most applications the effective output resistance is less than 1 Ω. This again illustrates why the internal parameters of the operational amplifier can be neglected in all but the most critical applications.

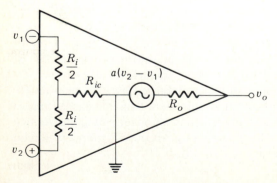

Figure 7-20 Operational amplifier equivalent circuit.

Gain, Bandwidth, and Slew Rate

The two most important parameters of all operational amplifiers are the gain and the bandwidth over which the gain is sufficiently large that the output is given by the simple resistance ratio in Eq. (7-18). The error introduced by the finite gain of practical amplifiers can be determined by introducing the feedback ratio.

$$\beta = \frac{R}{R + R_f} \quad = \frac{4.86}{4.86+47.45} \; = \; 9.3 \times 10^{-2} \atop \sim 10 \times 10^{-2} \quad (7\text{-}29)$$

into Eq. (7-17).

$$v_o\left[1 + \frac{1}{a}\left(1 + \frac{R_f}{R}\right)\right] = v_o\left(1 + \frac{1}{a\beta}\right) = -\frac{R_f}{R}\,v_i$$

$$v_o = -\frac{R_f}{R}\left(\frac{a\beta}{1 + a\beta}\right)v_i \qquad (7\text{-}30)$$

According to Eq. (7-30) a large value of $a\beta$, called the *loop gain*, ensures that errors are minimized.

As discussed previously, however, it is important that a becomes less than unity before the phase shift between input and output reaches $-360°$ in order that the feedback amplifier remain stable. Very frequently this is accomplished by an *internal compensation* network in the operational amplifier circuit that in effect makes the entire amplifier appear to have the gain and phase characteristics of a single-stage amplifier, since a single-stage feedback amplifier is unconditionally stable. Alternatively, *external compensation* may be employed, but this requires careful design to ensure stability in any given application.

Examples of the *open-loop* (that is, without feedback) gain versus frequency characteristics of internally compensated and externally compensated operational amplifiers are shown in Fig. 7-21. The gain of the internally compensated amplifier decreases with frequency according to Eq. (6-47), while the externally compensated unit does not. Note that if $\beta = 10^{-2}$, the internally compensated amplifier has a *closed-loop* (with feedback) gain of 100 and a bandwidth extending to 10^4 Hz. The externally compensated amplifier, by contrast, can achieve a gain of 10^3 over a frequency range extending to nearly 10^6 Hz.

Comparing these characteristics, the internal compensation approach may appear to be unduly restrictive, but this is not necessarily so. It is important to establish absolute stability for complicated feedback networks which may be difficult to analyze, and the internally stabilized amplifier provides this assurance. The externally compensated amplifier in Fig. 7-21 is very likely to be unstable when $\beta = 1$, for example, whereas the internally compensated unit is not. The unity feedback ratio is an important application of operational amplifiers, the so-called *voltage follower*, which is discussed in a later section. On the other hand, the externally compensated amplifier is inherently more flexible in that the external compensation may be adjusted to provide single-stage absolute stability as well as to optimize the frequency-response curve for specific feedback configurations.

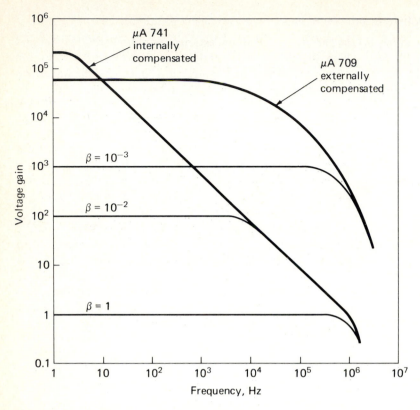

Figure 7-21 Open-loop frequency characteristics of internally and externally compensated operational amplifiers and typical gain characteristics with feedback.

Table 7-1 Properties of operational amplifiers

Type	Open-loop voltage gain	Open-loop bandwidth, unity gain, MHz	Slew rate, V/μs	Input offset, 10^{-3} V	Offset drift, μV/°C
General purpose:					
μA741	2×10^5	1.5	0.7	0.8	15
LM101A	3×10^5	1.0	1.0	0.7	3
TL070	2×10^5	3.0	13	3	10
LF351A	3×10^5	4.0	13	1	10
High speed:					
HA2525	1×10^5	20	100	5	20
OP-17	8×10^4	30	62	1	10
Special purpose:					
μA741E	2×10^5	1.5	0.7	0.075	1.3
HA2905	5×10^8	2.0	2.5	0.02	0.2

It is common practice to specify the frequency-response characteristics of operational amplifiers in terms of the open-loop gain at low frequencies and the open-loop bandwidth at unity gain. These parameters describe the bounds within which useful performance is possible. The properties of the general-purpose operational amplifier types listed in Table 7-1 show that gains are of the order of 10^5 and bandwidths range to several megacycles. These figures are consistent with the frequency-response characteristic of the internally compensated amplifier in Fig. 7-21.

The first two general-purpose units in the table are bipolar devices and the other two are FET amplifiers. Note that the FET types exhibit a somewhat greater bandwidth, as well as, of course, a much larger input resistance. Special-purpose, high-speed operational amplifiers exemplified by the bipolar HA2525 and the FET type OP-17 extend the unity gain frequency to tens of megacycles.

Very frequently the signal amplitudes applied to operational amplifiers range from cut off to saturation. In this case the ability of the amplifier to respond to rapid signal changes is limited by the ability of the internal amplifier circuitry to charge and discharge stray capacitances and the compensation capacitors. The *slew rate* specifies the maximum rate of change of the output signal that can be produced. According to Table 7-1, FET units are considerably faster than bipolar units, although very great slew rates can be achieved in high-speed devices of either kind.

Input Offset and Drift

Even with careful design and fabrication techniques, it is difficult to achieve exact balance in practical operational amplifiers and a small output voltage usually exists even with the input terminals short-circuited. This defect is specified in terms of the *input offset voltage*, which is the voltage difference needed between the input terminals to produce a zero output signal. As shown in Table 7-1, bipolar devices, which have input offset voltages less than 10^{-3} V, are better balanced than are FET amplifiers. The spurious signal represented by the input offset voltage can be troublesome in some applications, as, for example, in the integrator circuit, since, according to Eq. (7-22), a constant signal results in an ever-increasing output voltage which eventually drives the amplifier into saturation. In such critical applications it proves possible to supply the input offset voltage needed to zero the output by external circuitry including an adjustable potentiometer.

Even with a balanced input, however, the amplifier is still subject to *offset drift*, which is a change in the input offset voltage with temperature. This effect can be significant since a slow change in offset voltage is indistinguishable from a slowly varying signal. The last column in Table 7-1 shows that temperature changes of several tens of degrees may cause drift signals comparable with the input offset voltage.

Special processing techniques can result in very small input offset and drift as exemplified by type μA741E in Table 7-1. These superior properties are attained by careful and individual adjustment of the integrated circuit during fabrication to achieve improved balance. Chopper-stabilized amplifiers, as in type H2905, reach even lower offset and drift values because of the continuous balancing action of the chopper amplifier. In this connection, very large dc gains are usable, since difficulties with offset and drift are minimal.

In addition to the parameters listed in Table 7-1, a number of others are commonly used to specify practical operational amplifiers. The *input bias current*, which is of the order of 10^{-8} A for bipolar devices and much less, 10^{-11} A, for FET input units, is important in certain applications. Some others are input current offset, common-mode rejection ratio, internal noise, and power-supply voltages. For a given specific application any one of these, including cost, may prove more important than those tabulated. It is common practice for manufacturer's data sheets to describe quite completely the operating parameters and conditions pertaining to each device type.

OPERATIONAL AMPLIFIER APPLICATIONS

The properties of operational amplifiers make very effective single-ended or difference amplifiers, as described in previous sections. The operational feedback connection is exceedingly versatile, however, and can be used in a wide variety of other important applications as well, such that the integrated-circuit operational amplifier is the single most important linear circuit device.

Integrator-Differentiator Circuit

A practical integrator and differentiator circuit is illustrated in Fig. 7-22. The integrated-circuit operational amplifier used here is similar to that in Fig. 7-18. The switch in the input circuit interchanges the positions of the integrating

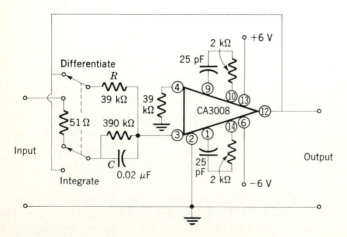

Figure 7-22 Practical integrator-differentiator circuit.

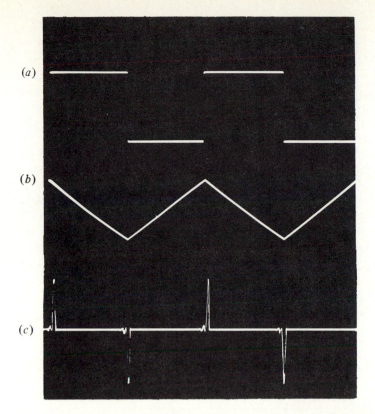

Figure 7-23 (*a*) Square-wave input (*b*) integrated and (*c*) differentiated by operational-amplifier circuit of Fig. 7-22.

capacitor and resistor, as may be noted by tracing through the circuit and comparing with Figs. 7-13 and 7-14. Thus the same circuit can be used either to integrate or differentiate the input signal. Typical experimental output waveforms for the case of a square-wave input signal are shown in Fig. 7-23.

The purpose of the 390-kΩ resistor shunting the integrating capacitor in Fig. 7-22 is to eliminate the effect of long-time integration of the input offset voltage. In effect the resistor keeps the capacitor discharged for slow changes in signal; therefore the integrator does not integrate properly at frequencies below about $f = 1/2\pi RC = 20$ Hz. The two series combinations of a small capacitor and resistor are phase compensation networks. As previously discussed, these networks ensure that the overall phase shifts do not produce positive feedback at the high-frequency limit of the amplifier bandpass.

Logarithmic Amplifier

Nonlinear feedback components in operational-amplifier circuits are used to develop special input-output characteristics. Consider, for example, the grounded-base transistor feedback element, Fig. 7-24, that results in an amplifier in which

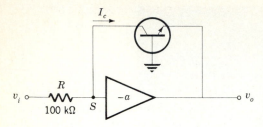

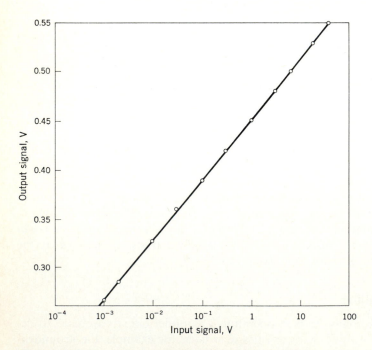

Figure 7-24 Elementary logarithmic amplifier.

the output voltage is the logarithm of the input signal. Such a logarithmic amplifier operates over a very wide range of input-signal magnitudes without satu-
rating.

The logarithmic input-output characteristic comes about in the following
way. The collector current in a transistor is just the emitter current times the
current gain, according to Fig. 4-11, and the emitter current is given by the
rectifier equation. Ignoring the unity in Eq. (4-7) since the exponential term
dominates, the collector current is

$$I_c = \alpha I_0\, e^{eV/kT} \tag{7-31}$$

In the operational-feedback connection appropriate for Fig. 7-24, the collector
current may be expressed in terms of the input signal

$$\frac{v_i}{R} = \alpha I_0\, e^{ev_o/kT} \tag{7-32}$$

Figure 7-25 Input-output characteristic of logarithmic amplifier.

where the output signal v_o is the base-to-emitter voltage. Rearranging and taking the logarithm of both sides,

$$v_o = \frac{kT}{e} \ln \frac{v_i}{\alpha I_0 R} \tag{7-33}$$

Experimental input-output characteristics, Fig. 7-25, illustrate the logarithmic properties predicted by Eq. (7-33) and the extremely wide range of input signals over which the amplifier operates. In practical circuits it is usually necessary to supplement the elementary circuit in Fig. 7-24 by including a collector bias supply and to compensate for offset voltages using the noninverting input in order to obtain the logarithmic characteristic over a wide range of input voltages.

In addition to its unique signal-handling capability, the logarithmic amplifier can be used to multiply two arbitrary input signals. The outputs of two logarithmic amplifiers are summed and then the output of the combination is passed through an inverse logarithmic amplifier. The result is a signal proportional to the product of the two input signals.

Active Filter

Consider the operational amplifier in Fig. 7-26 in which both the inverting and noninverting inputs have the operational-feedback connection. The difference-amplifier input together with Kirchhoff's voltage equation around the outer loop and around the upper loop yield three equations,

$$v_o = a(v_1 - v_2)$$
$$v_1 - i_1 r_1 + i_2 r_2 - v_2 = 0 \tag{7-34}$$
$$v_1 - i_1 r_1 - v_o = 0$$

Inserting the second equation into the first and equating to v_o from the third,

$$v_o = a(v_1 - v_2) = a(i_1 r_1 - i_2 r_2) = v_1 - i_1 r_1$$

or
$$i_1 r_1 \left(1 + \frac{1}{a}\right) - \frac{v_1}{a} = i_2 r_2 \tag{7-35}$$

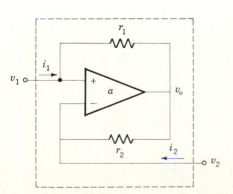

Figure 7-26 Operational amplifier connected as negative impedance converter.

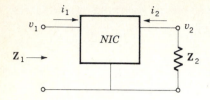

Figure 7-27 Block diagram of negative impedance converter.

Since a is very large,

$$i_1 r_1 = i_2 r_2$$

$$i_1 = \frac{r_2}{r_1} i_2$$

(7-36)

and, from the second relation in Eq. (7-34),

$$v_1 = v_2$$

(7-37)

Consider now the apparent input impedance of the circuit when an impedance $\mathbf{Z}_2$ is connected to the output terminals. To focus on the impedance transformation, use the block diagram, Fig. 7-27, in which the portion in the dashed rectangle in Fig. 7-26 is labeled *NIC*, for reasons justified below. According to Fig. 7-27, the impedance $\mathbf{Z}_2$ is given by

$$\mathbf{Z}_2 = \frac{v_2}{-i_2}$$

(7-38)

The input impedance is

$$\mathbf{Z}_1 = \frac{v_1}{i_1} = \frac{v_2}{r_2 i_2/r_1}$$

$$\mathbf{Z}_1 = -\frac{r_1}{r_2} \mathbf{Z}_2$$

(7-39)

where Eqs. (7-36) and (7-37) have been used. According to Eq. (7-39) the circuit has the interesting property that the apparent input impedance is the negative of the impedance connected to the output terminals. That is, the circuit is a *negative impedance converter*, or *NIC*. If, for example, the output impedance is capacitive, the input impedance is inductive. Furthermore, the magnitude of the inductance depends upon the scale factor r_1/r_2.

A simple application of this useful property is the *active filter*, Fig. 7-28, in which capacitor C_2 at the output terminals is converted to an inductance to form a resonant circuit with C_1. The input-output characteristic of this circuit is just

$$\frac{v_o}{v_i} = \frac{v_2}{i_1 \mathbf{Z}_1 + v_1} = \frac{1}{1 + (i_1/v_1)\mathbf{Z}_1}$$

(7-40)

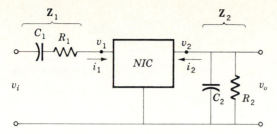

Figure 7-28 Active filter.

where Eq. (7-37) has been used. Inserting Eq. (7-39),

$$\left(\frac{v_o}{v_i}\right)^{-1} = 1 - \frac{r_2}{r_1}\frac{Z_1}{Z_2} \tag{7-41}$$

The ratio Z_1/Z_2 is

$$\frac{Z_1}{Z_2} = \left(R_1 - \frac{j}{\omega C_1}\right)\frac{R_2 - j/\omega C_2}{-jR_2/\omega C_2} = \left(\frac{R_1}{R_2} + \frac{C_2}{C_1}\right) + j\omega\left(R_1 C_2 - \frac{1}{\omega^2 C_1 R_2}\right) \tag{7-42}$$

The circuit is in resonance when the imaginary part vanishes,

$$\omega_0^2 = \frac{1}{R_1 C_1 R_2 C_2} \tag{7-43}$$

Inserting Eqs. (7-42) and (7-43) into Eq. (7-41) results in

$$\left(\frac{v_o}{v_1}\right)^{-1} = \left[1 - \frac{r_2}{r_1}\left(\frac{R_1}{R_2} + \frac{C_1}{C_2}\right)\right] - j\frac{r_2}{r_1}\sqrt{\frac{R_1 C_2}{R_2 C_1}}\left(\frac{\omega}{\omega_0} - \frac{\omega_0}{\omega}\right) \tag{7-44}$$

This expression should be compared with Eq. (5-49) for an RLC circuit. In the simple case where $R_1 = R_2$ and $C_1 = C_2$, the input-output characteristic is

$$\left(\frac{v_o}{v_i}\right)^{-1} = \left(1 - 2\frac{r_2}{r_1}\right) - j\frac{r_2}{r_1}\left(\frac{\omega}{\omega_0} - \frac{\omega_0}{\omega}\right) \tag{7-45}$$

Note that the equivalent Q of the circuit is just r_2/r_1, which means that the Q can be adjusted independently of the resonant frequency. Alternatively, it is possible to adjust the resonant frequency by changing either R_1 (and R_2) or C_1 (and C_2) without changing Q. The output signal may exceed the input voltage at resonance if the real term in Eq. (7-45) is less than unity, so that an active filter may also have gain.

Comparator

Operational amplifiers are useful in comparing the magnitudes of two signals. If one signal is applied to the inverting input and the other to the noninverting input, the output of the amplifier is zero only when the two signals are equal and the output is saturated either positively or negatively when the inputs are unequal.

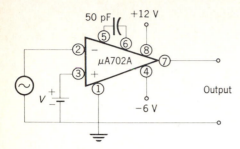

Figure 7-29 Operational-amplifier comparator.

A practical *comparator* circuit, Fig. 7-29, employing an integrated-circuit operational amplifier yields a zero output voltage whenever the instantaneous input signal is equal to V. The large common-mode rejection ratio and excellent input-stage balance characteristic of operational amplifiers are important in this application. The 50-pF capacitor is a simple phase compensation network.

Sample-and-Hold Amplifier

It often proves useful to sample a transient voltage signal at a given instant and to retain the sampled value for a time interval long enough to accomplish some necessary measurement. The so-called *sample-and-hold* amplifier, Fig. 7-30, employs an operational amplifier together with a low-leakage capacitor and an electronic switch to accomplish this action. A short sampling pulse closes the fast-acting switch, which allows the capacitor, C, to charge to the instantaneous voltage of the input signal. This value is presented to the output terminals through the operational amplifier connected as a voltage follower.

The electronic switch is essentially a MOSFET with the sampling pulse applied to the gate, although the integrated-circuit version in Fig. 7-30 includes additional components to optimize performance. The type 4066 switch, for example, has a closed resistance of $R_c = 100 \ \Omega$ and an open resistance of $R_o = 10^9 \ \Omega$ and can be used with either switch terminal as the input. A small value of R_c is important, since this resistance determines the minimum sampling duration. Suppose, for example, it is necessary to achieve 1 percent accuracy, then, according to Eq. (2-59), the time, T_s, required to charge the capacitor to 99 percent of the input signal is

$$0.99 \ v_i = v_i \ (1 - e^{-T_s/R_c C}) \tag{7-46}$$

$$-\frac{T_s}{R_c C} = \ln (0.99 - 1) = -4.61$$

$$T_s = 4.61 \ R_c C \tag{7-47}$$

According to Eq. (7-47), the sample-and-hold amplifier of Fig. 7-30 can capture the input signal over sample durations as short as $T_s = 4.61 \times 10^2 \times 10^{-7} = 50 \ \mu s$.

The operational-amplifier voltage follower has unity gain according to Eq. (7-5), since the feedback ratio is unity. The input resistance, R_i, is large, about

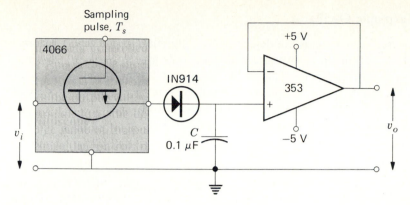

Figure 7-30 Sample-and-hold circuit uses integrated-circuit electronic switch and voltage follower.

10^{12} Ω, because of the voltage-follower connection and because the type 353 has FET input transistors. This is desirable to ensure that the charge on C is retained as long as possible. Again, assuming 1 percent accuracy is desired, the retention time is $4.61 \times 10^{12} \times 10^{-7} = 4.61 \times 10^5$ s, or about 130 h. Actually, this extreme value is not achieved in practice due to stray leakage currents, principally in the capacitor itself.

Note that the diode effectively isolates the capacitor from the resistance of the open switch, which otherwise would reduce the calculated retention time to less than 10 min. Also, the value of C represents a compromise between rapid acquisition and correspondingly short retention (small C) or vice versa. It can be selected to emphasize whichever feature is more important in any given application.

ANALOG COMPUTERS

Simulation

Operational-feedback circuits can be assembled into *analog computers* in which voltage signals are analogous to variables in physical systems. Many physical systems can be described by mathematical equations based on the laws of nature. Analog computers are used to solve these equations and thereby to display the behavior of the system.

The voltages and other circuit parameters of an analog computer correspond to the variables and properties of the real system. The parameters of the computer are easily altered and adjusted, however, so that the behavior of the system may be examined over a wide range of conditions. The computer is, in fact, looked upon as *simulating* the physical system, and its response to external stimuli corresponds to the actual system to the extent that the mathematical description is accurate.

Damped Harmonic Oscillator

A great many physical systems are described by differential equations, for example, mechanical motions in accordance with Newton's laws of motion. Consider the *damped harmonic oscillator*. This is the mechanical vibration of a body of mass m on the end of a spring with force constant k in the presence of viscous damping described by the damping constant b and driven by an arbitrary force $F(t)$. The differential equation for the position x of the body is

$$m\frac{d^2x}{dt^2} + b\frac{dx}{dt} + kx = F(t) \tag{7-48}$$

Rearranging,

$$\frac{d^2x}{dt^2} = -\frac{b}{m}\frac{dx}{dt} - \frac{k}{m}x + \frac{1}{m}F(t) \tag{7-49}$$

The design of the analog computer in Fig. 7-31 to solve this equation begins by assuming a voltage signal corresponding to d^2x/dt^2 is available. This is integrated to yield $-dx/dt$, where for convenience the RC time constant in Eq. (7-22) is made equal to unity. Next, $-dx/dt$ is integrated again to obtain x. A fraction b/m of $-dx/dt$ is obtained from a voltage-divider potentiometer across the output of the first integrator; this is inverted and added to the fraction k/m of x from the

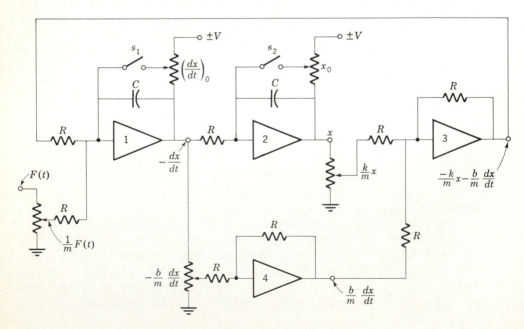

Figure 7-31 Analog computer for vibrating-mass problem.

output of the second integrator and to a voltage signal corresponding to $(1/m)F(t)$ (Fig. 7-31). The sum is equal to d^2x/dt^2, according to Eq. (7-49), and is returned to the input, where the second-derivative signal was assumed to be originally. This circuit, therefore, continuously solves the original differential equation Eq. (7-48) and voltages corresponding to x and, if desired, dx/dt can be measured at appropriate points in the circuit.

It is necessary to set voltages corresponding to dx/dt, and x for their initial values at the time when the solution begins, as in solving any differential equation. This is most effectively accomplished by opening switches s_1 and s_2 at $t = 0$. The voltages across the integrating capacitors represent the velocity and displacement at all times, as can be seen by the fact that the summing points are at ground potential. These switches must be opened simultaneously with the beginning of $F(t)$ and in practice this is most often done with electronic switches such as diodes.

Note that by simply adjusting the several potentiometers in Fig. 7-31 the physical behavior of the system can be examined over a wide range of masses, spring constants, damping, and initial conditions. Thus the analog-computer simulation provides an insight into system performance that would be much more difficult to attain by measurements on actual physical systems.

SUGGESTIONS FOR FURTHER READING

Jerald Graeme: "Applications of Operational Amplifiers," McGraw-Hill Book Company, New York, 1973.

Walter G. Jung: "IC OpAmp Cookbook," Howard W. Sams and Company, Inc., Indianapolis, Ind., 1980.

Z. H. Meiksin and Philip C. Thackray: "Electronic Design with Off-the-Shelf Integrated Circuits," Prentice-Hall, Inc., Englewood Cliffs, N.J., 1984.

David F. Stout and Milton Kaufman: "Handbook of Microcircuit Design and Application," McGraw-Hill Book Company, New York, 1980.

EXERCISES

7-1 Plot the frequency variation of β for the bridged-T feedback amplifier, Fig. 7-4, and determine the response curve of the feedback amplifier. Assume $a = 10^4$.

7-2 Analyze the source follower, Fig. 6-8, as a feedback amplifier and develop expressions for the gain, and for the input and output impedances.

    Answer: $a/(1 + a)$; $(R_g + R_2)(1 + a)$; $R_2/(1 + a)$

7-3 Calculate the input and output impedances of the transistor feedback amplifier in Fig. 7-2. The h parameters are $h_{ie} = 3600\ \Omega$, $h_{fe} = 150$.

    Answer: $2.25 \times 10^6\ \Omega$, $1.6\ \Omega$

7-4 Analyze the circuit of Fig. 6-16 as an amplifier with current feedback if the emitter bypass capacitor is omitted. Calculate β and the gain.

    Answer: 1, 3.3

7-5 Derive an expression for the input impedance of a feedback amplifier with both voltage and current feedback. Use this expression to show that both current and voltage negative feedback increase the input impedance. Find the expression for the input impedance when the output impedance is zero.

7-6 State the necessary frequency-response properties for a feedback amplifier to be stable if the feedback ratio is independent of frequency.

7-7 State three benefits of negative feedback. What is the main disadvantage?

7-8 Explain how negative feedback increases the input impedance of an amplifier. Why is this desirable?

7-9* Derive an expression for the input impedance of the operational amplifier in Fig. 7-15a. Determine by inspection the input impedance of the same unit connected as a difference operational amplifier, Fig. 7-15b.

    Answer: $R_i/2$; $2R$

7-10 Calculate the voltage and current feedback ratios β_v and β_i in the output stages of the operational-feedback amplifier of Fig. 7-16.

    Answer: 10^{-2}, 1

7-11 Using the principles employed in Fig. 7-24, sketch a circuit for which the output voltage is either the square or the cube of the input signal.

7-12* Sketch the input-output characteristic of the so-called *precision diode* circuit, Fig. 7-32, if the gain is 1000. Contrast the performance of this circuit with the simple diode circuit, Fig. 3-6, for millivolt input signals. Use the silicon diode characteristic in Fig. 3-2.

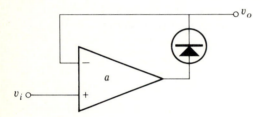

Figure 7-32 Precision diode circuit.

7-13 Determine the equivalent input inductance of the circuit in Fig. 7-33 at the frequency for which the Q is infinite.

    Answer: $R_1 R_2 C$

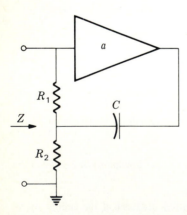

Figure 7-33 Equivalent inductance circuit analyzed in Exercise 7-13.

7-14* High input impedance and excellent common-mode rejection ratio are achieved in the *instrumentation amplifier* circuit, Fig. 7-34. If R_1 is set at 10 Ω and R_2 is set at 200 Ω, what is the overall gain of the amplifier? What is the purpose of R_1? Of R_2?

    Answer: 1000; adjust gain; adjust CMRR

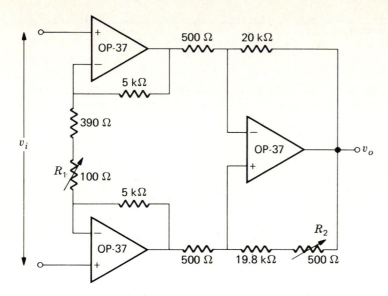

Figure 7-34 Instrumentation amplifier.

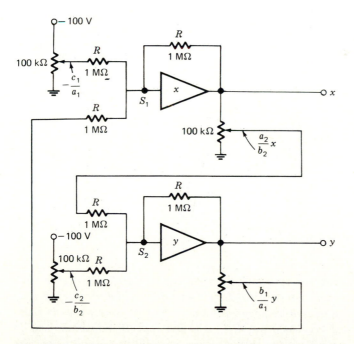

Figure 7-35 Analog-computer circuit solves simultaneous linear equations.

7-15 Write down the two simultaneous equations in x and y that the analog-computer circuit in Fig. 7-35 is set up to solve.

    Answer: $a_1 x + b_1 y = c_1$; $a_2 x + b_2 y = c_2$

EIGHT

OSCILLATORS

Electronic circuits can generate ac signals of a variety of waveforms over a wide range of frequencies. In fact, transistor and vacuum tube oscillators are the only convenient way of generating high-frequency voltages. They are widely used in radio and TV transmitters and receivers, for dielectric and induction heating, and in electronic instruments for timing and testing purposes. An oscillator, in effect, converts power delivered by the dc supply voltages into ac power having the desired characteristics. In addition to the frequency and waveform of the oscillations, the conversion efficiency and frequency stability are important in the design of oscillator circuits.

POSITIVE FEEDBACK

Oscillation is achieved through positive feedback which produces an output signal without any input signal. According to Eq. (7-4), the gain of a feedback amplifier is given by

$$a' = \frac{a}{1 - a\beta} \tag{8-1}$$

If circuit conditions are arranged so that

$$a\beta = 1 \tag{8-2}$$

Eq. (8-1) indicates that the gain becomes infinite. The physical interpretation is that an output signal exists even when the input signal is zero. For sinusoidal oscillations the feedback network is designed so that Eq. (8-2), called the *Barkhausen criterion*, is satisfied at only one frequency, and the circuit oscillates at that frequency. The Barkhausen criterion requires that the overall phase shift of the feedback signal be 360°, and this is the significant factor in determining the frequency of oscillation. In addition, the amplifier gain must be large enough to assure that the $a\beta$ product is equal to unity, in order for the oscillations to persist.

The amplitude of oscillation is determined indirectly by Eq. (8-2). The gain of any amplifier is reduced at large-scale amplitudes upon the onset of cutoff or saturation conditions. Accordingly, the quiescent amplitude is such that the absolute value of gain is $1/\beta$. Since the feedback network is most often a passive circuit, the amplitude depends primarily upon amplifier characteristics.

It is not necessary to supply an input signal in order to initiate oscillations. Random-noise voltages or transients accompanying application of the supply voltages are sufficient to start the feedback process. Since the amplitude of the feedback signal depends upon amplifier gain, the rapidity with which the oscillations reach the steady-state magnitude increases when the gain is large. It is usually desirable for the small-signal gain to be significantly larger than required by the Barkhausen criterion. This produces strong oscillations unaffected by minor circuit changes. On the other hand, if the gain is very great, nonsinusoidal oscillations may result from nonlinearities accompanying large-signal amplitudes.

RC OSCILLATORS

Phase-Shift Oscillator

A simple but useful oscillator circuit employing a conventional amplifier stage and an *RC* feedback network is the *phase-shift oscillator*, Fig. 8-1. The grounded-emitter stage has an inherent phase shift of 180°, and the three cascaded *RC* circuits shift the phase an additional 180° in order to satisfy the Barkhausen criterion. At some particular frequency the phase shift in each *RC* section is 60°,

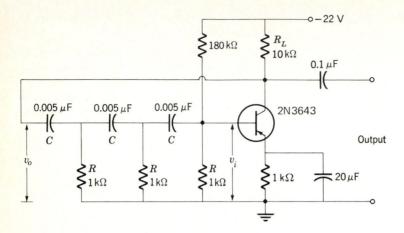

Figure 8-1 Phase-shift oscillator.

so the total phase shift in the feedback network is 180° and the circuit oscillates at this frequency, provided the amplification is great enough. Note that the maximum phase shift in one RC section is limited to 90°. This means that a two-section feedback network is not possible because an infinite gain would be required to overcome the attenuation in the feedback network at a total phase shift of 180°. Conversely, there is no particular advantage to having more than three RC sections in the feedback network, although it is possible to design such an oscillator.

The phase-shift oscillator is analyzed by first ignoring the loading effect of the amplifier upon the network. Considering that a voltage v_o is applied to the feedback network, the signal v_i applied to the transistor can be calculated by straightforward network analysis. The result is

$$\beta = \frac{v_i}{v_o} = \frac{1}{1 - 5/(\omega RC)^2 + j[1/(\omega RC)^3 - 6/\omega RC]} \tag{8-3}$$

In order for the phase shift of the feedback network to be 180°, the imaginary part of Eq. (8-3) must vanish, or

$$\frac{1}{(\omega_0 RC)^3} = \frac{6}{\omega_0 RC} \tag{8-4}$$

The oscillation frequency is found by solving for f_0,

$$f_0 = \frac{1}{2\pi\sqrt{6}} \frac{1}{RC} \tag{8-5}$$

Inserting Eq. (8-5) into Eq. (8-3), $\beta = 1/(1 - 5 \times 6) = -1/29$, which means that the gain must be

$$a = \frac{1}{\beta} = -29 \tag{8-6}$$

in order to satisfy the Barkhausen criterion. According to this result the amplification must be at least 29 or the circuit cannot oscillate.

Actually the amplifier gain must be somewhat larger than 29 in order to assure stable oscillations in the face of circuit losses and component aging effects. The amplitude of oscillations increases until limited to a value of 29 by nonlinearities in the transistor. Most often, the onset of cutoff at the peak of the wave is the limiting factor, and a peak signal amplitude nearly equal to the quiescent collector potential is expected.

The simplicity of the phase-shift oscillator makes it attractive for noncritical applications, particularly at medium and low frequencies down to about 1 Hz. The frequency stability is not as good as can be obtained with other RC oscillators, however. In addition, in order to change frequency it is necessary to vary all three capacitors (or the three resistors), and this is inconvenient.

Wien Bridge Oscillator

The frequency-selective properties of the Wien bridge discussed in Chap. 5 are very appropriate for the feedback network of an oscillator. Actually, the *Wien bridge oscillator*, Fig. 8-2, is widely used for variable-frequency laboratory instruments called *signal generators*. In Fig. 8-2 a conventional integrated-circuit amplifier provides positive feedback at the resonant frequency of the bridge, where, according to Eq. (5-68), the phase shift in the feedback network is zero. The characteristic frequency is, from Eq. (5-64),

$$f_0 = \frac{1}{2\pi RC} \tag{8-7}$$

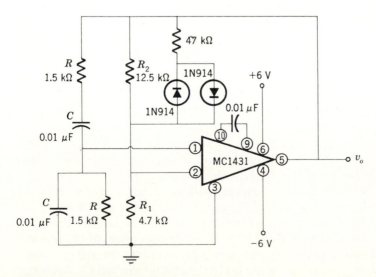

Figure 8-2 Integrated-circuit Wien bridge oscillator.

The output voltage of the Wien bridge is zero at exact balance when $R_2 = 2R_1$ (refer to Fig. 5-15), so it is necessary to unbalance the bridge slightly by adjusting the ratio R_2/R_1. This provides sufficient feedback voltage to maintain stable oscillations, since the feedback ratio is determined by the relative values of R_1 and R_2. The combination of the two 1N914 diodes and the 47-kΩ resistors shunting R_2 provides amplitude stability by automatically adjusting the feedback ratio. Suppose the output signal is small so that the signal voltage across the diodes is less than the characteristic diode forward-voltage drop, approximately 0.6 V. Then neither diode conducts and the shunt arm is open-circuited. If the output signal is large enough for the diodes to conduct, the 47-kΩ resistor is, in effect, connected in parallel with R_2 and feedback is reduced. Therefore, the output amplitude is stabilized at that level at which the diodes just conduct at the peaks of the waveform.

The excellent frequency stability of the Wien bridge oscillator compared with the phase-shift oscillator is a result of the rapid change of phase of the feedback voltage with frequency. The feedback ratio is, from Eq. (5-68),

$$\beta = \frac{v_o}{v_i} = \frac{1}{3 + j(\omega/\omega_0 - \omega_0/\omega)} - \frac{1}{1 + R_2/R_1} \tag{8-8}$$

The phase shift is found by rationalizing Eq. (8-8) and computing the tangent of the phase angle. For frequencies near to the resonant frequency, the result is

$$\tan \theta \simeq \frac{1 + R_2/R_1}{9 - 3(1 + R_2/R_1)} \left(\frac{\omega}{\omega_0} - \frac{\omega_0}{\omega} \right) \tag{8-9}$$

If, for example, $R_2 = 1.9R_1$, the coefficient of the frequency term in Eq. (8-9) is equal to 9.7. By comparison, the corresponding expression for the phase-shift oscillator is, after rationalizing Eq. (8-3),

$$\tan \theta \simeq -\frac{\sqrt{6}}{5} \left(\frac{\omega}{\omega_0} - \frac{\omega_0}{\omega} \right) \tag{8-10}$$

where Eq. (8-5) has been used to introduce ω_0. The numerical factor in Eq. (8-10) is only 0.49, which is smaller by a factor of 20 than in the case of the Wien bridge.

This comparison is illustrated more clearly in Fig. 8-3, where the phase angle of the feedback signal is plotted as a function of ω/ω_0 for both oscillators. The phase angle changes much more rapidly with frequency in the Wien bridge feedback circuit. This means that the oscillation frequency is quite stable since only feedback signals with a near-zero phase angle are effective.

The excellent frequency stability together with the relative ease in changing frequency (only the two capacitors need be variable) makes the Wien bridge oscillator popular. In most practical circuits the variable capacitors provide a frequency range of about 10 to 1. In addition, fixed decade values of the resistors are selected by a switch so that a wide frequency range can be covered in a single instrument. A typical commercial Wien bridge oscillator can have a frequency range extending from 5 Hz to 1 MHz in decade steps.

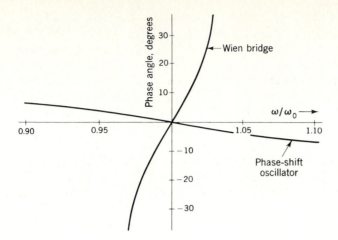

Figure 8-3 Phase-shift characteristics near resonant frequency for two RC oscillators.

RESONANT CIRCUIT OSCILLATORS

LC Oscillators

Resonant LC circuits are often used in the feedback network of oscillators to select the frequency of oscillation. Consider the so-called *Hartley oscillator*, Fig. 8-4, in which a parallel resonant circuit is connected between gate and drain. The inductance is tapped so that, in effect, the portion L_1 is part of the drain load while the remainder L_2 is in the gate circuit. The resonant frequency involves the series inductance of L_1 and L_2, so that, as in Chap. 5,

$$\omega_0 = \frac{1}{\sqrt{(L_1 + L_2)C}} \tag{8-11}$$

The feedback ratio is determined by first calculating the feedback voltage across L_2,

$$v_i = ij\omega L_2 = \frac{v_o\,j\omega L_2}{j\omega L_2 + 1/j\omega C} \tag{8-12}$$

Introducing the resonance condition $\omega_0(L_1 + L_2) = 1/\omega_0 C$,

$$v_i = v_o\,\frac{\omega_0 L_2}{-\omega_0 L_1} \tag{8-13}$$

so that the feedback ratio is just

$$\beta = -\frac{L_2}{L_1} \tag{8-14}$$

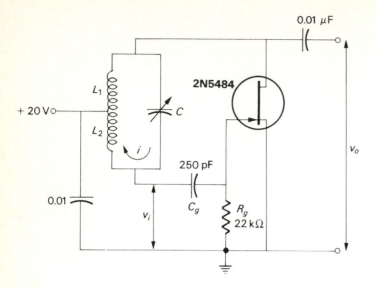

Figure 8-4 Hartley oscillator.

Note that the ratio is negative, which means that the additional 180° phase shift of the amplifier produces positive feedback as required. Equation (8-14) and the Barkhausen criterion specify the gain of the amplifier needed to sustain oscillations.

The bias conditions in the Hartley oscillator are worthy of note: Operating bias is supplied by the $R_g C_g$ combination in the gate circuit. When oscillations start, the gate bias is zero. As oscillations build up, the gate-source diode rectifies the feedback signal, thereby charging C_g to nearly the peak value of the input signal. The $R_g C_g$ time constant is much longer than the period of oscillation so the voltage across C_g is constant and represents the necessary dc gate bias. In effect, the gate is clamped at ground potential (compare Figs. 3-24 and 3-25). The gate bias therefore automatically adjusts itself to the amplitude of the feedback signal and this action stabilizes the amplitude of oscillation.

This form of bias means that the transistor is cut off during most of the cycle, a condition labeled *class C* operation. The gate voltage produces drain current at the peak of the feedback voltage cycle. The large impedance of the resonant circuit allows only the fundamental component of the output signal to have any appreciable amplitude, however, so that the output waveform is sinusoidal. The *Colpitts* oscillator is similar to the Hartley circuit except that the feedback ratio is determined by replacing the capacitor C with two series capacitors, rather than splitting the inductance.

Another way to develop feedback voltage is by including a secondary winding, or *tickler* winding, coupled to the inductance. Consider, for example, the grounded-base oscillator, Fig. 8-5a. Mutual inductance between L and the tickler winding induces a feedback signal of the proper amplitude and phase to sustain

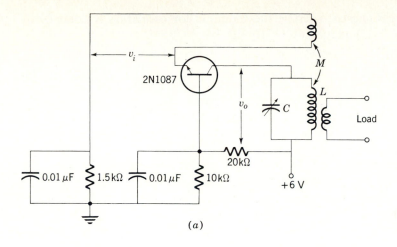

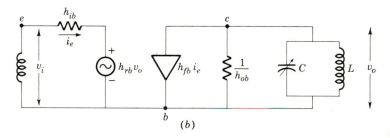

Figure 8-5 (a) Grounded-base tickler oscillator and (b) equivalent circuit.

oscillations. From the definition of mutual inductance in Chap. 5, the feedback voltage is just the ratio of the mutual inductance M between coils to the total inductance L times the voltage across L. Therefore

$$v_i = \frac{M}{L} v_o \tag{8-15}$$

or

$$\beta = \frac{v_i}{v_o} = \frac{M}{L} \tag{8-16}$$

The appropriate equivalent circuit of the oscillator, Fig. 8-5b, is the complete hybrid equivalent circuit discussed in Chap. 6. The approximate version is not applicable because the impedance of the parallel resonant load is very great at resonance and $1/h_{ob}$ cannot be neglected. According to Fig. 8-5b, the gain of the grounded-base amplifier is

$$a = \frac{v_o}{v_i} = \frac{v_o}{h_{ib} i_e + h_{rb} v_o} = \frac{1}{h_{ib} i_e/v_o + h_{rb}} \cong \frac{1}{h_{rb}} \tag{8-17}$$

Applying the Barkhausen criterion, the condition for oscillation is obtained by comparing Eqs. (8-17) and (8-16),

$$h_{rb} = \frac{M}{L} \tag{8-18}$$

In a good-quality transistor h_{rb} is of the order of 10^{-4}, which means that the mutual inductance may be quite small. Accordingly, a tickler winding consisting of only a few turns is sufficient.

A major advantage of the tickler feedback circuit is that the amplitude of the feedback voltage can be easily adjusted by choosing the number of turns on the feedback winding. The proper phase is obtained by interchanging the leads of the winding, if necessary, to obtain positive feedback. Thus, although Fig. 8-5 is a grounded-base circuit, the grounded-emitter and grounded-collector configurations are equally useful.

Crystal Oscillators

The frequency stability of LC oscillators is determined primarily by the Q factor of the resonant circuit. The resonance curve is sharply peaked and the rate of change of phase with frequency is rapid when the Q is large, and both factors contribute to frequency stability of the oscillator. In this connection, any equivalent resistance connected in parallel with the resonant circuit lowers the effective Q. Therefore, the loading effect on the resonant circuit should be minimized to improve frequency stability. According to the discussion in Chap. 6, the output impedance of the grounded-base amplifier is larger than for any other configuration. This is why the grounded-base oscillator with the resonant combination in the collector circuit is the most satisfactory transistor oscillator configuration. Correspondingly, FET oscillators most often have the resonant circuit in the gate circuit. In either case, practical values of Q from 100 to 500 can be obtained, and quite stable oscillations result.

Many applications require a higher order of frequency stability than can be obtained with LC resonant circuits, and *crystal oscillators* are widely used to fill this need. Certain crystalline materials, most notably quartz, exhibit piezoelectric properties; that is, they deform mechanically when subjected to an electric field. Piezoelectricity also implies that the inverse is also true: when the crystal is forcibly deformed, an electric potential is developed between opposing faces of the crystal. As a result of this piezoelectric property, a thin plate of quartz provided with conducting electrodes vibrates mechanically when the electrodes are connected to an alternating voltage source. The vibrations, in turn, produce electrical signals which interact with the voltage source. The vibrations and electrical signals are a maximum at the natural mechanical resonant frequency of the crystal.

The equation for the motion of a vibrating body has already been written in

Eq. (7-48),

$$m \frac{dx^2}{dt^2} + b \frac{dx}{dt} + kx = F(t) \qquad (8\text{-}19)$$

where in the present case m is the mass of the crystal, b is the internal mechanical loss coefficient, and k is the elastic constant of the crystal. This expression is identical in form with that for the current in a series resonant circuit, Eq. (5-20),

$$L \frac{d^2 i}{dt^2} + R \frac{di}{dt} + \frac{1}{C} i = F(t) \qquad (8\text{-}20)$$

Comparing Eqs. (8-19) and (8-20) it can be seen that the vibrating mass is analogous to inductance, mechanical losses are equivalent to resistance, and the elasticity corresponds to the reciprocal of capacitance. Because of the identical form of the two equations, mechanical resonance is expected, and it is useful to define a mechanical Q factor by analogy with Eq. (5-47),

$$Q = \frac{\omega m}{b} \qquad (8\text{-}21)$$

It turns out that the internal losses in quartz crystals are very small and that Q values reaching as high as 100,000 can be achieved. Furthermore, the elastic constants are such that resonant frequencies ranging from 10 kHz to several tens of megahertz are possible, depending upon the mechanical size and shape of the crystal.

The piezoelectric properties of quartz result in electrode potentials corresponding to the mechanical vibrations. This suggests that the electrical characteristics can be represented by an equivalent circuit. Comparing Eq. (8-19) and (8-20), the appropriate circuit is a series combination of a resistance, inductance, and capacitance. To this must be added the electric capacitance resulting from the parallel-plate capacitance of the electrodes with the crystal as a dielectric. Therefore, the complete equivalent circuit of a quartz crystal is the series-parallel combination shown in Fig. 8-6a. In this equivalent circuit L, C, and R are related to the properties of the quartz crystal and C' is the electrostatic capacitance of the electrodes. Appropriate values for a 90-kHz crystal are $L = 137$ H, $C = 0.0235$ pF, $R = 15,000$ Ω, and $C' = 3.5$ pF. The conventional circuit symbol for a crystal is a parallel-plate capacitor with the crystal between the plates, Fig. 8-6b.

The series-parallel equivalent circuit of a quartz crystal shows that there is both a series resonant frequency (zero impedance) and a parallel resonant frequency (infinite impedance). The frequency of series resonance is $\omega_S = 1/\sqrt{LC}$. The parallel resonance occurs when the reactance of C' equals the net inductive reactance of the combination of L and C, $\omega_P = \sqrt{1/L(1/C + 1/C')}$. Accordingly, the parallel resonant frequency is always greater than the series resonant frequency, although, since $C' \gg C$, the two are very close. The reactance is capacitive both below and above the resonant frequencies, Fig. 8-7.

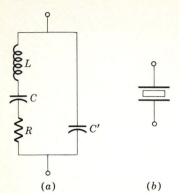

(a)   (b)

Figure 8-6 (a) Equivalent circuit of quartz crystal and (b) circuit symbol.

Extremely stable oscillator circuits are possible with the large Q of a quartz resonator, and a variety of circuits have been designed. Either the series or parallel resonance frequency may be used, although parallel resonance is more common. Consider, for example, the *Pierce oscillator*, Fig. 8-8, in which the crystal is connected between base and collector. This circuit is the Colpitts oscillator, with the crystal replacing the resonant circuit and the feedback ratio determined by the relative values of C_1 and C_2. The inductance RFC (for *radio-frequency choke*) in the collector lead is a useful way of applying collector potential without shorting the collector to ground at the signal frequency. It can be replaced by a 10,000-Ω resistor with some loss in circuit performance. A parallel resonant LC combination can also be used with attendant gain in circuit performance. In the latter case the resonant circuit is merely a convenient collector load impedance and does not determine the oscillation frequency.

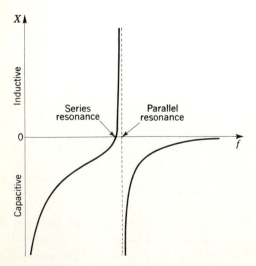

Figure 8-7 Impedance of quartz crystal.

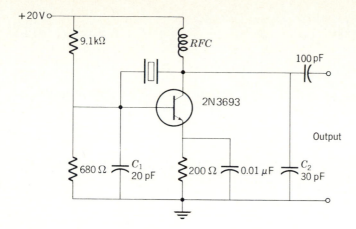

Figure 8-8 Quartz-crystal oscillator.

Quartz is almost universally used in crystal oscillators because it is hard, reasonably strong, and has a small temperature coefficient of expansion. Suitable orientation of the plane faces with respect to the crystalline structure makes the resonant frequency independent of temperature over a reasonable range. As a result, frequency stability of the order of 100 ppm can be achieved. Even greater accuracy is obtained by placing the crystal in a temperature-controlled oven and evacuating the crystal holder to reduce air damping forces on the vibrating crystal. It is also common practice to stabilize the temperature of the remainder of the circuit and to employ a regulated power supply for the oscillator. Amplifier stages are used to isolate the oscillator from variations in load. Such carefully designed crystal oscillators provide extremely precise time standards which can be accurate to 1 part in 100 million.

NEGATIVE RESISTANCE OSCILLATORS

Stability Analysis

The current-voltage characteristics of several devices, most notably the tunnel diode and the unijunction transistor, exhibit *negative resistance* properties. That is, over a portion of the characteristic the current decreases as the applied voltage increases. The physical mechanisms leading to negative resistance properties which are discussed in Chap. 4 prove to be very useful in a variety of oscillator circuits.

Analysis of negative resistance oscillation is best carried out by examining the complex impedance of the ac small-signal linear equivalent circuit of the oscillator. Consider, for example, the series resonant circuit in Fig. 8-9 where r

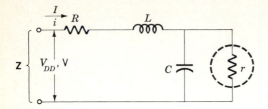

Figure 8-9 Equivalent circuit of negative resistance oscillator.

represents the negative resistance of a tunnel diode at the operating point. The impedance of the circuit is

$$Z = R + j\omega L + \frac{1}{1/r + j\omega C} = R + j\omega L + \frac{r}{1 + jr\omega C}$$

$$= R + \frac{r}{1 + (\omega rC)^2} + j\omega\left[L - \frac{r^2 C}{1 + (\omega rC)^2}\right] \tag{8-22}$$

As usual, the circuit current is given by Ohm's law

$$i = \frac{v}{Z} \tag{8-23}$$

Equation (8-23) predicts an infinite current under conditions for which the impedance vanishes. This situation is analogous to Eq. (8-2) and implies that the circuit oscillates. Analysis of circuit oscillation is effected, therefore, simply by searching for zeros of the impedance function.

The impedance vanishes when both the real and imaginary parts of Eq. (8-22) are equal to zero,

$$R + \frac{r}{1 + (\omega rC)^2} = 0 \quad \text{and} \quad L - \frac{r^2 C}{1 + (\omega rC)^2} = 0 \tag{8-24}$$

These relations are each solved for the angular frequency

$$\omega^2 = -\left(\frac{1}{rC}\right)^2\left(1 + \frac{r}{R}\right) \quad \text{and} \quad \omega^2 = \frac{1}{LC} - \left(\frac{1}{rC}\right)^2 = \omega_0^2 - \left(\frac{1}{rC}\right)^2 \tag{8-25}$$

where ω_0 is the resonant frequency. Since both expressions must yield the same value for the oscillation frequency,

$$-\left(\frac{1}{rC}\right)^2\left(1 + \frac{r}{R}\right) = \frac{1}{LC} - \left(\frac{1}{rC}\right)^2 \tag{8-26}$$

Solving for the value of negative resistance necessary to sustain oscillations,

$$r = -\frac{L}{RC} \tag{8-27}$$

Inserting this value into Eq. (8-25), the frequency of oscillation is

$$\omega^2 = \omega_0^2 - \left(\frac{R}{L}\right)^2 \qquad (8\text{-}28)$$

This expression is just the natural frequency of a resonant circuit with resistance. In effect, the diode negative resistance cancels out the positive resistance in the circuit and the circuit rings continuously at its natural frequency. Note that the absolute value of r must be greater than the circuit resistance R in order for the frequency calculated from Eq. (8-25) to have a real value.

The oscillatory condition, $|r| > R$, actually has several different modes of operation that depend upon the ac load line. If, for example, the inductance L is large, the frequency calculated from Eq. (8-25) is imaginary even though the magnitude of the diode negative resistance is proper for oscillation. This condition leads to nonsinusoidal *relaxation* oscillations which are treated later in this chapter. On the other hand, if the capacitance C is large, the frequency is zero, according to Eq. (8-25). This means that the circuit oscillations are completely damped out.

The relation between circuit conditions for sinusoidal oscillation, relaxation oscillation, and damped behavior may be appreciated from the current-voltage characteristic, Fig. 8-10, and the ac load lines. According to this analysis sinu-

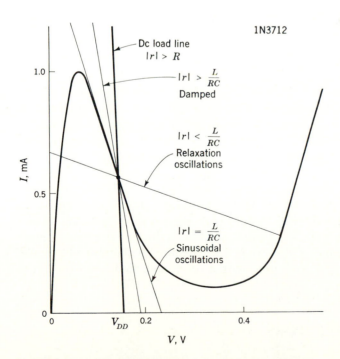

Figure 8-10 Combination of dc and ac load lines determine oscillating mode of tunnel-diode oscillator.

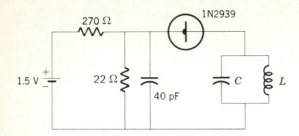

Figure 8-11 Tunnel-diode oscillator.

soidal oscillations at the natural frequency of the circuit are obtained when the magnitude of the negative resistance at the operating point is properly matched to circuit parameters, Eq. (8-27). A large value of inductance and small capacitance lead to relaxation oscillations and large C and small L produce complete damping. This latter condition is useful in experimentally tracing out the entire current-voltage characteristic which, clearly, cannot be accomplished with the circuit oscillating.

Tunnel-Diode Oscillator

The circuit of a practical tunnel-diode oscillator is shown in Fig. 8-11. In this configuration the diode is effectively in parallel with a parallel resonant circuit but the principle of operation is identical with the analysis of the preceding section. The resistive voltage divider biases the diode to an operating point on the negative resistance portion of the current-voltage characteristic.

It may appear that the condition for oscillation, Eq. (8-27), is so restrictive that sinusoidal oscillations are difficult to attain. This is not the case since r is the slope of the IV characteristic at the operating point and therefore varies considerably with operating points ranging from the peak to the valley. Furthermore, the circuit tends to adjust itself to the proper value since the amplitude of oscillation changes to yield an average value of r over a complete cycle which is in agreement with Eq. (8-27).

Tunnel diodes are excellent high-frequency devices and oscillation frequencies as large as 10^{11} Hz have been achieved. Actually this virtue often proves troublesome in experimental circumstances since even tiny stray capacitances and inductances are sufficient to cause oscillations at very high frequencies. Tunnel diodes operate at very small power levels. Correspondingly, however, the maximum signal amplitude and ac power output of tunnel-diode oscillators are limited.

RELAXATION OSCILLATORS

Oscillator circuits considered to this point can be analyzed in terms of linear elements. Circuits which employ highly nonlinear active elements are termed *relaxation oscillators* for reasons which become clear in the following discussion. Very often relaxation oscillators are based upon negative resistance properties of the active element. Although, as discussed in the previous section, it is possible to generate sinusoidal waveforms by means of negative resistance characteristics, relaxation oscillators characteristically produce nonsinusoidal signals.

Sawtooth Generators

Consider the unijunction relaxation-oscillator circuit in Fig. 8-12. The operation of this circuit can be understood after examining the current-voltage characteristic of a unijunction transistor, Fig. 8-13. This characteristic exhibits a negative resistance region between the peak voltage V_p and the valley point V_v, as discussed in Chap. 4.

   The unijunction oscillator operates as follows: Capacitor C charges through resistor R and the voltage across C increases exponentially until the potential V_p is attained. At this point the emitter junction becomes biased in the forward direction and the capacitor rapidly discharges through the emitter junction. When the capacitor potential drops to a low value (essentially equal to V_v), the emitter junction is again reverse-biased and the capacitor begins to recharge. The waveform at output terminal 1, Fig. 8-14, is a series of RC charging curves with a peak-to-peak amplitude equal to $V_p - V_v$. Note that following every charging period the circuit "relaxes" back to the starting point. This is the origin of the terminology for this type of oscillator. The output waveform at terminal 2 is a series of sharp positive pulses, Fig. 8-14.

   The period of oscillation is found from the expression for the capacitor voltage in a simple RC circuit, Eq. (2-59). The voltage reaches V_v at t_1,

$$V_v = V(1 - e^{-t_1/RC}) \tag{8-29}$$

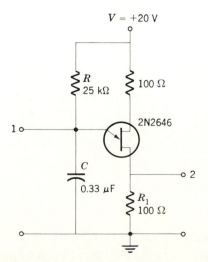

Figure 8-12 Unijunction relaxation oscillator.

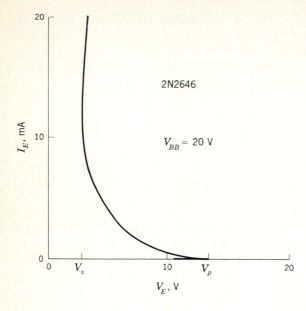

Figure 8-13 Current-voltage characteristic of unijunction transistor.

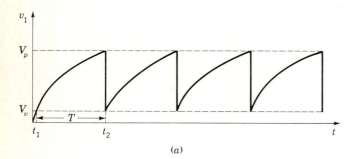

(a)

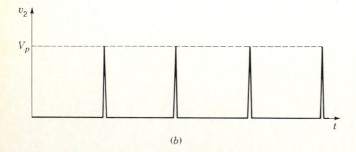

(b)

Figure 8-14 Output waveforms of unijunction oscillator at (a) output terminal 1 and (b) output terminal 2.

Solving for t_1

$$t_1 = -RC \ln (1 - V_v/V) \tag{8-30}$$

Similarly, the voltage reaches V_p at t_2, so that

$$t_2 = -RC \ln (1 - V_p/V) \tag{8-31}$$

The period of the oscillation is just $t_2 - t_1$, or

$$T = RC \ln \frac{V - V_v}{V - V_p} \tag{8-32}$$

According to Eq. (8-32) the oscillation frequency depends upon the properties of the unijunction transistor and magnitude of the supply voltage, as well as upon the circuit time constant. This equation assumes that the discharge time is zero, which is not true in practice. Because of the finite discharge time, unijunction relaxation oscillators are limited to frequencies less than about 100 kHz.

The frequency stability of relaxation oscillators is inherently, and characteristically, very poor. The initiation of the discharge is a probabilistic phenomenon, so that the discharge does not always start exactly at the potential V_p. Furthermore, the rate of change of the capacitor voltage is comparatively slow, which means that the exact instant at which the discharge is initiated can vary slightly from cycle to cycle. In many applications, however, the poor frequency stability is an advantage since the oscillations can be *triggered*, or *synchronized*, by an external signal.

Suppose, for example, that a synchronizing signal comprising a series of sharp voltage pulses is applied across the capacitor, Fig. 8-15. The sharp voltage rise of one of the pulses causes the capacitor voltage to exceed V_p at some point and the transistor is triggered into conduction. This repeats on successive cycles with the result that the frequency of oscillation is synchronized to the period of the pulse signal. Note that the period of the relaxation oscillator is a multiple or submultiple of the period of the pulses. In this way a relaxation oscillator acts as a frequency multiplier or divider. Frequency division by factors as large as 10 or so is easily possible.

The current-voltage characteristic of an SCR, Fig. 3-19, is similar in form to that of the unijunction transistor. However, the breakdown voltage is easily

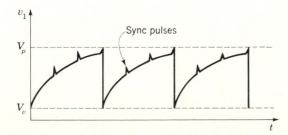

Figure 8-15 Synchronizing frequency of a relaxation oscillator.

adjusted by varying the gate current. This means that the amplitude and frequency of oscillation can be changed easily in an SCR relaxation oscillator. Furthermore, synchronizing signals can be introduced into the gate circuit to take advantage of the inherent amplification of the SCR.

A useful relaxation oscillator employing an SCR is shown in Fig. 8-16. The gate potential is determined by the variable voltage divider comprising R_1 and R_2. Initially, when C is uncharged, the cathode of the SCR is at a potential of 20 V with respect to ground, so the gate is biased in the reverse direction. As the capacitor voltage increases, the SCR remains switched off until the cathode potential becomes slightly less positive than the gate. At this point the SCR switches on, the capacitor is discharged, and the cycle repeats. The purpose of the 5-Ω resistor in series with the cathode is to limit the discharge current to a safe value for the SCR. This is necessary because the internal resistance of the SCR in the on state is so small that the peak current can be destructive.

A linear sawtooth waveform is produced by the constant-current collector characteristic of the transistor. Since the charging current is constant, the capacitor voltage increases linearly with time according to

$$\frac{q}{C} = \frac{It}{C} = \beta \frac{I_b}{C} t = \frac{\beta V_{dc}}{CR_b} t \tag{8-33}$$

where β is the current gain of the transistor, V_{dc} is the supply potential, and t is the time measured from the instant that the capacitor voltage is equal to zero. Thus the period of the waveform can be adjusted by changing the value of R_b. Furthermore, the peak amplitude of the sawtooth can be nearly as large as V_{dc}, which is advantageous in most applications.

The charging current can also be controlled by introducing a control voltage, V_c, through R_b in place of V_{dc}. Since the period of the sawtooth, $T = 1/f$, is just the time for the capacitor voltage to reach the potential, V_T, at which the SCR

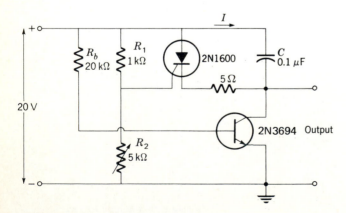

Figure 8-16 Relaxation sawtooth generator.

switches on, the frequency of oscillation is, from Eq. (8-33),

$$f = \frac{\beta}{CR_bV_T} V_c \qquad (8\text{-}34)$$

According to Eq. (8-34), the oscillator frequency depends linearly upon the control voltage. This is one version of a *voltage controlled oscillator*, or *VCO*. Applications of VCOs are considered in Chap. 10.

Multivibrators

Two-transistor feedback circuits called *multivibrators* are important relaxation oscillators. Additionally, suitably modified versions have exceedingly versatile properties other than oscillation.

It is easiest to begin with a study of the so-called *astable* multivibrator, Fig. 8-17, the terminology for which will become clear later. This circuit can be recognized as a two-stage RC amplifier with the output coupled back to the input. The feedback ratio is unity and positive because of the 180° phase shift in each stage; thus, the circuit oscillates. Because of the very strong feedback signals, the transistors are driven into either cutoff or saturation, and nonsinusoidal oscillations are generated.

Suppose that at some instant the feedback voltage drives Q_1 into cutoff; this implies that Q_2 is conducting because of the relative phase shift between the two stages. The voltage drop across the collector resistor of Q_2 puts the collector at nearly ground potential, and C_1 charges through R_1 toward the collector supply potential. When the voltage across C_1 increases sufficiently to bias the emitter junction of Q_1 in the forward direction, Q_1 begins to conduct. The collector voltage of Q_1 drops and Q_2 is driven into cutoff through coupling capacitor C_2. Now, C_2 charges through R_2 until Q_2 becomes forward-biased and the cycle repeats.

Notice that the circuit alternates between a state in which Q_1 is conducting and simultaneously Q_2 is cut off and a state in which Q_1 is cut off and Q_2 is conducting. The transition between these two states is rapid because of the strong

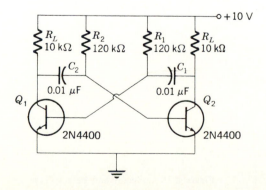

Figure 8-17 Astable multivibrator.

feedback signals. The time in each state depends upon the coupling capacitor and bias resistor time constant. Since each transistor is driven alternately into cutoff and saturation, the voltage waveform at either collector is essentially a square wave with a peak amplitude equal to the collector supply voltage.

This picture is confirmed by actual collector voltage waveforms, Fig. 8-18. The triangular base voltage waveforms illustrate the alternate charging and discharging of the coupling capacitors. Of particular interest is the very low voltage drop across the transistors when saturated ($\cong 0.1$ V) and the fact that the transition from one state to another is initiated when the base voltage of the cutoff transistor just slightly exceeds zero.

A small voltage spike, or *overshoot*, is present in the base voltage waveform at the transition from one state to the other. At each transition, one coupling capacitor charges quickly through R_L and the charging current results in an overshoot voltage drop across the corresponding emitter junction resistance. The charging current through R_L also prevents the collector voltage from rising immediately to V_{cc}, and this is the source of the rounding of the collector waveform pulse edge shown in Fig. 8-18. Overshoot introduces only a minor departure from a square collector waveform, however.

The pulse width in a multivibrator depends upon the time constant of C_1 (or C_2) charging through R_1 (or R_2). A value for t_w can be written down immediately from the expression for the period of a relaxation oscillator, Eq. (8-32). Com-

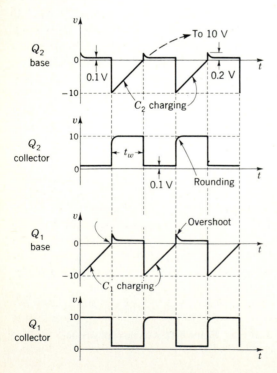

Figure 8-18 Collector and base voltage waveforms in transistor multivibrator.

paring the waveforms in Fig. 8-18 with those of Fig. 8-14,

$$t_w = R_1 C_1 \ln \frac{V_{cc} - (-V_{cc})}{V_{cc} - 0}$$

$$t_w = R_1 C_1 \ln 2 \tag{8-35}$$

When $R_1 = R_2$ and $C_1 = C_2$ the waveform is a square wave of frequency $f = 1/2t_w$. It is equally possible to generate asymmetrical waveforms by choosing nonequal values for, say, the coupling capacitors. The result is alternating pulses of widths given by Eq. (8-35) for each time constant.

The speed with which a multivibrator makes the transition from one state to the other, and, hence, the upper limit to the oscillation frequency, is governed by the same considerations that apply to the high-frequency response of amplifiers. This is so because during the transition the circuit really operates as a feedback amplifier. Integrated-circuit techniques, such as those discussed in Chap. 9, make it possible to achieve oscillation frequencies in the 100-MHz range.

Practical multivibrators are very effectively implemented using integrated-circuit devices. In Fig. 8-19a, for example, dual amplifiers of an integrated circuit are cross-coupled to produce an astable multivibrator analogous to the two-transistor version. In this circuit diagram the small circle at the apex of each amplifier triangle symbol is used to designate an inverting amplifier, which is necessary in order to achieve positive feedback. The actual circuit based on a type 914 integrated circuit, Fig. 8-19b, is quite similar to the transistor arrangement depicted in Fig. 8-17.

If one RC coupling network of an astable multivibrator is replaced by dc coupling, Fig. 8-20, the circuit remains stably in the state with $A2$ conducting and $A1$ cut off. This is determined by the strong positive input voltage applied to $A2$ through resistor R. A positive input pulse turns on $A1$, is amplified and inverted and turns off $A2$, and this, in turn, drives $A1$ into saturation. Thus the regenerative feedback rapidly causes a change of state initiated by the input signal. This state lasts while C charges through R until the input of $A2$ just becomes positive, at which time the circuit reverts to the stable state. As in the astable multivibrator, the transition between states is very rapid because of the strong positive feedback.

This *monostable*, or *one-shot*, multivibrator produces a single output pulse for every positive input signal sufficient to trigger the transition. It is often used to generate uniform pulses from irregular input signals. Also, the output waveform may be differentiated to yield a negative pulse delayed from the input pulse by the time the circuit remains in the unstable state. Thus the combination of a monostable multivibrator together with a differentiator produces a uniform time delay between input and output pulses. In either application the duration of the output pulse or the length of the time delay is given by Eq. (8-35).

A completely dc-coupled multivibrator, Fig. 8-21, can remain stably in either of its two states and can be caused to make transitions from one state to the other by suitable input signals. Because of its two stable states, a multivibrator with dc coupling is called a *binary*, and is also commonly referred to as a *flip-flop*.

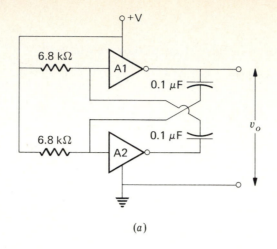

(a)

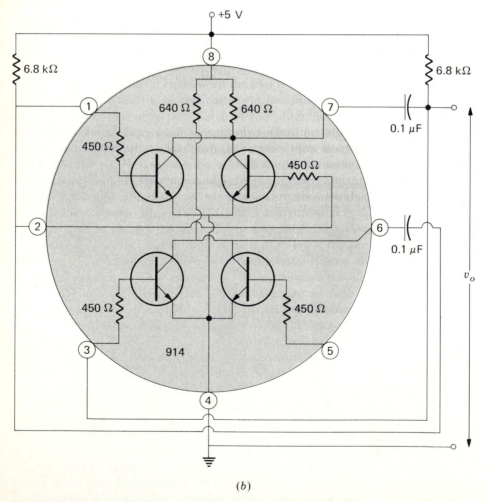

(b)

Figure 8-19 (a) Integrated-circuit astable multivibrator and (b) actual circuit based on type 914 dual inverter.

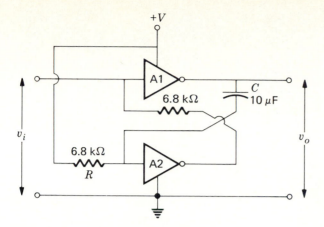

Figure 8-20 Monostable multivibrator uses both ac and dc coupling between stages.

This terminology distinguishes the binary from the monostable multivibrator with its one stable state and from the astable multivibrator, which is a free-running relaxation oscillator.

Suppose in the binary of Fig. 8-21 that $A1$ is saturated; this means that $A2$ is cutoff because of the cross-coupled feedback connection. Also, because $A1$ is saturated, the output voltage is zero. A negative input pulse can initiate a transition to the other stable state, but a positive input signal has no effect because $A1$ is already saturated. In the other stable state, that is, with $A1$ cut off and a positive output signal, only a positive input pulse can cause the reverse transition. Note that the binary is, in effect, a memory circuit in that it remembers the polarity of the last input pulse. That is, a zero output voltage indicates that the last input pulse was positive and vice versa. More elaborate and versatile versions of the binary are considered in Chap. 9.

A related binary circuit called the *Schmitt trigger* (after its inventor) is obtained by achieving dc feedback through a common resistor, Fig. 8-22. This

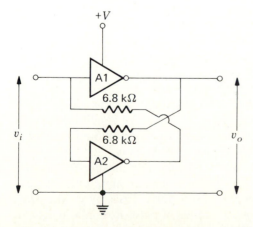

Figure 8-21 Binary flip-flop is dc-coupled multivibrator.

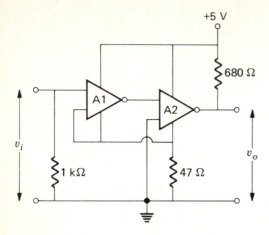

Figure 8-22 Schmitt-trigger circuit.

circuit has two stable states and the magnitude of the input voltage determines which of the two is possible. Suppose that the input signal is zero so that $A1$ is cut off; correspondingly, $A2$ is saturated and the output signal is zero. The voltage drop across the common feedback resistor which keeps $A1$ cut off is determined by the saturation current of $A2$. For example, if the circuit uses a type 914 integrated circuit (refer to Fig. 8-19), the saturation current is approximately given by the supply potential divided by the resistance of the parallel combination of the 640-Ω collector resistor and the 680-Ω external resistor. The corresponding voltage drop across the feedback resistor is 0.62 V.

Suppose now the input voltage is increased from zero. Nothing happens until the input voltage just exceeds 0.62 V, at which point the net input signal to $A1$ becomes positive and regenerative feedback flips the circuit into the other state. This condition, with $A1$ saturated and $A2$ cut off (thus a positive output signal), continues so long as the input voltage is maintained. As the input signal is decreased, the circuit does not regain its original state until a value smaller than 0.62 V is reached. The reason for this is that the voltage drop across the feedback resistor is now determined by the saturation current in $A1$. This is approximately 0.34 V (since there is no external resistor in the $A1$ circuit). As the input voltage is decreased below 0.34 V, regeneration quickly causes the circuit to revert to the original state and the output signal drops to zero.

An interesting application of the Schmitt trigger is as a pulse regenerator or *squaring circuit*, illustrated in Fig. 8-23. The input signal, a series of degraded pulse waveforms, is converted to a square pulse output as the circuit is triggered back and forth between its two states by the input waveform. Note that even a sinusoidal signal may be converted to a square wave in this way.

The Schmitt trigger is also useful as a *pulse-height discriminator* to measure the voltage pulse amplitudes of an input signal. Each time an input pulse exceeds the trigger threshold the circuit generates an output pulse. Thus, by varying, say, a dc voltage in series with the input signal, the range of pulse sizes in a signal can

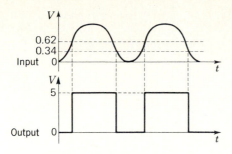

Figure 8-23 Waveforms in Schmitt-trigger squaring circuit.

be determined. Note that the output pulses are all of uniform amplitude, independent of the input trigger pulse. This is useful in that subsequent circuits need not be capable of handling a wide range of pulse sizes.

WAVEFORM GENERATORS

Nonsinusoidal waveforms can be generated by relaxation oscillators and multivibrator circuits. The output signals from these generators can be further modified through the use of diode clippers and clamps to select a portion or set the dc level of a waveform. Also, integrating and differentiating circuits can be used to modify waveforms in specific ways. A number of additional waveform generators are described in this section.

Diode Pump

Consider the action of the so-called *diode-pump* circuit, Fig. 8-24, in response to negative-going signal pulses. On each negative pulse, diode D_1 conducts and C_1 charges to the peak value of the input pulse. Since the input pulse biases D_2 in reverse, no charge reaches C_2. Thus the charge on C_1 is

$$Q_1 = C_1 v_i \tag{8-36}$$

During the time that the input voltage is zero, the voltage on C_1 biases D_2 in the forward direction and C_2 becomes charged. Since the two capacitors are now

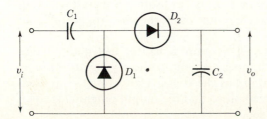

Figure 8-24 Diode pump.

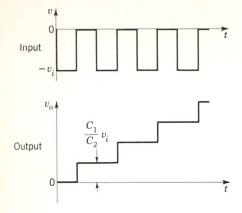

Figure 8-25 Staircase waveform produced by diode pump from negative square-wave input.

effectively connected in parallel, the charge on C_2 is

$$Q_2 = C_2 v_o = C_2 \frac{Q_1}{C_1 + C_2} = \frac{Q_1}{1 + C_1/C_2} \tag{8-37}$$

If $C_2 \gg C_1$, $Q_2 = Q_1$ and the charge has effectively been pumped from the source to C_1 and thence to C_2. Note that under this condition C_1 is essentially discharged and the process can be repeated on the next cycle.

The output voltage after the first cycle is

$$v_o = \frac{Q_2}{C_2} = \frac{v_i}{1 + C_2/C_1} \cong \frac{C_1}{C_2} v_i \tag{8-38}$$

where Eqs. (8-36) and (8-37) have been used. On each subsequent cycle a voltage increment given by Eq. (8-38) appears across C_2 as long as the total voltage remains small compared with the input pulse amplitude. The result is a *staircase* waveform, Fig. 8-25. In most applications the voltage across C_2 is rapidly discharged by an auxiliary circuit after a finite number of steps and the staircase waveform then repeats.

Ramps

A voltage that increases linearly with time, called a ramp, is widely used for the horizontal sweep voltage in oscilloscopes. Recurring ramps such as those produced by relaxation oscillators are referred to as sawtooth waves. Other ramp generators are designed to produce a single ramp waveform when triggered by an external signal.

Figure 8-26 is the circuit diagram of a ramp generator called the *Miller sweep*, which yields a linear ramp with a peak amplitude nearly equal to the dc supply potential. The quiescent condition has the trigger transistor Q_1 saturated and the Miller sweep transistor Q_2 turned off so that the output potential is equal to V_{cc} and the capacitor C is fully charged.

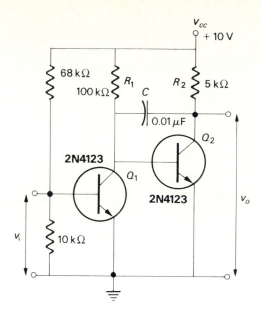

Figure 8-26 Miller sweep ramp generator.

A negative input pulse turns Q_1 off and allows Q_2 to operate as an amplifier. Thus C charges through R_1 and Q_2 and the output voltage decreases toward zero. The decrease is linear because of the feedback between collector and base provided by the capacitor. At the end of the input pulse Q_1 becomes saturated, Q_2 is turned off, and the capacitor charges rapidly to V_{cc} through R_2. The circuit is then ready for another input pulse.

Actually the Miller sweep circuit may be thought of as an operational-amplifier integrator, Fig. 7-13, in which Q_2 is the amplifier. The input signal is the constant voltage V_{cc}, which is applied through R_1. Since the integral of a constant value increases linearly, the output signal is a ramp. The function of Q_1 is to start and stop the integration interval.

Circuits similar to the Miller sweep are often used in *triggered-sweep* oscilloscopes, which are particularly useful for examining waveforms of transient signals. A typical block diagram of the sweep circuit for such an oscilloscope, Fig. 8-27, starts with a Schmitt trigger activated by the input signal applied to the vertical deflection amplifier. The output of the Schmitt trigger is differentiated and clipped to produce a sharp negative pulse suitable for triggering a monostable multivibrator. The multivibrator generates a square pulse input to the Miller integrator. The output ramp is clamped at a voltage level such that the sweep starts at the desired place on the oscilloscope screen.

A number of adjustable controls are included to increase the flexibility of the circuit. For example, bias on the Schmitt trigger can be adjusted to permit the circuit to trigger on input signals of different amplitudes (*trigger amplitude*, in Fig. 8-27). Similarly, the bias of the multivibrator is variable (*sweep stability*) so that it may also be operated as a synchronized astable multivibrator to produce recur-

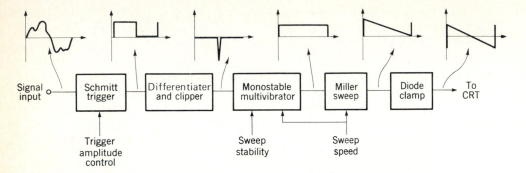

Figure 8-27 Block diagram and typical waveforms of triggered-sweep circuit used in oscilloscopes.

ring sweeps. The sweep speed is altered by changing the charging capacitor (and/or resistor) in the Miller sweep circuit. At the same time, it is necessary to change the time constant of the multivibrator (by switching capacitors) so that the sweep duration is commensurate with the sweep speed.

Pulses

Pulse waveforms of endless variety can be generated using combinations of multi-vibrator circuits. Consider the two cascaded monostable multivibrators in Fig. 8-28. A negative trigger pulse causes a positive pulse with a duration T_1. The negative trailing edge of this pulse triggers a second monostable multivibrator and the output is a square pulse of duration T_2. Thus, this combination generates one standard-shaped output pulse delayed by a specified interval following each input trigger pulse.

A simple binary triggered by a repeating waveform produces an output square wave with half-cycles as precisely regular as the frequency of the input

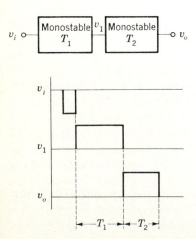

Figure 8-28 Cascaded monostable multivibrators.

signal. This is so because the binary changes state on every negative swing of the input wave and the precision of the square-wave output depends only upon the precision of the successive negative swings of the input signal.

Binary triggering may be accomplished by pulses fed directly to only one or the other transistor base. In this case a negative pulse induces a transition only when the transistor is conducting. Such *unsymmetrical* triggering is useful when a binary is triggered from two separate sources, one applied to each base. The output of this binary is a square pulse with a duration equal to the time interval between input pulses.

SUGGESTIONS FOR FURTHER READING

Paul Horowitz and Winfield Hill: "The Art of Electronics," Cambridge University Press, New York, 1980.

J. D. Ryder: "Electronic Fundamentals and Applications," 4/e, Prentice-Hall, Inc., Englewood Cliffs, N. J., 1970.

Joseph A. Walston and John R. Miller (eds.): "Transistor Circuit Design," McGraw-Hill Book Company, New York, 1963.

EXERCISES

8-1 Derive Eq. (8-3) for the feedback network of a phase-shift oscillator. *Hint*: Use the technique of loop currents and solve for the current in the last resistor by determinants.

8-2 Draw the *h*-parameter equivalent circuit of the phase-shift oscillator, Fig. 8-1, and derive an expression for the feedback ratio in terms of the *h* parameters and the input impedance of the feedback network, Z_i.

  $Answer: \beta = -(h_{ie}/h_{fe})(1/R_L + 1/Z_i)$

8-3 Using the results of Exercise 8-2, show that the circuit of Fig. 8-1 does oscillate. The *h* parameters of the 2N1414 are $h_{ie} = 1260 \ \Omega$, $h_{fe} = 60$. What is the frequency of oscillation?

  $Answer: 1.28 \times 10^4 \ Hz, 1/\beta = -121$

8-4* Plot the signal voltage at terminal 2 and the reciprocal of the feedback ratio as a function of the output signal for the integrated-circuit Wien bridge oscillator in Fig. 8-2. What is the stabilized amplitude of the output signal?

  Answer: 1.7 V peak-to-peak

8-5 State the two requirements for oscillation as embodied in the Barkhausen criterion.

8-6 Develop an expression analogous to Eq. (8-10) for the phase of the feedback voltage as a function of frequency in a tickler *LC* oscillator. Show that a resonant circuit with a large *Q* is desirable for maximum frequency stability.

  $Answer: \tan \theta = Q_0(\omega_0/\omega - \omega/\omega_0)$

8-7 Obtain an expression for the phase angle of the impedance of a quartz crystal near the parallel resonant frequency. Using component values given in the text, plot the phase angle as a function of frequency and compare with a plot of an *LC* resonant circuit having a *Q* of 200. Use the results of Exercise 8-6.

8-8 Neglecting the resistance in the equivalent circuit of a quartz crystal, determine the equivalent impedance as a function of frequency. Using the component values given in the text, plot Fig. 8-7.

8-9 Using the current-voltage characteristics of Fig. 8-10, estimate the maximum coil resistance that the inductance of the tunnel-diode oscillator in Fig. 8-11 can have. What is the frequency of oscillation if $C = 10$ pF and $L = 45 \, \mu$H?

   Answer: 160 Ω, 7.5 MHz

8-10 Find the frequencies and amplitudes of the output signals from the SCR relaxation oscillator, Fig. 8-16, for the two extremes of R_2, if $\beta = 20$. Plot the output-voltage waveforms.

   Answer: 1.18×10^4 Hz, 6.67×10^5 Hz, 3.3 V, 20 V

8-11 Astable multivibrators are often used as light flashers to mark road barricades, construction work, etc. Qualitatively describe the operation of a typical circuit, Fig. 8-29, starting from the time when Q_1 and Q_2 are off and Q_1 is starting to turn on.

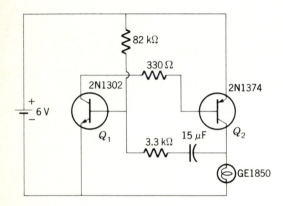

Figure 8-29 Light-flasher circuit analyzed in Exercise 8-11.

8-12 Sketch the waveform at the output of the circuit in Fig. 8-30 with the 50-Ω potentiometer at the center. Repeat for the potentiometer set one-fourth of the way from the $+20$-V terminal. Assume the 2N491 characteristics are the same as those of the 2N2646 in Fig. 8-13.

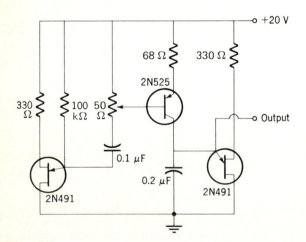

Figure 8-30

8-13 Describe the condition under which relaxation oscillations rather than sine-wave oscillations are likely in a bipolar transistor oscillator.

8-14 Determine the frequency of oscillation and minimum amplifier gain necessary for stable oscillations using the feedback network in Fig. 8-31. What is the only single-stage, bipolar-transistor configuration possible?

Answer: $1/RC$; $+3$; grounded-base

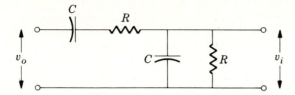

Figure 8-31 Feedback network analyzed in Exercise 8-14.

8-15* Determine the period of the operational-amplifier relaxation oscillator in Fig. 8-32. What is the output waveform?

Answer: $2RC \ln 3$; square wave

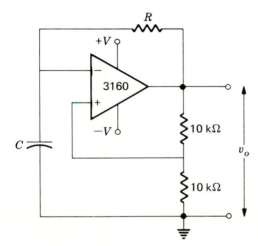

Figure 8-32 Operational-amplifier relaxation oscillator.

DIGITAL ELECTRONICS

Numerical digits can be represented by electric signals which have only two possible magnitudes, say, zero and some finite value. In this situation only the existence of one signal state or the other is significant, and the actual magnitude is relatively unimportant. Therefore, digital circuits need have only two stable conditions represented by a transistor fully conducting, or completely cut off. Such circuits are inherently more reliable than those which must handle a continuous range of signal levels. Any given number can be represented by a digital waveform, so accuracy is not limited by the stability of circuit parameters. Digital signals are manipulated by circuits according to specific logic statements which make possible exceedingly flexible and powerful processing of information.

DIGITAL LOGIC

Binary Numbers

Voltage signals are used to represent the digits of numbers in order that mathematical logic operations such as addition and subtraction can be carried out by electronic circuits. The stability advantages of digital circuits are best realized if the signal waveforms have only two amplitudes, called state 0 and state 1, or *off* and *on*. This is so because a transistor, for example, need then only be either completely cut off or fully conducting. Since only two digits are available, the *binary* number system is used, rather than the more familiar *decimal* system based on the use of 10 digits. Although the binary system is less familiar, there is no difference in principle between the two.

Consider, for example, the meaning of the decimal number 528. This array of decimal digits is a shorthand notation for the increasing powers of ten in the number. Thus,

$$528 : (5 \times 10^2) + (2 \times 10^1) + (8 \times 10^0)$$

$$: \quad 500 \quad + \quad 20 \quad + \quad 8 \quad = 528$$

Similarly, the two digits in the binary system, 0 and 1, tell the number of increasing powers of two in the number. The binary number 10110, for example, means

$$10110 : (1 \times 2^4) + (0 \times 2^3) + (1 \times 2^2) + (1 \times 2^1) + (0 \times 2^0)$$

$$: \quad 16 \quad + \quad 0 \quad + \quad 4 \quad + \quad 2 \quad + \quad 0 \quad = 22$$

That is, the binary number 10110 represents the same quantity as the decimal number 22. Table 9-1 contains the decimal-binary equivalents for the numbers from 0 through 15.

Arithmetic manipulations with binary numbers employ mathematical logic quite familiar from decimal numbers. Addition and subtraction operations involve digits that are carried and borrowed in an analogous fashion. It is useful to state the mathematical logic contained in the addition of two binary numbers A and B by means of a *truth table*, Table 9-2, which accounts for all of the possible combinations of A and B. A digit is carried, as in the fourth entry in Table 9-2, by shifting it to the next higher position, that is, to the left. For example, the sum of 0110 and 0101 is written

$$
\begin{array}{rr}
0110 & 6 \\
+0101 & +\ 5 \\
\hline
1011 & 11 \\
\end{array}
$$

Binary numbers can be represented by voltage waveforms with regularly spaced pulses of uniform amplitude. Conventionally, the pulses corresponding to

Table 9-1 Binary and decimal numbers

Decimal number	Binary number
0	0000
1	0001
2	0010
3	0011
4	0100
5	0101
6	0110
7	0111
8	1000
9	1001
10	1010
11	1011
12	1100
13	1101
14	1110
15	1111

increasing powers of 2 appear in time sequence beginning with 2^0. The waveforms of the binary numbers 1011 and 0011 are illustrated in Fig. 9-1, together with their sum, 1110. Note that the pulse lengths of both 0 and 1 digits are equal and that in actual practice the time interval between successive digits may be reduced to zero. One digit, be it a 0 or a 1, is termed a *bit*, which is a contraction for *binary digit*.

It should be noted in passing that the waveform of the sum bears no direct relationship to the waveforms of the two numbers. That is, the waveform of the sum is not obtained simply by adding the voltages of the waveforms representing the individual binary numbers. This points up the fact that waveforms in digital circuits are a coded representation of numbers and the voltages of the waveforms have no significance in themselves. Further examples of digital arithmetic are considered in Appendix 2.

Table 9-2 Truth table for addition

A	B	Sum	Carry
0	0	0	0
0	1	1	0
1	0	1	0
1	1	0	1

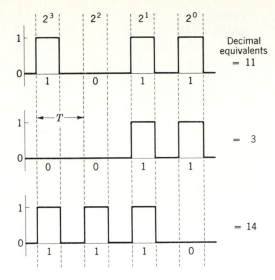

Figure 9-1 Waveforms of the binary numbers 1011 and 0011 and their sum, 1110.

Logic Gates

The various operations performed on digital waveforms are accomplished using circuits termed *logic gates* which have two or more inputs and one output. The output waveform of a logic gate depends upon the input waveforms and the input-output characteristic of the circuit described in terms of mathematical logic.

Consider the logic statement "if A is true and B is true, then T is true," which is written symbolically, as

$$A \cdot B = T \tag{9-1}$$

This is called the AND concept and is signified by the dot between the two quantities A and B in Eq. (9-1). Conventionally the condition *true* is identified with logic state 1 and the opposite condition, *false*, is identified with logic state 0 in digital circuits. In this fashion a truth table for the AND operation may be devised as in Table 9-3. Extending the truth table for more than two input quantities is easily accomplished by applying the logic in Table 9-3 several times to each pair of inputs. According to Table 9-3 an output is obtained (that is, T is true) only when A AND B are true.

A simple three-input diode AND gate that accomplishes the logic in Table 9-3 is illustrated in Fig. 9-2a. The diodes are biased in the low-resistance forward direction and the output voltage is zero, which means that the output condition is state 0. If positive input voltage signals somewhat greater than V_c are applied simultaneously to all three inputs, the diodes become reverse-biased and the output voltage rises to V_c, or state 1. Note, however, that if even one input

Table 9-3 Truth table for AND		
A	B	T
0	0	0
0	1	0
1	0	0
1	1	1

Table 9-4 Truth table for OR		
A	B	T
0	0	0
0	1	1
1	0	1
1	1	1

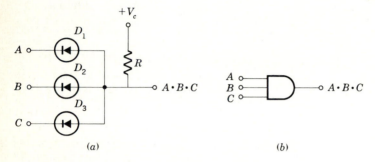

(a)   (b)

Figure 9-2 (a) Diode AND gate and (b) circuit symbol.

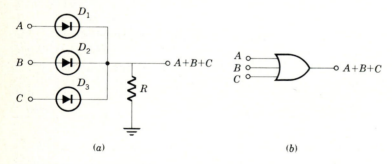

(a)   (b)

Figure 9-3 (a) Diode OR gate and (b) circuit symbol.

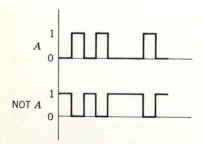

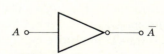

Figure 9-4 Waveforms of binary numbers A and NOT A.

Figure 9-5 Inverter symbol for NOT operation.

remains connected to ground, corresponding to state 0, the corresponding diode is under forward bias and the output signal remains at state 0. This action is logically described by saying that an output signal is obtained only when A AND B AND C are state 1.

The operation of an AND gate should not be confused with the mathematical operation of addition. That is, the output of an AND gate is not the sum of the input signals, as may be verified by noting the differences between the corresponding truth tables. AND gates are so extensively used in digital electronic circuits that it is convenient to use the special symbol in Fig. 9-2b.

The logical-OR statement is "if A or B is true, then T is true," written as

$$A + B = T \tag{9-2}$$

where the $+$ symbol indicates the OR concept. The corresponding truth table, Table 9-4, shows that an output is obtained whenever any input is present.

A diode OR gate and its circuit symbol are shown in Fig. 9-3a and b, respectively. Note that a positive signal at any input biases the corresponding diode in the forward direction and appears at the output. At the same time the other diodes are reverse-biased so that signals fed back to the other inputs are negligible. In effect the OR gate is a simple mixing circuit that brings several input signals to a common output with minimum interaction between signal sources.

In binary logic the NOT operation inverts the polarity of an input signal, as indicated schematically in Fig. 9-4. Since a digital waveform has only two states, 0 and 1, if at any given instant the signal is in state 1 it may equally be said to be in state NOT 0. The symbol for NOT is a bar over the quantity,

$$\text{NOT } A = \bar{A} \tag{9-3}$$

and the truth table is quite simple, Table 9-5. A simple grounded-emitter amplifier produces the logic in Table 9-5 as a result of the 180° phase shift between input and output. Accordingly, the appropriate circuit symbol for a NOT or *inverter* gate is similar to that previously used for an amplifier, Fig. 9-5. The small circle at the apex of the triangle (sometimes placed at the input) specifically indicates inversion.

Table 9-5 Truth table for NOT

A	$\bar{A}$
0	1
1	0

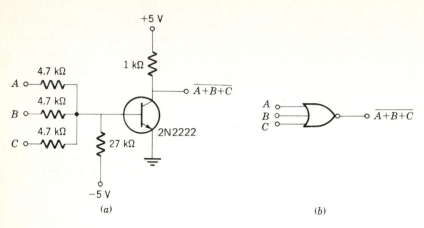

Figure 9-6 (a) Simple NOR gate performs combined operation NOT-OR and (b) circuit symbol.

It is common practice to combine the NOT inversion with other logic-gate functions. This results in amplification of the signals in the logic gate itself and consequently preserves the signal amplitude throughout a network of multiple logic gates. Consider the bipolar transistor amplifier in Fig. 9-6a with three equivalent inputs so that the waveform at each input may appear at the output terminal. This is the same logic as in Table 9-4 and therefore this circuit is a NOT-OR, or NOR, gate. The circuit symbol, Fig. 9-6b, includes the small circle to indicate inversion. A NOT-AND, or NAND, gate is equally possible and a typical circuit is illustrated in a later section. The circuit symbol for a NAND gate is shown in Fig. 9-7. Truth tables for NOR and NAND gates are simply obtained by applying the NOT operation to the last columns of the truth tables for OR, Table 9-4, and for AND, Table 9-3.

One further simple logic statement is the exclusive-OR function. The everyday usage of *or* is ambiguous for it can mean "one or the other or both" or "one or the other and not both." The logical-OR function previously defined is the "one or the other or both" situation according to the truth table, Table 9-4. It could correctly be called the inclusive-OR function. The exclusive-OR statement is

$$A \oplus B = A \cdot \bar{B} + B \cdot \bar{A} \tag{9-4}$$

which can be implemented by the array of logic gates in Fig. 9-8. Here again, it is useful to examine the truth table, Table 9-6, which is constructed with the aid

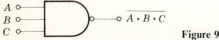

Figure 9-7 NAND gate.

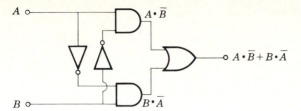

$A \cdot \bar{B}$

$A \cdot \bar{B} + B \cdot \bar{A}$

$B \cdot \bar{A}$

Figure 9-8 Exclusive-OR logic.

of Tables 9-3 to 9-5. According to the last column in Table 9-6, true output from the exclusive-OR circuit is produced only when either A or B is true, but not both, as expected.

Boolean Algebra

Analysis of logic networks is greatly aided by the logical algebra developed in the last century by George Boole, an English mathematician. The theorems of Boolean algebra are used to simplify digital logic networks in much the same fashion that mathematical logic is used to manipulate ordinary algebraic expressions. One major difference, of course, is that the variables in Boolean expressions can assume only one of two possible values.

A list of theorems in Boolean algebra is given in Table 9-7. Theorems 1 through 4 may be recognized as OR logic and are proved using the OR truth table, Table 9-4. Similarly, Theorems 5 through 8 are based on the AND concept in Table 9-3. Theorem 9 is a formal definition of the NOT function. The commutation, association, distribution, and absorption theorems are relatively straightforward and may be proved by means of truth tables.

De Morgan's theorems, numbers 18 and 19, are particularly interesting as they show a useful relationship between the AND and OR functions. Theorem 18 is proved in Table 9-8 by first finding $\overline{(A + B)}$ and then separately finding $\bar{A} \cdot \bar{B}$. The result is that the two truth tables are equal and the theorem is proved. A similar process may be used to establish Theorem 19. The basic duality of Boolean algebra is expressed by these theorems and can also be noted in pairs of other expressions in Table 9-7. Compare, for example the two association theorems, and also Theorem 1 with Theorem 6. In each case the OR and AND functions are dually related.

Table 9-6 Truth table for exclusive-OR

A	B	$\bar{A}$	$\bar{B}$	$A \cdot \bar{B}$	$B \cdot \bar{A}$	$A \cdot \bar{B} + B \cdot \bar{A}$
0	0	1	1	0	0	0
0	1	1	0	0	1	1
1	0	0	1	1	0	1
1	1	0	0	0	0	0

Table 9-7 Boolean algebra theorems

OR function	1	$0 + A = A$
	2	$1 + A = 1$
	3	$A + A = A$
	4	$A + \bar{A} = 1$
AND function	5	$0 \cdot A = 0$
	6	$1 \cdot A = A$
	7	$A \cdot A = A$
	8	$A \cdot \bar{A} = 0$
NOT function	9	$(\bar{\bar{A}}) = A$
Commutation	10	$A + B = B + A$
	11	$A \cdot B = B \cdot A$
Association	12	$A + (B + C) = (A + B) + C$
	13	$A \cdot (B \cdot C) = (A \cdot B) \cdot C$
Distribution	14	$A \cdot (B + C) = A \cdot B + A \cdot C$
	15	$(A + B) \cdot (A + C) = A + B \cdot C$
Absorption	16	$A + A \cdot B = A$
	17	$A \cdot (A + B) = A$
De Morgan's theorems	18	$\overline{(A + B)} = \bar{A} \cdot \bar{B}$
	19	$\overline{A \cdot B} = \bar{A} + \bar{B}$

An elementary example of the use of Boolean algebra to simplify logic networks is provided by the circuit in Fig. 9-9 which contains three NOR gates. The outputs of the two input gates are $\bar{A}$ and $\bar{B}$, respectively, so that the output of the total circuit may be written down and subsequently simplified by means of De Morgan's Theorem 18

$$(\overline{\bar{A} + \bar{B}}) = \bar{\bar{A}} \cdot \bar{\bar{B}} = A \cdot B \qquad (9\text{-}5)$$

In arriving at Eq. (9-5), Theorem 9 is also used. According to this result, the combination of three NOR gates may be simplified to one simple AND gate.

A second example of logic-network reduction is the exclusive-OR logic network illustrated in Fig. 9-8 and repeated in Fig. 9-10a. Observe that the output logic may be written in the form

$$A \cdot \bar{B} + B \cdot \bar{A} = \overline{\overline{(A \cdot \bar{B}) + (B \cdot \bar{A})}} = \overline{\overline{(A \cdot \bar{B})} \cdot \overline{(B \cdot \bar{A})}} \qquad (9\text{-}6)$$

Table 9-8 Proof of De Morgan theorem 18

A	B	$A + B$	$\overline{(A + B)}$	$\bar{A}$	$\bar{B}$	$\bar{A} \cdot \bar{B}$
0	0	0	1	1	1	1
0	1	1	0	1	0	0
1	0	1	0	0	1	0
1	1	1	0	0	0	0

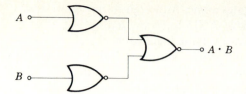

Figure 9-9 Three NOR gates produce AND gate.

where Theorems 9 and 18 have been used. The last logic statement may be accomplished by replacing the output OR gate with a NAND gate, if the inputs are $(A \cdot \bar{B})$ and $(B \cdot \bar{A})$. These logic functions are easily produced by replacing the two AND gates with NAND gates. The result is the logic network in Fig. 9-10b. This network may be further simplified to Fig. 9-10c by noting that a NAND gate can replace the two inverters. This is proved by determining the output of the upper NAND gate,

$$\overline{A \cdot (\overline{A \cdot B})} = \overline{A \cdot (\bar{A} + \bar{B})} = \overline{(A \cdot \bar{A}) + (A \cdot \bar{B})} \qquad \text{Theorems 19 and 14}$$

$$= \overline{0 + (A \cdot \bar{B})} = \overline{\bar{0} \cdot (\overline{A \cdot \bar{B}})} \qquad \text{Theorems 8 and 18}$$

$$= \overline{1 \cdot (\overline{A \cdot \bar{B}})} = (\overline{A \cdot \bar{B}}) \qquad \text{Theorem 6}$$

$$(9\text{-}7)$$

A similar analysis shows that the output of the lower NAND gate is $(\overline{B \cdot \bar{A}})$. Therefore the network in Fig. 9-10c is exactly logically equivalent to the exclusive-OR logic in Fig. 9-10a.

The results of the two previous paragraphs illustrate an important feature of Boolean logic networks. This is that any logic function may be accomplished

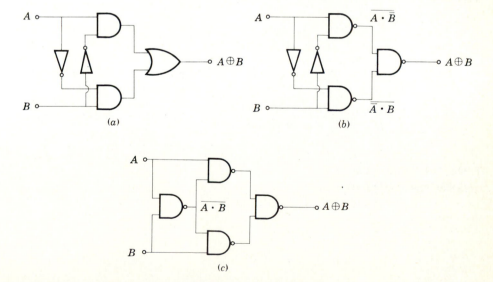

Figure 9-10 Reduction of exclusive-OR logic.

with NOR gates alone, or with NAND gates alone. Although it may seem wasteful to employ three NOR gates to achieve AND logic as in Fig. 9-9, it is an appreciable convenience to be able to assemble large logic networks of thousands of gates from the same simple subunit. This greatly simplifies design, production, and maintenance considerations. Furthermore, it turns out that, as discussed in a later section, the operating characteristics of either the NAND or NOR gate of a particular design are more favorable than the other type of gate. Therefore, considerable advantage is realized in using only the more favorable form. The choice of which to use is entirely a matter of convenience and the logic statements may be converted from one form to the other by means of the theorems of Boolean algebra in Table 9-7.

LOGIC CIRCUITS

Logic Signals

In the preceding discussions it has been assumed that signal waveforms corresponding to state 1 are at a small positive potential while state 0 is represented by ground potential. This is in keeping with much current practice in which the voltage of a state 1 bit is $+5$ V to $+15$V and that of a state 0 bit is near 0 V. This choice is completely arbitrary, of course, since Boolean algebra does not depend on the specific choice of signals used but only upon the presence of two distinct states.

It is interesting to note the effect of interchanging the designation of logic levels upon actual gate circuits. That is, suppose state 1 is represented by 0 V and state 0 by $+5$ V. This is called *negative logic* because state 1 is more negative than state 0. Consider the action of the diode AND gate in Fig. 9-2a in negative logic. The diodes are biased in the forward direction so that the output is state 1 when any one of the inputs is at ground potential, or state 1. The output is positive, state 0, only when all inputs are positive, or state 0. The truth table for this logic is written in Table 9-9 for the case of only two inputs. Table 9-9 is identical to Table 9-4 for the OR function. According to this analysis, an AND gate in negative logic becomes an OR gate in positive logic. This also means that a negative logic OR gate is a positive logic AND gate.

Table 9-9 Truth table for negative logic AND gate

A	B	T
0	0	0
0	1	1
1	0	1
1	1	1

It is useful to think of an inverter gate as a logic level exchanger since a state 1 input yields a state 0 output, according to Table 9-5. Thus, for example, a logical AND function can be produced using a positive logic OR gate by first inverting input signals to negative logic. The negative logic signals are acted upon by the positive logic OR gate to give negative AND logic. Subsequently, the signals may be converted back to positive logic by means of another inverter gate. The steps in this process can be symbolized in the following fashion:

1. Positive logic signals A, B
2. Negative logic signals $\bar{A}, \bar{B}$
3. Positive OR gate $\bar{A} + \bar{B}$
4. Theorem 19 $\overline{A \cdot B}$
5. Positive logic signal $A \cdot B$

This illustrates how the combination of positive and negative logic may be analyzed by means of the standard theorems of Boolean algebra.

There are two fundamentally different ways of transmitting signals representing the several bits of a complete binary number. The first, already encountered in connection with Fig. 9-1, employs a time sequence of pulses corresponding to increasing powers of 2 in the number. This is called *serial* representation and is indicated schematically in Fig. 9-11a. An alternate method, *parallel* representation, illustrated in Fig. 9-11b, is equally possible. In parallel transmission, pulses representing the increasing powers of 2 in the number all appear simultaneously on corresponding wires.

Both serial and parallel representation of numbers are commonly used in the same digital circuitry, as is examined in subsequent sections. Clearly, serial transmission has some advantages since only one signal path is required, while a path for each bit is necessary in the case of parallel transmission. On the other hand, a time interval of only one bit length is needed to transmit an entire digital number

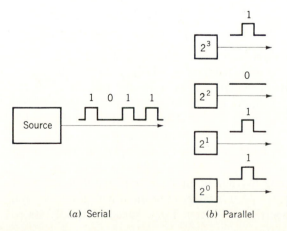

(a) Serial

(b) Parallel

Figure 9-11 (a) Serial and (b) parallel representation of binary numbers.

when the parallel representation is used, whereas the time required is much longer for the serial representation and the time interval also depends upon the number of digits in the number. The method selected in any given instance depends upon which feature is more advantageous.

Gate Circuits

Several different designs of electronic circuits are used as logic gates. The simplest, illustrated in Figs. 9-2*a* and 9-3*a*, use diodes and, accordingly, are known as *diode logic*, or DL, gates. These simple circuits perform the desired logic function satisfactorily but are not easily cascaded to carry out complicated logical functions. The reason for this is the inevitable signal loss through the circuit and the impedance mismatch between the output of one gate and the input of the next.

Much more satisfactory in these respects is *resistor-transistor logic*, RTL, already shown in Fig. 9-6*a* as a NOR gate. Actually this design is used only in the form of a NOR gate, but this is sufficient to carry out all logic functions, as pointed out previously. The transistor provides gain so that logic signals are not degraded in traversing the gate, and the transistor also significantly reduces the effect of load impedance upon the input circuit. One useful measure of this effect is the *fan-out* characteristic. Fan-out refers to the number of identical logic gates that can be connected to the output without seriously impairing circuit operation. Clearly a large fan-out is advantageous since complex logic networks are then possible. A fan-out of about 5 is characteristic of RTL gates. Larger values are achieved if the bipolar transistor is replaced by a FET, because of the high input-impedance characteristic of a FET.

Although such discrete-component logic gates can be used for specific applications, it is universal practice to employ integrated-circuit gates and both bipolar and MOSFET versions yield a range of useful characteristics. The combination of a diode AND gate and an inverter, Fig. 9-12*a* results in a NAND gate. Fan-out in this *diode-transistor logic*, DTL, gate can be as large as 10, and, although an analogous NOR gate is possible, DTL logic normally employs only NAND gates.

The versatility of integrated-circuit fabrication permits *transistor-transistor logic*, TTL, in which a multiple-emitter transistor is the logic element, Fig. 9-12*b*. The operation of the circuit may be looked upon as a DTL gate with the diode anodes in common. When any one of the emitters is at ground potential, state 0, the collector of Q_1 is essentially at ground and the output transistor Q_2 is cut off since no current flows in its emitter-base junction. With the output transistor cut off the output is at state 1. Only if all three emitters are at state 1 do both transistors conduct and the output go to state 0. This is logically equivalent to a NAND gate.

TTL logic has a fan-out as great as 15, and eight or more input emitters are not unusual. Furthermore, operating speed is faster than the other gate types and the so-called noise-immunity properties are better. Logic speed is usually stated

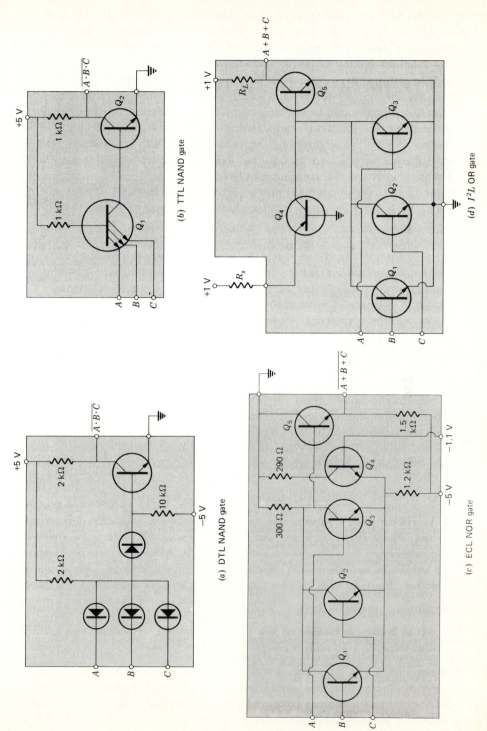

Figure 9-12 Bipolar-transistor integrated-circuit gates.

285

in terms of the *propagation delay* of the gate, that is, the time between introducing a change of logic state at the input and achieving a stable change at the output. The propagation delay is about 10×10^{-9} s (called nanoseconds, or ns) for TTL gates compared to values of 50 ns for DTL and RTL gates.

The noise immunity of logic gates is an important parameter that specifies how well a gate remains in a given logic state in the face of spurious signals due, for example, to changes in supply voltages, etc. Noise immunity is the difference between the minimum effective input signal in state 1 and the maximum state 0 output signal. A large value implies the gate is most likely to remain in a given logic state. Noise immunity exceeds 1 V for TTL gates, is about 0.75 V for DTL gates, and may be as small as 0.2 V for RTL gates.

Exceedingly small propagation delays are achieved in *emitter-coupled logic*, or *ECL*, gates, Fig. 9-12c, in which the input transistors Q_1, Q_2, and Q_3 are coupled to Q_4 through a common-emitter resistor. When the input transistors are all cut off (state 0), the base of Q_5 is at 0 V and the NOR output signal is just the built-in emitter-base voltage, or state 1. If one or more inputs are raised to state 1, the corresponding input transistor conducts, causing Q_5 to cut off and put the output at state 0.

The bias on Q_4 is set so that the gate transistors are not saturated when in state 1, which materially reduces the time required for them to be subsequently turned off again. This feature, together with the relatively small voltage swing between state 0 and state 1, leads to the small propagation delay of ECL gates. It should be remarked that in ECL state 0 is at -1.5 V, while state 1 is -0.75 V. While this is positive logic, the voltage levels are not compatible with the logic levels of other gate circuits.

So-called *integrated-injection logic*, or I^2L, gates, Fig. 9-12d, prove useful in VLSI bipolar circuits because the physical size of the gate is small and power dissipation is much less than for other bipolar types. The base injection current in Q_5 is set by the collector current in Q_4, which, in turn, is given by

$$I = \frac{V_{cc} - 0.6}{R_s} \tag{9-8}$$

where the forward voltage drop across the emitter junction of Q_4 is taken to be 0.6 V, as usual. If any input turns on Q_1, Q_2, or Q_3, the Q_4 base injection current is shunted to ground and Q_5 is cut off, causing the output to go to state 1. This is just OR logic.

A single resistor R_s external to an entire integrated-circuit chip is used to set the injection currents for all gates on the chip. This permits the propagation delay to be adjusted over a considerable range. The I^2L configuration also allows interconnections between gates without collector load resistors. Thus, for example, the input transistors in Fig. 9-12d would each actually be a $Q_4 Q_5$ pair and the direct connection between the collectors and the Q_5 input is evident. Similarly, the Q_5 collector can be directly connected to a subsequent gate circuit, depending somewhat upon the logic configuration desired. This feature further minimizes space and power requirements.

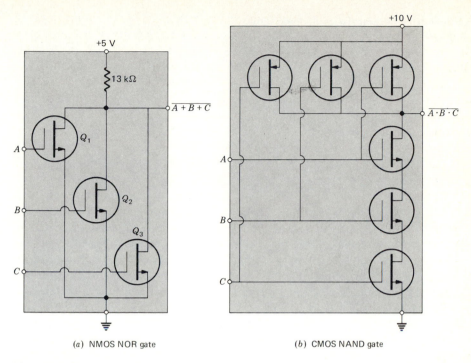

(a) NMOS NOR gate   (b) CMOS NAND gate

Figure 9-13 Integrated-circuit MOSFET gates.

The properties of both n-channel and p-channel MOSFET transistors are particularly appropriate for VLSI circuits because of their inherently small physical size and minimal power dissipation. An NMOS NOR gate, Fig. 9-13a involves only the three input MOSFETs and a common collector resistor. If any input gate is at state 1, the corresponding MOSFET conducts and the output is state 0, which is NOR logic. Obviously an analogous PMOS gate using p-channel MOSFETs is equally possible.

A particularly useful gate in digital circuits uses a combination of n-channel and p-channel MOSFETs. Such *complementary-MOS*, or CMOS, gates dissipate zero power regardless of the logic state, in distinction to all other gate circuits. If all inputs to a CMOS NAND gate, Fig. 9-13b are at state 0, the n-channel MOSFETs are cut off while the p-channel MOSFETs are conducting, so that the output is state 1. Any state 1 input turns on the corresponding n-channel input MOSFET and simultaneously cuts off the corresponding p-channel MOSFET load, but all three inputs must be state 1 for the output to go to state 0, and this is NAND logic. Note that in state 1 the n-channel input MOSFETs are cut off, while in state 0 the p-channel load MOSFETs are cut off; in either case no current passes and the static power dissipation is zero. Power is required to switch from one logic state to the other, however, so this very considerable advantage of CMOS circuitry is somewhat mitigated in actual practice.

Table 9-10 Characteristics of logic gates

Gate type	Propagation delay, ns/gate	Power dissipation, mW/gate	Size, gates/mm^2
DTL	50	10	10
PMOS	10	2	100
CMOS	1	0.3	50
NMOS	1	1	150
I^2L	10–100	0.1–1.0	100
TTL	5	5	15
ECL	0.5	30	30

Digital integrated circuits composed of many hundreds of thousands of gates per chip are only possible using basic gate circuits which are physically small so that many gates can be included on single chips of practical size. Gate circuits must also dissipate minimum power so that operating temperatures are not excessive as the gate density on the chip is increased. In addition, it is desirable for propagation delays to be small so that signals can be processed through hundreds of logic steps without undue time delay. The practical gate circuits described above meet these requirements to varying degrees, as illustrated in Table 9-10.

In general, small propagation delays imply high power dissipation, so that ECL gates are useful when speed is important and MOSFET gates are most appropriate when power dissipation is to be minimized. The NMOS and PMOS configurations also tend to be physically small. The I^2L circuit provides a range of easily adjustable operating parameters and CMOS gates offer very low power dissipation under static or slow-speed operating conditions. Obviously, other considerations such as signal-voltage swing, power-supply voltages, noise immunity, and cost can also be important for any given application. The properties in Table 9-10 refer to gates fabricated from silicon, by far the most universal integrated-circuit semiconductor. More recently, gallium arsenide gates have been produced with very small propagation delays, 0.1 ns, and low power dissipation, 1.0 mW/gate.

ADD Gates

It is pertinent to examine the addition of binary numbers since addition is an arithmetic operation basic to numerical computation. Multiplication, for example, can then be accomplished simply by repeated additions (see Appendix 2). Similarly, division can be carried out by the inverse of addition, that is, subtraction. Logic gates are used to achieve binary addition in the following two-step way. First, the digits in each column are added and then the carry digits are added to the column representing the next higher power of 2. This is identical to conventional addition as illustrated in a previous section.

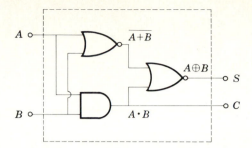

Figure 9-14 Half-adder circuit.

The truth table for the addition of bits, Table 9-2, shows that the sum S of two bits is just exclusive-OR logic, Table 9-6, while the carry C is AND logic, Table 9-3. Therefore, the addition of bit A to bit B is stated by the following logic

$$S = A \oplus B \tag{9-9}$$

$$C = A \cdot B \tag{9-10}$$

A network to accomplish this logic is called a *half-adder*, and one form is illustrated in Fig. 9-14. Note that Eq. (9-10) is immediately evident, while the sum output is

$$S = \overline{\overline{(A + B)} + (A \cdot B)} = \overline{(A + B)} \cdot \overline{(A \cdot B)} \qquad \text{Theorem 18}$$

$$= (A + B) \cdot (\bar{A} + \bar{B}) \qquad \text{Theorems 9}$$

$$\text{and 19}$$

$$= A \cdot \bar{A} + A \cdot \bar{B} + B \cdot \bar{A} + B \cdot \bar{B} \qquad \text{Theorem 14}$$

$$= A \cdot \bar{B} + B \cdot \bar{A} \qquad \text{Theorem 8}$$

which corresponds to Eq. (9-9).

The half-adder accomplishes the first step in the addition of numbers, that is, the addition of bits. Two half-adders are combined into a *full adder*, Fig. 9-15, in order to add carry bits to the bit sum. The OR gate is needed to include the possibility that adding the carry bit to the sum generates a new carry bit.

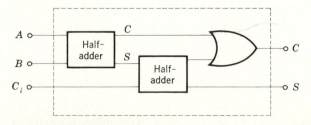

Figure 9-15 Full adder.

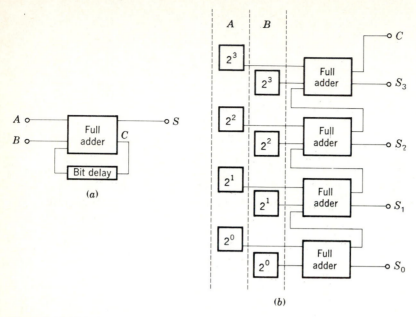

Figure 9-16 (a) Serial and (b) parallel addition.

The way in which a full adder is used to add binary numbers depends upon whether the numbers to be added are available in serial or parallel representation. In serial representation, Fig. 9-16a, the waveforms of the two numbers are introduced to a full-adder circuit, and each bit sum is added in sequence. Any carry bit produced is returned to the carry input, delayed by a time interval equal to the time between successive bits. This delay has the effect of adding the carry bit to the next column of bits in the binary number. A convenient 1-bit delay might be the combination of two cascaded binaries discussed in Chap. 8.

Addition in parallel representation is similar but a full adder is required for each digit, Fig. 9-16b. In this case the carry bit from the least significant bit addition is introduced into the full adder of the next most significant bit, etc. Although parallel addition may appear to require very elaborate circuitry since some 28 logic gates are required to add the two 4-digit numbers in Fig. 9-16b, fabrication processes in integrated circuits are readily capable of producing elaborate networks. Thus a complete integrated-circuit full-adder gate is physically no larger than the container needed to provide the necessary terminals.

INFORMATION REGISTERS

Flip-Flops

If two inverting gates are cross-coupled, as in the case of the two NAND gates in Fig. 9-17, the combination has two stable states. Suppose, for example, the S input (for *set*) is state 1 and the R input (for *reset*, or *clear*) is state 0. If the upper

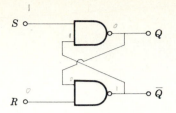

Figure 9-17 Cross-coupled NAND gate binary.

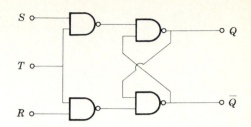

Figure 9-18 Gated NAND flip-flop.

NAND gate has a state 1 at the other input, its output, Q, must be state 0. The two state 0 inputs at the lower NAND gate make the $\overline{Q}$ output state 1, as originally assumed at the upper gate input. That is, Q and $\overline{Q}$ always have opposite logic states. Note that, if S is state 0 and R state 1, the output Q becomes state 1 and $\overline{Q}$ becomes state 0. Actually, this cross-coupled logic gate is quite equivalent to the binary multivibrator discussed in Chap. 8.

The truth table for a NAND binary is given in Table 9-11. When S is state 0 the flip-flop is said to be *set* since $Q = 1$. Similarly, a state 0 input to R *resets*, or *clears*, the flip-flop by ·putting Q at state 0. The binary is a memory circuit since it indicates which input was at state 0 last. Simultaneous state 1 inputs to both R and S leave the circuit in its original state, while two state 0 inputs give an indeterminate result, since the logic of the circuit is not satisfied and the final state after these input signals are removed is only a matter of chance. A similar binary results from cross-coupled NOR gates; the corresponding truth table is related but not identical to Table 9-11.

If input gates are added to the NAND flip-flop, Fig. 9-18, the circuit responds to input logic signals only when the T, *trigger*, or *clock* input is in state 1. The reason for this is that when the clock input, T, is at state 0, the input gates must both have state 1 outputs (independent of the signals at R and S). According to Table 9-11, the binary stays in its original state under this condition. When the T input is in state 1, logic signals at R and S can cause transitions in the circuit, as described by Table 9-12. Note that this logic is reversed compared to the simple NAND flip-flop because of inversion in the input gates. If the signals applied to T are pulses, the transitions determined by the R and S inputs are *clocked* in synchronism with the T input signals.

Table 9-11 Truth table for NAND binary

R	S	Q
0	0	Indeterminate
0	1	0
1	0	1
1	1	Q

Table 9-12 Truth table for gated NAND flip-flop

R	S	Q
0	0	Q
0	1	1
1	0	0
1	1	Indeterminate

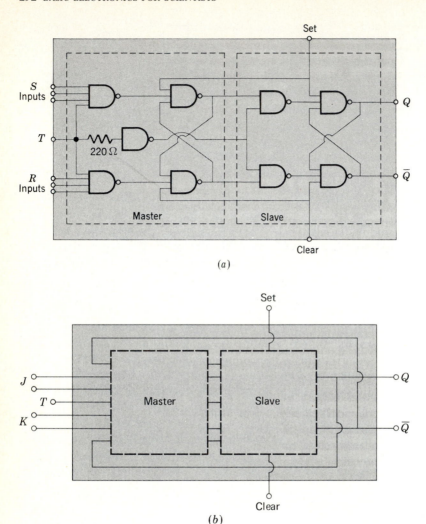

Figure 9-19 (*a*) *RS* master-slave flip-flop and (*b*) *JK* modification which toggles.

It is often useful to have the binary alternate states in response to successive pulses at the T input. This action is much like that of a mechanical toggle switch, and for this reason the clock input is also called the *toggle* terminal. The logic network of a *RS master-slave* flip-flop in Fig. 9-19 illustrates how two cascaded flip-flops are connected to achieve toggle action.

Note that the circuit is two gated flip-flops controlled by the same clock input. The slave flip-flop is gated by an inverted clock signal, however, and because of the 220-Ω resistor in the clock line is activated at a slightly different time than the master flip-flop. When the clock is at state 0, logic signals at the R and S inputs are ineffective, as in a regular gated flip-flop. The inverted clock signal is state 1 at the slave flip-flop, however, and the slave assumes the same

Table 9-13 Truth table for NAND *JK* master-slave flip-flop

J	K	Q
0	0	Q
0	1	0
1	0	1
1	1	$\bar{Q}$

Table 9-14 Truth table for *D* flip-flop

D	Q
0	0
1	1

state as the master. As the clock signal goes towards state 1, the input of the clock inverter gate reaches state 0 before the master input gates reach state 1. Therefore, the slave input gates isolate the slave flip-flop from the master, and the state of the slave flip-flop stores that of the master.

Next, the master input gates receive the state 1 clock signal and the logic signals at R and S determine the state of the master flip-flop according to Table 9-12. At the end of the clock pulse, the master input gates isolate the master flip-flop from the R and S inputs. With the clock back at state 0, the inverted clock signal at the slave flip-flop causes the slave to take up the state of the master flip-flop. Thus the output state reflects the logic signals at R and S, but the output state is not attained until the end of the clock pulse.

Note that multiple terminals are available at both the R and S inputs. According to Table 9-12, a state 0 at any S input puts Q at state 0, and a state 0 at any R input puts Q at state 1. Direct *set* and *clear* inputs are also provided. A state 0 signal at the *set* input overrides all other signals and produces Q = 1. Similarly, a state 0 signal at the *clear* input makes Q = 0 and the flip-flop is cleared.

Toggle action is achieved by connecting $\bar{Q}$ to one of the S inputs and Q to one of the R inputs, Fig. 9-19b. The remaining S terminals are renamed J inputs and the free R terminals are called K inputs. The configuration is known as a *JK master-slave* flip-flop. The new truth table, given in Table 9-13, is similar to that for a gated NAND flip-flop except that the circuit toggles, as indicated by the last entry. Stated in words, the logic of a *JK* flip-flop is: a state 0 at any *J* input prevents a state 1 at Q, a state 0 at any *K* input prevents a state 0 at Q, and the circuit toggles if both *J* and *K* inputs are at state 1.

Another useful modification of a gated flip-flop has a single input terminal connected to the S input and uses an inverter to connect $\bar{S}$ to the R input, Fig. 9-20. The appropriate truth table, Table 9-14, shows that the D flip-flop (for *data latch*) stores the logic signal at the input terminal during the last interval that the clock was in state 1. An application of this circuit is considered in the next section.

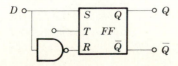

Figure 9-20 Data latch flip-flop.

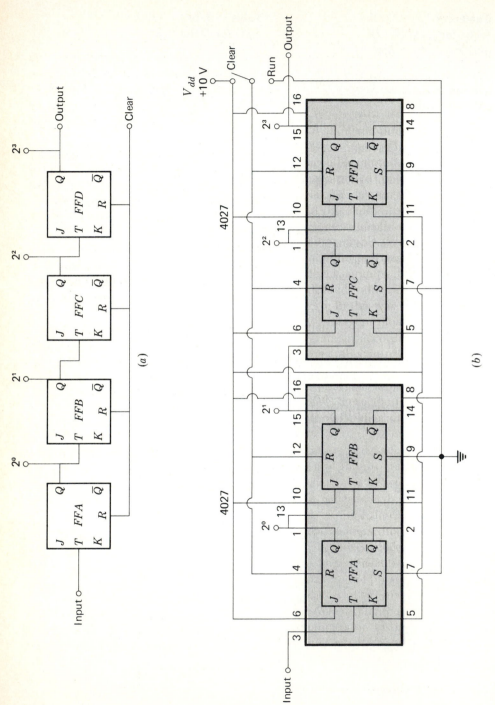

Figure 9-21 (a) Scale-of-16 binary counter and (b) actual circuit employing type 4027 dual JK flip-flops.

Counters

A single *JK* flip-flop produces one output pulse for every two pulses applied to the clock input. This is so because two pulses cause the circuit to shift from one stable state to the other and then back to the first state again. If the *Q* output of one binary is connected to the *T* input of a second binary, a total of four input pulses produces a single output pulse. That is, a string of cascaded binaries is an electronic *counter* which denumerates the number of input pulses. The state of each binary tells the total number of input pulses at any time.

Consider the four-binary counter in Fig. 9-21*a*. Suppose that initially all four flip-flops are cleared so that $Q = 0$ in each stage. The waveforms at the output of each flip-flop as a result of a regular series of input pulses are shown in Fig. 9-22. The first input pulse causes *FFA* to change state while the others remain in state 0.

At the second input pulse *FFA* returns from state 1 back to state 0 and its output signal triggers *FFB* into state 1. The net result of two input pulses is that *FFB* is in state 1 while the other three are in state 0. This process continues with further input pulses, as illustrated in Fig. 9-22, until at the sixteenth input pulse a single negative-going output signal is produced. The cascade combination of four binaries is, accordingly, called a *scale-of-16* counter.

A complete scale-of-16 counter employing two type 4027 dual *JK* flip flops is shown in Fig. 9-21*b*. Note that the *JK* inputs are at logic 1 as required by Table 9-13 for the circuit to toggle. The *reset* (*R*) and *set* (*S*) terminals are at logic 0. Placing the *R* terminals at logic 1 momentarily resets all flip-flops to state 0. This is the reverse of the set and clear logic of the *JK* flip-flop in Fig. 9-19.

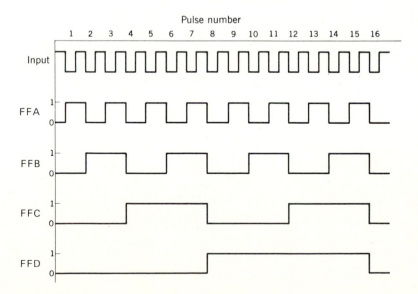

Figure 9-22 Waveforms in scale-of-16 counter.

A major early application of cascaded binaries was to reduce or scale down the pulse rate produced, for example, by nuclear radiation detectors, to the point where the output pulses could be recorded by an electromechanical register. The scale factor of such a *scaler* may be increased simply by adding additional stages. Cascaded flip-flops, called *registers*, are also widely used to store and manipulate digital logic signals and signals representing binary numbers.

The simple cascade configuration in Fig. 9-21 suffers from the propagation delay associated with the transition time of each stage. On the eighth and sixteenth input pulses, for example, all four flip-flops change state in sequence, so that a total time equal to the transition time of each stage times the number of stages is required for the last transition. At fast input pulse rates the first flip-flop may already have responded to succeeding input pulses before the last flip-flop settles in its final state. Thus the state of the counter at any given instant may not represent an exact number.

This difficulty is eliminated in the so-called *synchronous* binary counter, Fig. 9-23, in which all four flip-flops change state simultaneously. Note also that in this figure the input terminal is placed to the right so that a digital number stored in the register is presented with its most significant digit on the left as it is normally read. Each flip-flop is toggled by input pulses applied to all clock inputs. Interconnections between the flip-flops inhibit transitions until the appropriate count is reached. Note, for example, that FFA toggles on every input pulse, while FFD is inhibited until the eighth (and sixteenth) input pulse when all its J and K inputs are at state 1. The waveforms are identical to those for the nonsynchronous binary counter, Fig. 9-22, since the final state of each JK flip-flop is attained on the negative transition of the clock pulse. This waveform pattern may be verified by using the truth table for the JK flip-flop, Table 9-13.

It is often convenient to store each digit of a decimal number separately in binary form. Thus each digit of the decimal number can be represented by a four-flip-flop register counting from 0 to 9, and, for example, five such registers can store the decimal numbers from 0 to 99,999. Larger numbers are readily accommodated by adding an additional register for each decade.

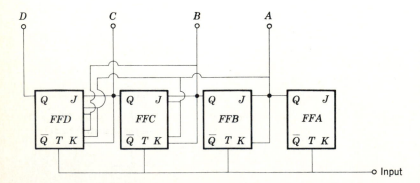

Figure 9-23 Synchronous binary counter.

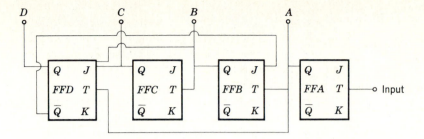

Figure 9-24 Asynchronous binary-coded decimal counter.

A four-flip-flop register is capable of counting to 16, but a total count of only 10 is desired in such a *binary-coded decimal*, or *BCD*, counter. This can easily be accomplished by interconnections between flip-flops, as in Fig. 9-24. Input pulses cause FFA to toggle and FFA, in turn, toggles FFB and FFC in cascade until the eighth count when the $\bar{Q}$ output from FFD is inhibiting. Similarly, FFD is inhibited by the Q outputs of FFB and FFC at its J inputs, one or the other of which is at state 0 until the sixth count. FFD flips on the eighth input pulse. The ninth pulse toggles FFA only and on the tenth count both FFA and FFD flip so that all Q's are at state 0 and the cycle is complete. This is best illustrated by a table showing the state of each flip-flop as a function of the input count, Table 9-15. Note that this pattern is identical to the waveforms of the binary counter, Fig. 9-22. That is, the counting pattern of this BCD counter is the same as the first ten states of a scale-of-16 counter. This is called the natural 8421 code and is the one most often used.

The cycle, or *modulus*, of a BCD counter is 10 since all flip-flops are cleared every 10 counts. Similarly, the modulus of a four-flip-flop binary counter is 16. Through appropriate logic connections, counters of any modulus are possible. Note, for example, that the combination of FFB, FFC, and FFD of the BCD counter is a *modulo*-5 counter if the input is at T of FFB. Alternatively, a simple binary counter may be converted to a counter of any desired modulus by using

Table 9-15 BCD counter states

Count	FFD	FFC	FFB	FFA
0	0	0	0	0
1	0	0	0	1
2	0	0	1	0
3	0	0	1	1
4	0	1	0	0
5	0	1	0	1
6	0	1	1	0
7	0	1	1	1
8	1	0	0	0
9	1	0	0	1

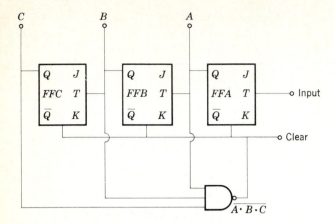

Figure 9-25 Modulo-7 counter using direct clearing.

the clear inputs and a logic circuit to clear all flip-flops at the proper count. This technique is illustrated in Fig. 9-25 for a direct-cleared modulo-7 counter. The circuit counts as a normal binary counter until the seventh input pulse when the Q outputs are all state 1 (refer to Table 9-15) for the first time. The output of the NAND gate goes to state 0, all the flip-flops are cleared, and the count cycle repeats.

It is often desirable to store the total count in a register, as for display purposes, while the counter proceeds to count a new signal waveform. One way to

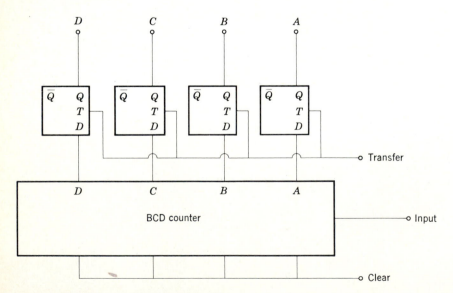

Figure 9-26 BCD counter and storage register.

accomplish this is by means of a *D* flip-flop, often called a *data latch*, connected to each counter flip-flop, Fig. 9-26. A state 1 pulse at the memory transfer terminal causes each *D* flip-flop to assume the state of its respective counter flip-flop, according to Table 9-14. Therefore, the number in the BCD counter at the time of the transfer pulse is stored in the data-latch register; subsequently the counter may be cleared to count again. The output of the data-latch register may be used, for example, to operate a visual display that remains constant until the next transfer pulse. Note also that the BCD counter and data-latch combination convert a serial representation into a parallel representation of the binary number.

Shift Registers

It is possible to move the information stored in a register along the chain of flip-flops without changing the waveform of the logic signal, and such a device is called a *shift register*. Consider the 4-bit shift register of *JK* flip-flops in Fig. 9-27*a* together with typical corresponding waveforms in Fig. 9-27*b*. The input

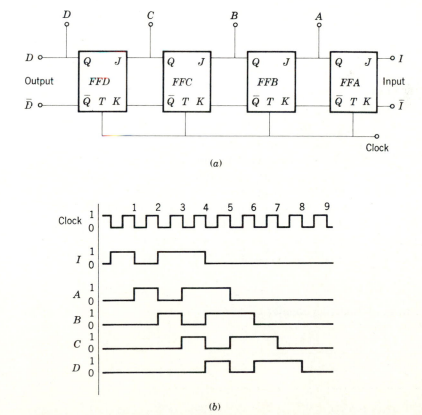

Figure 9-27 (*a*) 4-bit shift register and (*b*) typical waveforms.

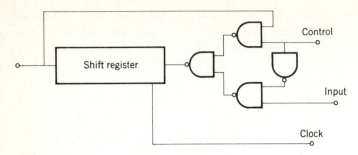

Figure 9-28 Circulating register.

signal, I, and $\bar{I}$ are applied to the J and K inputs, respectively. Whenever I is state 1, the same logic appears at Q of FFA at the trailing edge of the clock pulse. This, in turn, results in state 1 at Q of FFB at the cessation of the next clock pulse, and so on. That is, information presented to the input appears at each succeeding flip-flop output delayed by one additional clock pulse. Therefore, the information pattern is shifted intact along the register, as indicated by the typical waveforms in Fig. 9-27b.

If the output of a shift register is returned to the input, the logic signals circulate continuously. A useful arrangement of such a *circulating register* is shown in Fig. 9-28. With the control input at state 1, signals at the output of the register are reintroduced at the input. The signals reappear at the output every m pulses, where m is the number of flip-flops in the register. New digital signals can be introduced by a state 0 signal at the control terminal. Integrated-circuit circulating registers many thousands of bits long are available.

A related register with useful properties is obtained by connecting the output of the shift register in Fig. 9-27a to the input terminals and regarding the clock terminal as the input signal. First, a state 1 signal is placed in FFA and state 0 in all other flip-flops using the direct-set and direct-clear inputs on each flip-flop. According to the waveforms in Fig. 9-27b this single 1 circulates around the register in response to input pulses at the clock terminal. The successive states of this 4-bit *ring counter* are shown in Table 9-16. Note that this operation is, in

Table 9-16 States of 4-bit ring counter

Input pulse	FFD	FFC	FFB	FFA
1	0	0	0	1
2	0	0	1	0
3	0	1	0	0
4	1	0	0	0

effect, a modulo-4 counter, and that the total decimal count is readily discernible by, say, connecting an indicator to each flip-flop output. This useful feature compensates for the fact that only one bit per flip-flop may be accomplished.

VISUAL DISPLAYS

Often it proves useful to show the information stored in a register by means of a visual display. Simple indicators that tell the logic state of each flip-flop in the register may be sufficient, but more elaborate displays that present the stored information in terms of decimal digits are usually more convenient.

Single-Element Displays

The logic state of a flip-flop can be indicated directly by a small incandescent lamp, as illustrated in Fig. 9-29. When $Q = 1$ the transistor base is biased in the forward direction and the lamp lights. Conversely, the transistor is cut off and the lamp is dark when the flip-flop is in state $Q = 0$. The transistor switch is preferable, rather than, say, simply putting the lamp in series with one collector load of the flip-flop itself, because the current required to light the lamp, approximately 50 mA, could easily disrupt normal operation of the circuit.

    The state of each binary in a counter can be indicated by a separate lamp and the total count determined at any time by noting which lamps are lit. The lamps connected to each binary in the scale-of-16 counter in Fig. 9-21 are then labeled 1, 2, 4, and 8, respectively, according to the waveforms in Fig. 9-22. The total count is found by adding the numbers associated with those lamps which are lit. This simple display obviously presents information in the binary number system.

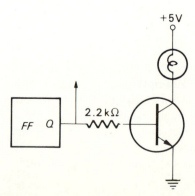

Figure 9-29 Incandescent-lamp circuit indicates flip-flop logic state.

Table 9-17 Binary to decimal code

Decimal number	Binary logic statement
0	$\bar{A} \cdot \bar{B} \cdot \bar{C} \cdot \bar{D}$
1	$A \cdot \bar{B} \cdot \bar{C} \cdot \bar{D}$
2	$\bar{A} \cdot B \cdot \bar{C} \cdot \bar{D}$
3	$A \cdot B \cdot \bar{C} \cdot \bar{D}$
4	$\bar{A} \cdot \bar{B} \cdot C \cdot \bar{D}$
5	$A \cdot \bar{B} \cdot C \cdot \bar{D}$
6	$\bar{A} \cdot B \cdot C \cdot \bar{D}$
7	$A \cdot B \cdot C \cdot \bar{D}$
8	$\bar{A} \cdot \bar{B} \cdot \bar{C} \cdot D$
9	$A \cdot \bar{B} \cdot \bar{C} \cdot D$

In order to provide visual indicators in the decimal number system it is first necessary to convert binary signals into signals corresponding to decimal digits. Consider, for example the BCD counter in Fig. 9-24. The logic statement (that is, the state of the counter) corresponding to each decimal number for this counter is shown in Table 9-17. This is just a restatement of the presentation in Table 9-15 which shows the state of each flip-flop resulting from successive input pulses.

The binary statements in Table 9-17 are all AND logic which can be accomplished by the array of diode AND gates in Fig. 9-30. Note that each binary BCD statement input signal causes one lamp representing the corresponding decimal number to light. This is illustrated most easily by isolating the circuit associated with the 0 lamp, Fig. 9-31. Note that if $\bar{B}$ and $\bar{C}$ and $\bar{D}$ are present, the diodes are all reverse-biased and the transistor base becomes forward-biased. If now, in addition, the transistor emitter is grounded (that is, $\bar{A}$), the lamp lights. A similar analysis can be made for each of the other BCD logic input statements. This circuit, in effect, takes the input information in a BCD code and converts it to output signals in decimal code and, accordingly, the array is called a *decoder*. The driver transistors are needed to provide the necessary lamp currents, as discussed in connection with Fig. 9-29.

In arriving at the configuration of this decoder, advantage has been taken of certain simplifications in BCD logic. Note, for example, that D and $\bar{D}$ only distinguish states 0 and 1 from 8 and 9 in Table 9-17; that is, $\bar{D}$ is really not needed to describe states 2 through 7. Correspondingly, no connection to the D and $\bar{D}$ lines is made for these digits in the decoder array. Also, A and $\bar{A}$ are introduced in connection with the driver transistors rather than as separate vertical logic lines, which, though equally possible, would result in a more complicated circuit.

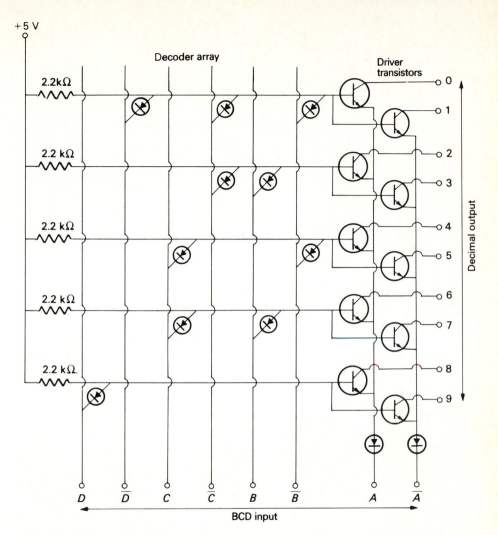

Figure 9-30 BCD-to-decimal decoder and driver.

Once the decoder logic is established by the logic statements, the operation of converting from binary to decimal numbers may be looked upon in the following way. Each 4-bit *binary word* which describes the state of the counter is decoded into a specific decimal number that is indicated by the lamp associated with that number. This display is obviously more convenient and informative than one which simply indicates the state of each flip-flop in the counter.

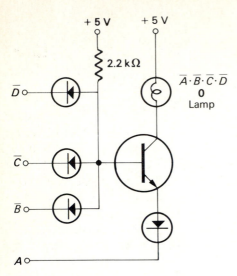

Figure 9-31 Isolated AND gate of BCD decoder for 0 lamp.

Seven-Segment Displays

A very useful numerical indicator uses the seven-segment display illustrated in Fig. 9-32. Each segment is lit by a small lamp and the appropriate combination of lit segments forms the numbers from 0 to 9. In particular, LED displays discussed in Chap. 4 are most useful because of their small size and long operating life. A typical LED seven-segment display is shown in Fig. 9-33.

A very popular seven-segment display is based on the property of certain materials known as liquid crystals to alter the polarization of transmitted light in response to an applied electric field. A liquid crystal is composed of large elongated

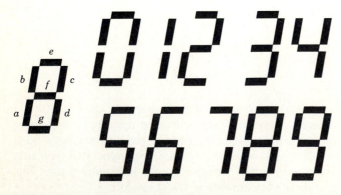

Figure 9-32 Seven-segment decimal display.

Figure 9-33 LED seven-segment display. (*Hewlett-Packard Corp.*)

molecules that are organized in an ordered structure which rotates the plane of polarization of light energy. In an electric field, the molecules line up parallel to the field and the rotation effect is lost. This unique property is used in a *liquid crystal display*, or *LCD*, by placing a liquid crystal layer between two glass plates each coated with transparent conducting electrodes (as in, for example, the seven-segment pattern) and two polarizing layers oriented such that their polarizations are at right angles, Fig. 9-34. The side away from the viewing side is also provided with a mirrored surface.

In the absence of an electric field, incident light is polarized by the top polarizer and then rotated 90° by the liquid crystal so that the light passes through the bottom polarizer. Upon reflection, the light energy passes back up through the stack in reverse order and emerges from the top polarizer. When a voltage signal is applied to the conducting electrode pattern, polarization rotation does not occur

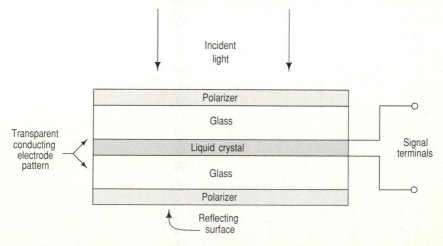

Figure 9-34 Structure of liquid crystal display.

in the liquid crystal layer and the crossed top and bottom polarizers absorb the light to make the conducting electrode area appear dark against the surrounding light background. The principal advantage of the LCD is that very little electric power is required to activate the display since the electrode pattern is essentially a small capacitor with the liquid crystal as dielectric.

Decoder Logic

The seven-segment display requires a decoder logic circuit to activate the proper segments corresponding to the BCD input word describing each decimal number. A truth table specifying the lit segments for each numeral is easily established, Table 9-18. This means, for example, that the logic necessary to light segment a, which is activated for decimal numbers 0, 2, 6, and 8, is

$$a = (\bar{A} \cdot \bar{B} \cdot \bar{C} \cdot \bar{D}) + (\bar{A} \cdot B \cdot \bar{C} \cdot \bar{D}) + (\bar{A} \cdot B \cdot C \cdot \bar{D}) + (\bar{A} \cdot \bar{B} \cdot \bar{C} \cdot D)$$

$$(9\text{-}11)$$

This expression may be simplified using Theorem 14 of Table 9-7 so that

$$a = (\bar{A} \cdot \bar{B}) \cdot (\bar{C} \cdot \bar{D} + \bar{C} \cdot D) + (\bar{A} \cdot B) \cdot (\bar{C} \cdot \bar{D} + C \cdot \bar{D})$$

$$= (\bar{A} \cdot \bar{B}) \cdot (\bar{C} \cdot \bar{D} + D) + (\bar{A} \cdot B) \cdot (\bar{D} \cdot \bar{C} + C)$$

$$= \bar{A} \cdot \bar{B} \cdot \bar{C} + \bar{A} \cdot B \cdot \bar{D}$$

$$(9\text{-}12)$$

A similar logic expression can be written for the other six segments. For certain segments such as d and e the logic statement for NOT lit is shorter since these segments are activated on all but one or two of the ten numerals. In each case the result is similar to Eq. (9-12) and can be implemented by AND and OR logic arrays analogous to Fig. 9-30. In this connection it should be noted that OR logic is achieved in such a DL array simply by reversing the forward direction of the diodes.

Table 9-18 Truth table for seven-segment decimal display

Decimal number	BCD word	Segment						
		a	b	c	d	e	f	g
0	$\bar{A} \cdot \bar{B} \cdot \bar{C} \cdot \bar{D}$	1	1	1	1	1	0	1
1	$A \cdot \bar{B} \cdot \bar{C} \cdot \bar{D}$	0	0	1	1	0	0	0
2	$\bar{A} \cdot B \cdot \bar{C} \cdot \bar{D}$	1	0	1	0	1	1	1
3	$A \cdot B \cdot \bar{C} \cdot \bar{D}$	0	0	1	1	1	1	1
4	$\bar{A} \cdot \bar{B} \cdot C \cdot \bar{D}$	0	1	1	1	0	1	0
5	$A \cdot \bar{B} \cdot C \cdot \bar{D}$	0	1	0	1	1	1	1
6	$\bar{A} \cdot B \cdot C \cdot \bar{D}$	1	1	0	1	1	1	1
7	$A \cdot B \cdot C \cdot \bar{D}$	0	0	1	1	1	0	0
8	$\bar{A} \cdot \bar{B} \cdot \bar{C} \cdot D$	1	1	1	1	1	1	1
9	$A \cdot \bar{B} \cdot \bar{C} \cdot D$	0	1	1	1	1	1	1

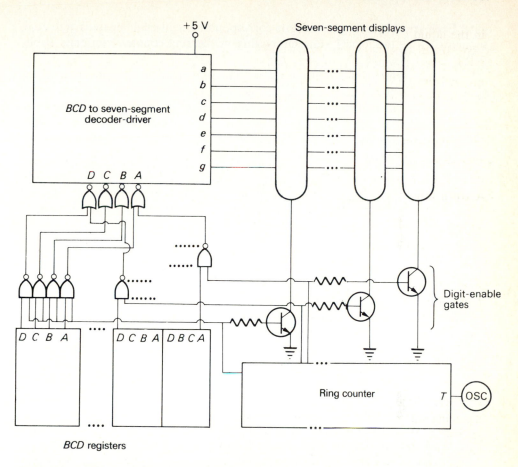

Figure 9-35 Multiplexing circuit for digital display.

Most often, decoder logic arrays are best implemented by integrated-circuit techniques because of the repetitive complexity involved. Thus a single integrated circuit decoder accepts the four input signals corresponding to a BCD word and provides the proper output signal at each of its seven output terminals.

Certain digital instruments may employ displays involving 6-, 8-, or even 12-decimal numbers. This means that an equivalent number of decoder-driver arrays are needed, which implies extensive circuitry. Alternatively, it proves possible to use a single decoder-driver in a multidigit display by rapidly switching to each digit in succession. In this case, persistence of vision makes it appear that the entire display is lit continuously.

The rapid-switching technique, called *multiplexing*, is illustrated in Fig. 9-35. All common segments in the display are connected together and to the appropriate output terminal of the decoder. The decoder input word is delivered from each of the BCD registers through a combination of NOR and NAND gates such that only one BCD register is connected at any instant. At the same instant only the corresponding decimal number is activated by the associated digit-enable gate. Each set of NAND gates and its corresponding digit-enable gate is activated sequentially by the logic 1 state circulating in the ring counter (compare Table 9-16).

The net result of this multiplexer is that each number is supplied with its proper seven-segment code and is lit once per cycle of the ring counter. The frequency of the oscillator driving the ring counter is made great enough to eliminate flicker in the display. In addition to a significant reduction in circuit complexity, multiplexing also lowers the overall power required to illuminate the display since, in effect, only one digit is lit at any time.

MEMORY CIRCUITS

Read-Only Memories

The display logic decoder arrays discussed in the previous section are each a simple example of what has come to be called a *read-only memory*, or *ROM*. A ROM associates a specific output binary number with each input binary number according to its fixed internal logic. The fixed relationship between input and output distinguishes the ROM from other memory circuits described subsequently.

Consider the circuit configuration of a 256-bit ROM arranged in a 32-word, 8-bit format, Fig. 9-36, which produces a specific 8-bit output corresponding to each of the 32 input states determined by the 5-bit input signal. As fabricated by integrated-circuit techniques, each word line is the base of a transistor having an emitter connected to all eight bit lines. A state 1 at the base terminal of a selected word transistor turns on output driver transistors, that is, results in a state 0 at the bit outputs. During final stages of fabrication, certain emitter connections between word transistors and bit lines are etched away to correspond to the desired logic pattern. Accordingly, those bit lines which are disconnected from a given word transistor result in a state 1 output when that word transistor is activated. The pattern of connected and absent transistors at the intersections of word lines and bit lines is the fashion by which ROM permanently remembers the relationship between input words and output signals.

The binary numbers from 00000 to 11111 specify 32 different words in Fig. 9-36, and an *address decoder* is used to select one of the 32 word transistors in accordance with each 5-bit input word. Thus the overall operation of this ROM is to store permanently 256 bits of information which are accessed by 5-bit binary words and which are presented as one of 32 8-bit binary words.

Note that the presence or absence of a transistor at the intersection between word lines and bit lines specifies the logic at that location. This is exactly analogous to the diodes used in the decoder-driver ROM, Fig. 9-30. Similarly, a MOSFET ROM is fabricated such that the presence or absence of a gate electrode at the intersection determines the logic. All three types of logic connections are used in practice. A typical ROM chip is shown in Fig. 9-37. ROMs capable of storing more than 2^{20} (1,048,576) bits are available.

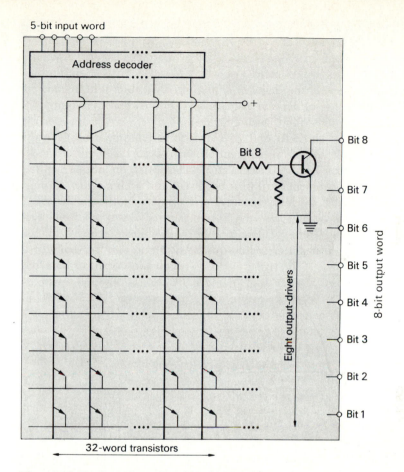

Figure 9-36 Circuit configuration of 256-bit ROM.

An important application of ROMs in digital circuits is to provide look-up tables for mathematical functions, such as trigonometric, exponential, square root, and logarithmic functions. In the case of a sine table, for example, the input signal corresponds to a given angle and the output represents the sine of that angle. A single 1024-bit ROM can store 128 angular increments corresponding to an angular resolution of 0.7 degree and can generate an 8-bit output signal that errs less than 0.1 percent from the handbook value of the sine. Obviously, accuracy is improved by increasing the bit storage capacity of the memory.

In certain applications, most notably in microprocessor circuits discussed in Chaps. 11 and 12, it proves useful to be able to enter the information in a ROM after fabrication. In such a *programmable ROM*, or *PROM*, the desired memory bits are stored by electrically altering the appropriate emitter connections, or their equivalent. Similarly, *erasable PROMs*, or *EPROMs*, are available in which information is stored as charge on stray capacitance at the gate electrodes of a MOSFET ROM without actually destroying the gate electrodes at the intersections. These

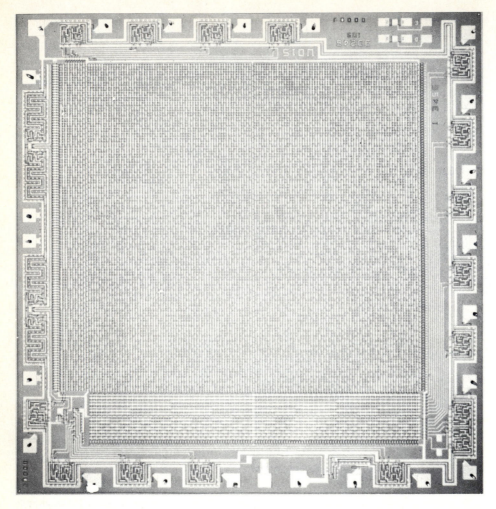

Figure 9-37 Photomicrograph of 65,536-bit ROM chip. (*University of Utah Research Institute.*)

bit patterns can be erased by irradiation with ultraviolet light to discharge the gate capacitors or other electrical signals (*electrically erasable PROM*, or *EEPROM*) and a new bit pattern introduced.

Shift-Register Memories

In many digital-circuit applications it proves useful to store digital signals temporarily for recall at a later time. This is, in effect, a memory into which information can be rapidly written and changed, as well as read out, as distinguished from the case of the ROM. Shift registers are convenient and effective memory circuits for this purpose. They are universally fabricated by integrated-circuit techniques in order to achieve a usefully large memory capacity. In this connection

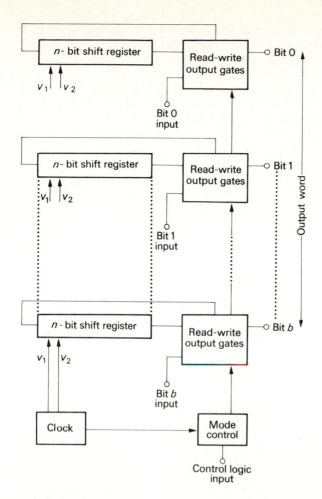

Figure 9-38 n-Word, b-bit shift register memory.

MOSFET circuits are particularly advantageous because of the small power dissipation per gate, as already noted in Table 9-10.

The organization of a shift-register memory based on circulating-register principles is shown in Fig. 9-38. Each register stores one bit of each digital word and the words are shifted along the registers in synchronism since the registers are all activated by the same clock signals. The read, write, and output gates, analogous to those in Fig. 9-28, determine whether the output signals are recirculated or lost, read as output and recirculated, or replaced by new input signals. These gates are activated in unison by mode control circuitry so that all bits of a word are written and read synchronously. In this connection the mode control circuit also performs address logic by keeping track of the location of the words as they circulate through the memory.

Memory capacity may be increased by paralleling additional shift registers to increase the number of bits in each word. Alternatively, the number of words may be increased by cascading additional shift registers. This has the disadvantage that the *access time*, that is, the time required to read out any given word, increases proportionally since the total time required for information to circulate through the memory depends upon the number of bits in each register.

Random-Access Memories

The access time in a shift-register memory depends upon the word address and upon the word storage capacity of the memory since information is only available sequentially at the shift-register outputs. In a *random-access memory*, or *RAM*, the access time is independent of the location of information in the memory and addressing logic permits immediate access to any information stored in the memory. A RAM is organized into word lines and bit lines much like a ROM except that the information is stored at each intersection by the state of a flip-flop memory cell.

Consider a single MOS memory cell, Fig. 9-39, in which four control FETs are used. When either the x-address line or the y-address line is at state 0, two of the four control transistors Q_3, Q_4, Q_7, and Q_8 are cut off and the memory cell is

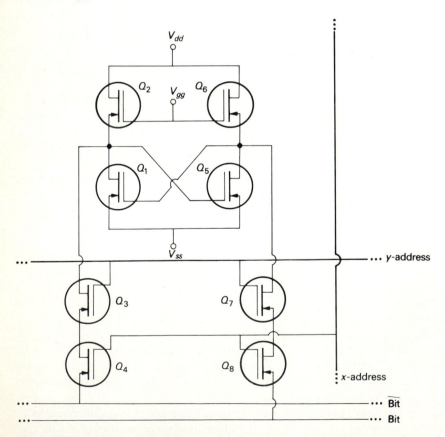

Figure 9-39 MOS memory cell for RAM.

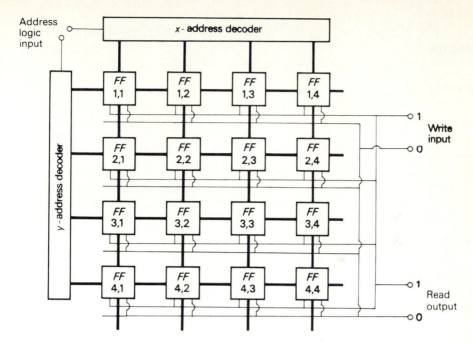

Figure 9-40 16-bit RAM.

electrically isolated from its surroundings. The cell is accessed by putting both the x-address line and the y-address line at state 1. This turns on all four control transistors so that logic levels at the bit and NOT bit lines correspond to the state of the binary. Alternatively, the state of the binary can be set by the logic levels presented externally to the bit and NOT bit lines.

A 16-bit RAM is shown in Fig. 9-40. Address decoders select the appropriate x and y lines corresponding to the address logic input and the state of the binary at the selected intersection appears at the output terminals. The information at any intersection is set by introducing the desired bit states to the write input.

The static memory cell is wasteful of chip space, since eight transistors are required per bit, and is also wasteful of power, since current is present in one or the other transistors in the binary. Power dissipation can be reduced by clocking the load transistor gates, V_{gg}, so that Q_2 and Q_6 are cut off except when the cell is addressed. This causes the current to be zero and memory information is retained by the charges on the stray capacitances at the gates of Q_1 and Q_5. In this case it is necessary to *refresh* the charges on the stray capacitances by periodically turning on Q_2 and Q_6 so that unavoidable leakage currents do not cause information to be lost.

Dynamic RAM, or DRAM, memory cells containing fewer transistors have been devised in order to increase the total memory capacity per chip. The ultimate in this direction is the single-transistor memory cell, Fig. 9-41, in which the logic

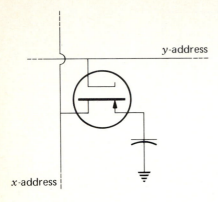

y-address

x-address

Figure 9-41 Single-transistor DRAM memory cell.

1 or 0 states are represented by the presence or absence of charge on the stray capacitance. A logic 1 bit is stored, for example, by addressing the cell to turn on the transistor by gate bias on the y-address line and providing a potential $-V_{cc}$ on the x-address line; the capacitor therefore charges to $-V_{cc}$. This charge must be periodically refreshed.

Read-out is accomplished by biasing the y-address line and sensing current in the x-address line. If the capacitor is at state 1, it discharges through the transistor and produces current in the x-address line. Conversely, if the capacitor is at state 0, no current is present. Note that this read-out is destructive in that a state 1 sensed in any given cell must be reintroduced if the information is to be available subsequently.

The additional address, refresh, sense, and restore circuitry required is much more extensive than in the case of a static memory cell. Since, however, this extra circuitry is common to all cells in a DRAM, it is not excessive for very large memory capacities. Dynamic RAM memories are available which can store as many as 2^{26} (67,108,864) bits on a single chip.

SUGGESTIONS FOR FURTHER READING

D. Childers and A. Durling: "Digital Filtering and Signal Processing." West Publishing Company, Inc., New York, 1975.

G. M. Glasford: "Digital Electronic Circuits," Prentice-Hall, Inc., Englewood Cliffs, N.J., 1988.

H. V. Malmstadt and C. G. Enke: "Digital Electronics for Scientists," W. A. Benjamin, Inc., New York, 1969.

Saul Ritterman: "Computer Circuit Concepts," McGraw-Hill Book Company, New York, 1986.

EXERCISES

9-1 Carry out the binary additions for $20 + 14$, $15 + 34$, and $56 + 25$. Check by forming the binary numbers of the decimal sums. Sketch the waveforms of the individual numbers and their sums.

Answer: 100010; 110001; 1010001

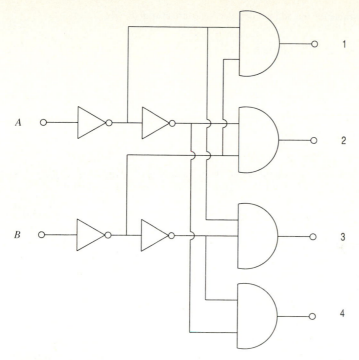

Figure 9-42 Address decoder circuit.

9-2 Carry out the binary subtractions for $25 - 14$, $35 - 16$, and $12 - 3$. Check by forming the binary numbers of the decimal subtractions. Sketch the waveforms of the individual numbers and their differences.

Answer: 1011; 10011; 1001

9-3 Sketch the waveforms of the binary numbers 1101 and 1001 and their sum. Compare the waveform of the sum with the output waveform of an AND gate with 1101 and 1001 as input signals.

9-4 Using De Morgan's theorem, simplify the logic expression for NOT exclusive-OR. Why is this sometimes called an *equality* comparator?

Answer: $A \cdot B + \bar{A} \cdot \bar{B}$

9-5 Devise a circuit to perform OR logic using only NOR gates. Repeat using only NAND gates.

9-6 Devise a full adder using only NOR gates. Repeat for NAND gates.

9-7 Prove the following logic statements:

$$A \oplus B = \overline{A \cdot B + \bar{A} \cdot \bar{B}}$$

$$A \oplus B = (A + B) \cdot (\bar{A} + \bar{B})$$

9-8 Evaluate the logic operations $0 \cdot 1$, $0 + 1$, 1, $1 + 1$.

Answer: 0, 1, 1, 1

9-9 Develop the truth table for the full adder in Fig. 9-15.

9-10 Develop the truth tables for a NOR binary and a gated NOR binary and contrast with Tables 9-11 and 9-12.

9-11* Derive the truth table for a *JK* NAND master-slave flip-flop, Table 9-13.

9-12* Determine the truth table for the address decoder logic circuit in Fig. 9-42.

Answer:

Digital address		Output			
B	A	1	2	3	4
0	0	1	0	0	0
0	1	0	1	0	0
1	0	0	0	1	0
1	1	0	0	0	1

9-13 Specify the states of the 4-bit ring counter similar to Table 9-16 if the Q output of FFD is connected to K of FFA and $\bar{Q}$ of FFD to J of FFA. Assume the initial state of this so-called *switch-tail* counter is all flip-flops cleared.

9-14 Determine the output of the BCD-to-decimal decoder in Fig. 9-30 to an input state specified by $\overline{AB}CD$. Repeat for $AB\bar{C}D$.

Answer: 4 and 8; 3 and 9

9-15 Cascade four flip-flops as shown in Fig. 9-43 and apply a square wave to terminal T. Put $FF1$ in state 1 and all others in state 0. Plot the waveforms at Q for all flip-flops for the first 10 pulses of the square wave. It is helpful to arrange the waveforms as in Fig. 9-22. Make a simple statement of what the circuit is doing.

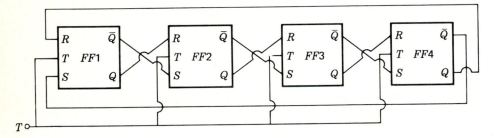

Figure 9-43 Cascaded clocked flip-flops.

TEN

ANALOG AND DIGITAL MEASUREMENTS

Electronic techniques are used to measure and control a wide variety of physical phenomena in both laboratory research and industrial applications. Indeed, it is common practice to develop an electric signal analogous to the phenomena of interest at the earliest opportunity so that subsequent processing can be carried out electronically. Circuits used to measure and process such analog signals and circuits for control purposes are investigated in this chapter.

Digital instruments use the logic circuits and techniques described in the previous chapter to carry out measurements and to process data. A significant advantage of digital over analog instruments is that digital data is decisive. The least significant bit must be either a 1 or a 0, which means precision is increased simply by employing additional digits. Because of these advantages, modern measurement techniques often digitize signals so that processing can be carried out by digital circuits.

TRANSDUCERS

Devices that produce electric signals in accordance with changes in some physical effect are called *transducers*. A typical example is a microphone which generates a voltage waveform corresponding to the variations of sound pressure impinging on a diaphragm. Other transducers, such as the strain gauge, operate by changing some circuit parameter (e.g., resistance) rather than by generating a voltage. A great variety of transducers have been developed so that it is often possible to generate a signal in more than one way. The deciding factor in any given situation is the accuracy with which the transducer output represents the physical parameter being measured.

Mechanical Transducers

A useful transducer for sensing mechanical motion or position measures the separation between a movable and a fixed plate by their mutual capacitance. Variations in capacitance accompanying a change in the plate separation are detected with a capacitance bridge. If it is only necessary to detect motions from an equilibrium position, simply connecting the capacitor in series with a resistor and a dc voltage is sufficient, as in the *condenser microphone*, Fig. 10-1. Variations in sound pressure cause the diaphragm to deflect, thus changing the capacitance C, and generating a current in the external circuit,

$$Q = CV \tag{10-1}$$

$$i = \frac{dQ}{dt} = V\frac{dC}{dt} \tag{10-2}$$

The output signal is

$$v_o = RV\frac{dC}{dt} \tag{10-3}$$

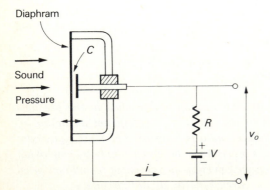

Figure 10-1 Sketch of condenser microphone.

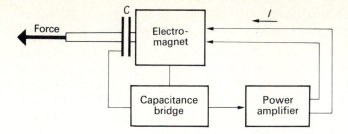

Figure 10-2 Capacitance transducer used in feedback system to measure mechanical force.

The capacitance transducer can be adapted to measure mechanical force, as in Fig. 10-2. Here the position of an arm to which the force is applied is detected by the capacitance between the arm and the pole face of an electromagnet. Any applied force displaces the arm, and the change in capacitance is detected by a capacitance bridge. The unbalance voltage of the bridge provides a signal for an amplifier feeding the coils of the electromagnet. The magnetic force pulls the arm back to the equilibrium position against the external force. Thus, the magnet current is a direct measure of the applied force.

A system that strives to return to the equilibrium, or *null*, position is very commonly used in measurement circuits. Two comments are worthy of note. First, the system is basically a feedback network and may be analyzed as such. Second, since the system returns to its equilibrium position it is not necessary for the transducer to have a linear relation between input action and output signal. As a matter of fact, it is not even essential that the transducer calibration be known, since it is not deflected in the equilibrium position.

Mechanical stress in structural members and materials is conventionally measured by means of a resistance *strain gauge*. One common strain gauge is a thin metal wire bent back and forth and fastened to a paper backing, as in Fig. 10-3. The gauge is cemented directly to the surface of the structural member under test

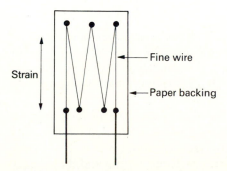

Figure 10-3 Resistance strain gauge.

so that changes in length are detected by changes in the resistance of the gauge. The magnitude of the change in resistance can be determined from

$$R = \rho \frac{L}{A} \tag{10-4}$$

where L is the total length, A is the cross-sectional area, and ρ is the resistivity of the wire. Taking the logarithm and the differential of both sides,

$$\log R = \log \rho + \log L - \log A$$

$$\frac{\Delta R}{R} = \frac{\Delta \rho}{\rho} + \frac{\Delta L}{L} - \frac{\Delta A}{A} \tag{10-5}$$

Introducing Poisson's ratio σ for the wire material,

$$\frac{\Delta A}{A} = -2\sigma \frac{\Delta L}{L} \tag{10-6}$$

into Eq. (10-5) yields

$$\frac{\Delta R}{R} = \frac{\Delta L}{L}(1 + 2\sigma) + \frac{\Delta \rho}{\rho} \tag{10-7}$$

Finally, the *gauge factor* $K = (\Delta R/R)/(\Delta L/L)$ is

$$K = 1 + 2\sigma + \frac{\Delta \rho}{\rho} \frac{L}{\Delta L}$$

$$K = 1 + 2\sigma + \frac{L}{\rho} \frac{d\rho}{dL} \tag{10-8}$$

According to Eq. (10-8) the change in resistance arises from changes in the wire dimensions because of mechanical strain plus the possibility of changes in resistivity. In metal-wire gauges, the third term is negligible and $K \cong 2$. The gauge resistance may be approximately 100 Ω and the strains of interest are of the order of 10^{-3}, so that the total resistance change is, from Eq. (10-7),

$$\Delta R = RK \frac{\Delta L}{L} = 100 \times 2 \times 10^{-3} = 0.2 \ \Omega$$

which is a fairly small value. Silicon strain gauges have much larger gauge factors, of the order of 150, as a result of change in resistivity with strain. The reason for this is that the number of current carriers in a semiconductor is sensitive to mechanical strain. Therefore, the conductivity changes significantly in response to elastic deformation.

Bridge circuits are universally used to measure the change in resistance of strain gauges because small resistance changes can be measured and temperature compensation is easily effected by including two identical strain gauges in adjacent arms of the bridge. One gauge is subjected to the mechanical strain and the other is

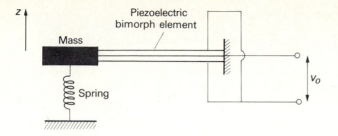

Figure 10-4 Elementary piezoelectric accelerometer.

isolated from the strain but positioned so that its temperature environment is the same as that of the strain detector. Thus, changes in temperature influence each gauge equally, and the bridge remains balanced.

Mechanical vibrations in structures and machines are determined with transducers called *accelerometers*. An accelerometer measures the forces on a small mass caused by accelerations of the object to which the transducer is fastened. The inertial force $F = ma$ is counterbalanced by the restoring force of a spring supporting the mass,

$$ma = kx \qquad (10\text{-}9)$$

where k is the spring constant. The deflection x is detected by the voltage generated in a piezoelectric crystalline material such as rochelle salts or barium titanate. As discussed in Chap. 8, piezoelectric materials generate a voltage proportional to the mechanical strain on the crystal. Usually, as illustrated in Fig. 10-4, the piezoelectric element is formed from two parallel slabs such that, for example, a vertical deflection of the mass puts the upper slab in compression and the lower slab in tension. The output terminals on the flat faces of the two slabs are connected in parallel. The result of this *bimorph* construction is a more sensitive transducer than can be obtained using only a single element.

Since the output voltage is proportional to the deflection x, Eq. (10-9) can be written

$$a = \frac{kc}{m} v_o \qquad (10\text{-}10)$$

where c is the piezoelectric constant. The useful frequency range of an accelerometer ranges up to the mechanical resonant frequency of the mass and the supporting spring combination. An upper frequency limit of 2 kHz is typical.

The output voltage of an accelerometer can be integrated electronically to develop a signal proportional to velocity. A second integration yields the displacement. Sensitive accelerometers form an important part of inertial guidance navigation systems in missiles and aircraft. Three mutually perpendicular accelerometers are arranged to measure the displacement of the vehicle from a known starting point.

Radiation Detectors

Light energy from infrared to ultraviolet wavelengths is detected and measured with various forms of *photocells*. A simple, though extensively used, photocell is a small piece of semiconductor provided with two electric contacts. The resistance of the semiconductor between contacts depends upon the intensity of light energy impinging upon the surface because absorbed photons produce electrons and holes that act as current carriers in the semiconductor. The response of a semiconductor photocell is zero for photon energies less than the width of the forbidden energy gap because such photons have insufficient energy to promote electrons from the valence band to the conduction band across the forbidden energy gap. The photoresponse is also small at short wavelengths (high photon energies) because photons are so heavily absorbed in the surface layers that most of the semiconductor is inactive. Therefore, the response is a maximum at photon energies corresponding to the width of the forbidden energy gap, as illustrated for the common photoconductor materials CdS, CdSe, and CdTe in Fig. 10-5. Note that CdS photocells are useful in the visible region of the spectrum, CdSe is most sensitive to red light, and CdTe responds to near-infrared radiation. Other semiconductors commonly used as infrared detectors are PbS and PbSe, which are useful at 3.0 μm and 4.5 μm, respectively, corresponding to forbidden energy gaps of 0.37 and 0.27 eV. In addition, InSb, with a forbidden energy gap of only 0.18 eV, responds to radiation at wavelengths as long as 7 μm.

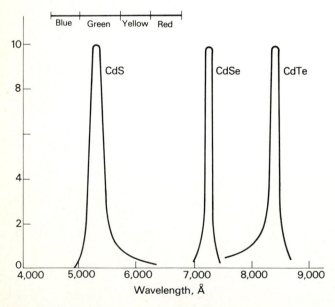

Figure 10-5 Wavelength response of CdS, CdSe, and CdTe photocells. Maximum response is at photon energies equal to width of forbidden energy gap.

A useful measure of the sensitivity of a photocell at the peak of the wavelength response curve is the *gain G*. The gain is the ratio of the number of current carriers generated by illumination to the number of absorbed photons per second. Expressed in terms of the photocurrent I the gain is

$$G = \frac{I/e}{F} \qquad (10\text{-}11)$$

where e is the electronic charge, and F is the rate of photon absorption. The photocurrent can be written in terms of an applied voltage V and the dimensions of the semiconductor using Ohm's law, Eq. (10-4), and an expression for the resistivity in terms of the carrier density n and the carrier *mobility*, μ, $\rho = ne\mu$,

$$G = \frac{1}{e}\frac{V}{R}\frac{1}{F} = \frac{1}{e}\frac{V}{\rho L/A}\frac{1}{F} = \frac{n\mu A}{L}\frac{V}{F} \qquad (10\text{-}12)$$

Equation (10-12) is further simplified by noting that the total number of carriers nAL is equal to the rate at which carriers are produced by photon absorption times the average time τ an electron spends in the conduction band before recombining with a hole in the valence band,

$$F\tau = nAL \qquad (10\text{-}13)$$

so that

$$G = \frac{\mu V \tau}{L^2} \qquad (10\text{-}14)$$

Finally, the transit time of a carrier between the contacts on the semiconductor is the distance between the electrodes L divided by the average carrier velocity. The carrier velocity is the mobility times the electric field,

$$T = \frac{L}{\mu(V/L)} = \frac{L^2}{\mu V} \qquad (10\text{-}15)$$

where T is the transit time. Inserting Eq. (10-15) into Eq. (10-14), the gain becomes simply

$$G = \frac{\tau}{T} \qquad (10\text{-}16)$$

According to Eq. (10-16), high sensitivity implies a sluggish response. That is, a large value of gain is achieved by increasing the time constant τ, and τ also determines the speed of response of the photoconductor to sudden changes in light intensity. Where speed is of secondary importance simple photoconductors are widely used because they can be quite sensitive. Most often, the change in resistance of the photocell accompanying incident illumination is measured by connecting the photocell in series with a dc voltage source and a load resistor. The voltage drop across the load resistor corresponds to the resistance changes in the semiconductor resulting from photon absorption.

Many semiconductor photocells having rapid response and good sensitivity are based on a reverse-biased *pn* junction. Photons absorbed in the region of the junction create electrons and holes that are swept to the opposite sides of the junction. The collected carriers are observed as an increase in reverse current.

Since the width of a *pn* junction is very small, the transit time T in Eq. (10-16) is also small, and a reasonable value of G can be attained even for fast response times. The *pn*-junction structure has the additional advantage that the reverse current in the dark is very small compared with the current in a simple piece of semiconductor at the same applied voltage. This means that the additional carriers produced by photon absorption can be detected with greater sensitivity.

Note that electrons and holes created in the region of the junction are collected even in the absence of an applied voltage because of the internal potential rise at a *pn* junction, Fig. 4-3. This means that an illuminated *pn* junction can also act as a battery. An open-circuit voltage equal to the internal potential step exists between terminals applied to the *n*-type and *p*-type regions and the short-circuit current depends upon the illumination intensity at the junction. This is the principle of the *solar battery* used as an energy source on space satellites. Silicon solar cells are most often used because the width of the forbidden energy gap in silicon corresponds favorably with the wavelength distribution of radiant energy from the sun. Similar photocells are also used in automatic-iris cameras and light-powered pocket calculators.

Highly ionizing radiation such as β rays and α rays are also detected by a *pn* junction. If an α particle strikes the junction, its energy is consumed in creating electrons and holes which are subsequently collected by the field at the junction. The output current pulse corresponding to this event is a measure of the energy of the incident particle. Therefore, a *pn-junction radiation detector* not only signals the presence of nuclear radiation, but also measures the energy of the individual particles. Most often, it is desirable to employ detectors having a wide *pn* junction to assure that the particle is completely absorbed in the junction region. Furthermore, the junction must be very near the surface of the semiconductor so that little energy is lost in creating carriers far from the junction where they cannot be collected.

CHEMFETs

If the metal gate electrode of a MOSFET is replaced by an ion-selective membrane, Fig. 10-6, ions absorbed in the membrane alter the electric field in the channel and change the drain current. Such *CHEMFET* sensors make exceedingly sensitive chemical ion detectors. The effect of the absorbed ions can be accounted for by adding the *Nernst potential* to Eq. (4-13) for the drain current

$$I_d = K \left(V_{rs} + \frac{RT}{nF} \ln a - V_k \right) V_{ds} \tag{10-17}$$

where V_{rs} is the voltage between the source and the reference electrode, R is the gas constant, T is the temperature, n is the charge on an absorbed ion, F is the Faraday constant, a is an activity coefficient which in simple cases is just the ion concentration, and V_k is a constant.

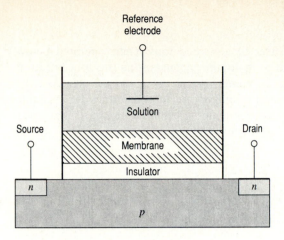

Figure 10-6 CHEMFET ion-sensitive detector.

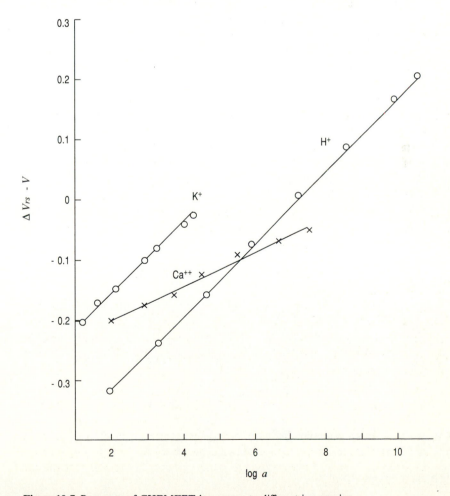

Figure 10-7 Response of CHEMFET ion sensor to different ion species.

The CHEMFET is most sensitive to absorbed ions when operated below pinchoff. Most often, V_{rs} is adjusted to maintain I_d constant as the concentration of ions in solution, and therefore in the membrane, changes and the change in V_{rs} is a direct measure of the ionic concentration. Polymeric membranes are available that are selective to only a single ion species so that CHEMFETs can be fabricated to measure only one type of ion. Typical experimental results, Fig. 10-7, are in agreement with Eq. (10-17).

FEEDBACK CONTROL CIRCUITS

Voltage Regulators

Special circuits are widely used to stabilize the output voltage of a rectifier power supply to a greater extent than is possible with the simple zener diode circuit described in Chap. 3. Such a *voltage regulator* is an important and useful example of measurement and control with electronic feedback circuits. The circuit measures the output voltage, compares it with the desired value, and adjusts conditions so that the difference between the two is zero. In this way control over the dc output voltage may be maintained in the face of changes in load and also in ac line voltage.

The output voltage of a power supply is conveniently controlled by introducing a power transistor in series with the rectifier-filter and the load, Fig. 10-8. It is useful to analyze this circuit as a feedback amplifier in which the reference voltage V is the input voltage and V_o is the output. Since the feedback ratio is $R_2/(R_1 + R_2)$, the output voltage is controlled by the amplifier and power transistor to be, from Eq. (7-3),

$$V_o = -\frac{R_1 + R_2}{R_2} V \qquad (10\text{-}18)$$

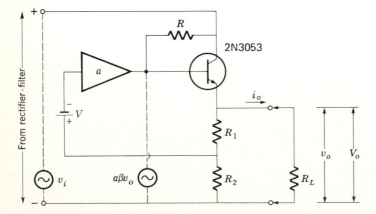

Figure 10-8 Series voltage regulator.

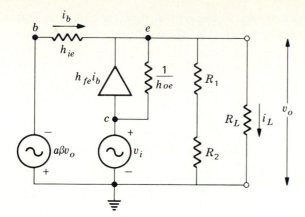

Figure 10-9 Equivalent circuit of voltage regulator in Fig. 10-8.

In effect, the voltage regulator compares a fraction of the dc output voltage with the reference voltage and adjusts the control transistor to maintain the difference equal to zero; hence V_o is held constant.

The stabilizing effect of the regulator is best determined by analyzing the control transistor separately from the dc amplifier. Variations in the supply voltage are assumed to arise from an ac generator v_i in Fig. 10-8 and result in a variation v_o in the output voltage, and a feedback signal $a\beta v_o$. The control transistor is an emitter-follower amplifier and the appropriate equivalent circuit is shown in Fig. 10-9. It is important to include the output conductance h_{oe} in the circuit since $1/h_{oe}$ is not negligible with respect to R_L for most bipolar power transistors.

Analysis begins by writing the voltage drops around the outer loop,

$$a\beta v_o + i_b h_{ie} + v_o = 0 \qquad (10\text{-}19)$$

The drops around the loop including $1/h_{oe}$ and R_L neglect the minor current drain in the feedback divider $R_1 R_2$,

$$v_i + \frac{1}{h_{oe}}[(1 + h_{fe})i_b - i_L] - v_o = 0 \qquad (10\text{-}20)$$

The base current is eliminated from Eq. (10-20) by substitution from Eq. (10-19),

$$v_i - \left[1 + \frac{(1 + h_{fe})(1 + a\beta)}{h_{oe} h_{ie}}\right]v_o - \frac{i_L}{h_{oe}} = 0 \qquad (10\text{-}21)$$

Solving for the output voltage,

$$v_o = \frac{h_{ie}}{h_{oe} h_{ie} + (1 + h_{fe})(1 + a\beta)}(h_{oe} v_i - i_L) \qquad (10\text{-}22)$$

This expression is the Thévenin equivalent of the regulated power supply in that the first term is the internal voltage generator and the coefficient of i_L represents the equivalent internal resistance.

For practical purposes, a sufficiently accurate form of Eq. (10-22) reduces to

$$v_o = \frac{h_{ie}}{a\beta h_{fe}}(h_{oe}v_i - i_L) \tag{10-23}$$

Typical values of the hybrid parameters for the control power transistor are $h_{ie} = 1000 \ \Omega$, $h_{fe} = 20$, and $h_{oe} = 10^{-3}$ mho; if $a = 50$ and $\beta = 0.3$, the effective internal resistance of the power supply is $1000/(50 \times 0.3 \times 20) = 3.3 \ \Omega$, a very small value. This means that variations in load current result in only small changes in load voltage. Additionally, output voltage changes are only $3.3 \times h_{oe} = 3.3 \times 10^{-3}$ of changes in the input voltage. This represents a reduction by a factor of 300. It is apparent that this simple regulator is effective against both input-voltage and output-load variations.

Note that the factor by which input voltage variations are reduced at the output terminals applies to ripple voltages as well as to other voltage changes. Therefore output voltage ripple is very small in a regulated power supply. Both the reduction factor and the effective internal resistance can be improved by increasing the gain of the dc amplifier, according to Eq. (10-22). In many critical applications the gain is made large enough that residual variations in the regulated voltage reflect the stability of the reference potential.

The series control transistor must be capable of dissipating the heat generated by the entire load current without overheating. Power transistors are often used, and it is feasible to employ two or more identical units connected in parallel, if this is necessary to carry the maximum load current.

Although the series regulator is most popular, it is also possible to connect the control transistor in parallel with the load, much as in the case of a simple zener-diode regulator, Fig. 10-10. The current through the transistor is controlled so that the voltage drop across the series resistor is the same independent of load-current changes. Effective voltage stabilization is obtained by such a *shunt regulator*, but the circuit suffers from the additional power loss in the series resistor. On the other hand, it is only necessary for the control transistor to carry a fraction of the full-load current, which is a considerable advantage in high-current power supplies.

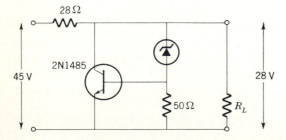

Figure 10-10 Shunt regulator.

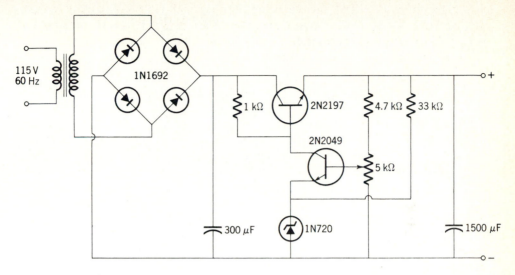

Figure 10-11 Complete 40-V 500-mA regulated power supply.

A complete regulated power supply is illustrated in Fig. 10-11. The dc amplifier is a single grounded-emitter transistor stage and the reference voltage is provided by an 18-V zener diode. The zener diode is placed in the emitter circuit of the amplifier, rather than in series with the feedback signal, so that one terminal of the diode can be grounded. The 33-kΩ resistor provides reverse potential for the diode. Note that the feedback ratio is adjustable. This enables the output voltage to be set at any desired value, within limits dictated by the operating conditions of the transistors. In this circuit, the output voltage can be adjusted over the range 40 to 50 V while maintaining good regulation. The 1500-μF capacitor across the output terminals provides a very low effective internal impedance for ac signals.

Very precisely regulated voltages are obtained with more elaborate versions of this circuit. The gain is increased by using a multistage dc amplifier. In particular, operational amplifiers prove convenient because of their large gain and dual inputs, which facilitate comparing the output voltage with the reference voltage. Such circuits can hold the output voltage to within 0.1 percent over the entire range of load current. In fact, stabilities as great as one part in 10^5 are achieved in special applications.

The separate components of an entire voltage regulator can be conveniently incorporated into a single integrated circuit, Fig. 10-12a. In this circuit the voltage drop across R_s, the so-called *shut-down resistor*, acts to open the circuit when the current capability of the regulator is exceeded, as, for example, by an inadvertent short circuit at the output terminals. Thus the regulator is self-protected against current overloads. A separate control transistor, Fig. 10-12b, may be included to increase the current capacity. In this fashion, the power transistor is effectively placed in the Darlington connection with the control transistor internal to the

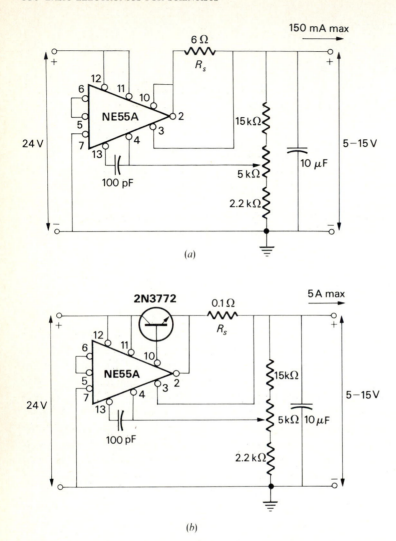

(a)

(b)

Figure 10-12 (a) Integrated-circuit voltage regulator and (b) with power transistor to increase current capability.

integrated circuit to achieve enhanced current gain. It proves convenient to mount the external control transistor on a heat sink to increase power dissipation.

These voltage-regulator circuits suffer from the power lost in the control transistor, $V_{ec} I_c$, which can be an appreciable fraction of the total power delivered by the power supply. This defect is remedied in so-called *switching regulators*, in which the control transistor is rapidly switched on and off by the control circuits. If the duration of the on time is t_1 and the off time is t_2, the average output voltage is just

$$V_o = \frac{t_1}{t_1 + t_2} V_i \tag{10-24}$$

where V_i is the dc input voltage to the regulator. In effect, the control circuit senses the output voltage and adjusts the ratio t_1/t_2 in accordance with Eq. (10-24) to mainain V_o at the desired value. A conventional LC filter (Chap. 3) passes the average value of the pulse waveform to the load.

Since the control transistor is either saturated ($V_{ec} \sim 0.1$ V) or completely cut off ($I_c = 0$), the power lost in the regulator is very small and the efficiency approaches 100 percent. Furthermore, the circuit may be arranged to use inductive transients in the LC filter to produce an output voltage which is actually greater than the dc input voltage. In this connection, more elaborate control circuitry may adjust $t_1 + t_2$, that is, the switching frequency, rather than, or in addition to, the ratio t_1/t_2 to effect control over the output voltage.

Phase-Locked Loop

A useful and versatile electrical feedback circuit is the so-called *phase-locked loop*, or *PLL*, in which the frequency of a voltage-controlled oscillator is adjusted by a voltage corresponding to the phase difference between the oscillator output and an external input signal. The feedback connection, Fig. 10-13, compares the oscillator output to the input signal in a phase detector such as an operational amplifier comparator (Exercise 10-8), or an exclusive-OR gate (Exercise 10-7), and the average or rms output is applied to the VCO. The PLL outputs are the VCO-control voltage and/or the VCO output itself, depending upon the specific application of the circuit.

Suppose, for example, that the input signal is a sine wave which is modulated in frequency. The VCO-control-voltage output is then a direct replica of the modulating waveform because the feedback loop continuously adjusts the VCO frequency to equal that of the input. Thus the PLL functions as a *frequency modulation detector* as, for example, in an FM radio.

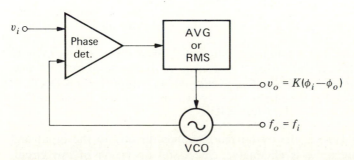

Figure 10-13 Feedback connection in phase-locked loop maintains VCO output in phase with input signal.

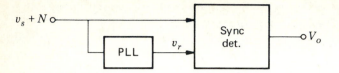

Figure 10-14 PLL with synchronous detector recovers sine-wave signals from noise.

A digital counter may be inserted in the feedback loop, causing the PLL to act as a frequency multiplier. In this case the VCO output is scaled down and compared in phase to the input waveform. Not only is the VCO output a multiple of the input signal by the scale factor of the counter, but it retains a constant phase and frequency relationship against changes of the input voltage. A wide range of output frequencies is obtained by selecting different scale factors.

An important application of the PLL principle is detection of sinusoidal signals in the presence of random noise using synchronous detection similar to the lock-in amplifier discussed in Chap. 6. In this circuit, Fig. 10-14, the PLL generates a reference signal phase-locked to the sinusoidal input. This reference signal is used in the synchronous detector to recover the desired signal and eliminate the noise. To show how this comes about, the random noise, N, is explicitly included in the input signal v_i together with the desired signal v_s, while the PLL output, v_r, is a sine wave at the frequency of v_s,

$$v_i = v_s + N = V_s \sin \omega t + N$$
$$v_r = V_r \sin \omega t \tag{10-25}$$

As in the synchronous detector of Fig. 6-39, and by analogy with Eq. (6-74), the current in the upper diode is

$$i_3 = a_1(v_s + N + v_r) + a_2(v_s + N + v_r)^2$$

The current i_4 in the lower diode is similar except that the polarity of v_i is reversed so that corresponding terms cancel in the output,

$$v_o = R(i_3 - i_4) = 2a_1 R(v_s + N) + 4a_2 R(v_s + N)v_r$$
$$= 2a_1 R(v_s + N) + 4a_2 RNv_r + 4a_2 RV_s V_r \sin^2 \omega t \tag{10-26}$$

Since N is random noise and v_s and v_r are sinusoidal, the average, or dc, value of the first two terms in Eq. (10-26) is zero. The average of the last term can be found using the identity $2 \sin^2 \omega t = 1 - \cos 2\omega t$, so that the dc output signal is just

$$V_o = (2a_2 RV_r)V_s \tag{10-27}$$

According to Eq. (10-27), the output corresponds directly to the input and the interfering random noise is eliminated. Successful operation of this circuit depends upon the phase-lock of the PLL with the input signal. Because of the fluctuating nature of random noise, lock can be maintained even though the random-noise level is much larger than the desired signal.

Servos

It is often useful to control mechanical position or motion in accordance with an electric signal. To assure that the desired motion takes place, a voltage feedback signal corresponding to the mechanical motion is developed. This voltage is compared with the input signal. When the two are equal, mechanical motion ceases since the mechanism has responded to the actuating signal. Feedback systems incorporating a mechanical link in the feedback network are called *servomechanisms*, or *servos*, from the Latin word for "slave," since the mechanical motion is the slave of the input signal. Servos are widely used in control applications, from laboratory chart recorders to automatic navigation systems.

Consider the servo system sketched in block-diagram form in Fig. 10-15. The amplifier drives a mechanical actuator as, for example, a dc motor. Suppose the motor moves a mechanical arm to a position x and the feedback voltage is proportional to the position of the arm and equal to kx. So long as the *error signal*, which is the difference between the input and feedback voltages, is not equal to zero, the amplified error signal continues to drive the motor. When the feedback signal equals the actuating signal, the error is zero, and motion ceases. Thus

$$v_i - v_f = v_i - kx_0 = 0 \tag{10-28}$$

so that

$$x_0 = \frac{v_i}{k} \tag{10-29}$$

According to Eq. (10-29), the equilibrium position of the mechanical motion x_0 is directly proportional to the input signal.

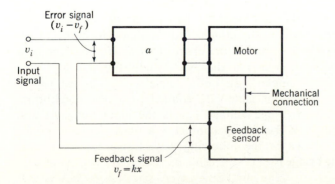

Figure 10-15 Block diagram of servo system.

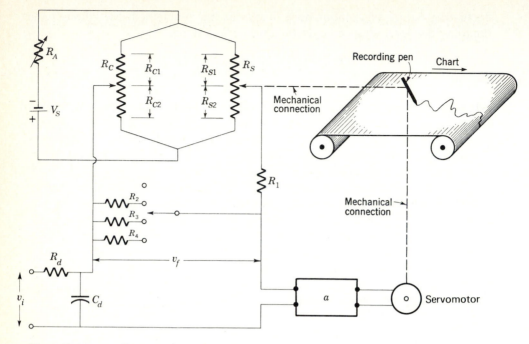

Figure 10-16 Recording-potentiometer servo.

A common laboratory instrument based on servo principles is the *recording potentiometer*. This is a self-balancing dc bridge in which the balance position is marked on a moving paper strip or chart producing a continuous record of the input voltage. The heart of the system, Fig. 10-16, is the slide wire R_S in which there is a calibrated current supplied by the battery V_S and variable resistor R_A.

The potentiometer in parallel with the slide-wire resistance forms a bridge circuit which is balanced when, from Eq. (1-46), $R_{C_1}/R_{C_2} = R_{S_1}/R_{S_2}$. Under this condition the feedback voltage v_f is zero. In the absence of an input signal, the adjustable contact on the slide wire is moved by the servomotor until balance is achieved. The recording pen mechanically attached to the slide-wire contact continuously records the balance position on the chart. This means that the zero position ($v_i = 0$) may be placed anywhere on the chart by adjusting R_C.

When an input voltage is applied, an initial signal $v_i - v_f$ is fed to the amplifier. The servomotor drives the slide-wire contact to a new equilibrium position where the unbalanced bridge voltage equals the input voltgage, and $v_i - v_f = 0$. The displacement of the sliding contact is a linear function of the input voltage, as can be shown in the following way. Using Eq. (1-51), the unbalance voltage from the bridge is

$$v_f = V\left(\frac{R_{C_1}}{R_{C_1} + R_{C_2}} - \frac{R_{S_1}}{R_{S_1} + R_{S_2}}\right) \tag{10-30}$$

where V is the voltage across the bridge. If the slide wire is uniform, and the zero position, $x = 0$, is taken to be at the center,

$$R_{S_1} = \frac{R_S}{2} - mx \quad \text{and} \quad R_{S_2} = \frac{R_S}{2} + mx \qquad (10\text{-}31)$$

where m is a constant of the slide wire. Introducing Eq. (10-31) into Eq. (10-30) and noting that $R_{C_1} = R_{C_2}$ since the zero position is at the center,

$$v_f = V\left(\frac{1}{2} - \frac{R_S/2 - mx}{R_S/2 - mx + R_S/2 + mx}\right) = \frac{V}{R_S} mx \qquad (10\text{-}32)$$

Therefore, according to Eq. (10-32),

$$x_0 = \frac{1}{mI_S} v_i \qquad (10\text{-}33)$$

where $I_S = V/R_S$ is the standardized current in the slide wire. Equation (10-33) shows that the deflection of the slider, and hence of the recording pen, is proportional to the input signal. It is common practice to select m and the standard current I_S such that the full chart width is an integral voltage such as 100 mV.

Resistors R_1 through R_4 in Fig. 10-16 constitute a voltage divider for changing the sensitivity of the potentiometer. When the divider is in the circuit, only a fraction of the unbalance voltage is available as the feedback signal and the sensitivity is increased correspondingly. The combination of R_d and C_d is a damping filter at the input to optimize the response time of the recorder.

The recording potentiometer is a very versatile instrument in the laboratory. It combines high sensitivity for dc signals and the inherently high input impedance of a potentiometer circuit with automatic operation and a permanent record. It is widely used as part of other instruments, such as optical spectrometers, x-ray diffractometers, and gas chromatographs, or as a separate unit.

ANALOG INSTRUMENTS

Oscilloscopes

The circuits and applications of the cathode-ray oscilloscope have been touched upon several times. The complete block diagram of a typical unit, Fig. 10-17, includes vertical and horizontal amplifiers coupled to the vertical and horizontal deflection plates of the cathode-ray tube, a triggered sawtooth, or ramp generator to provide a linear horizontal sweep, and suitable power-supply circuits. In the most common mode of operation, the waveform of interest is applied to the vertical input terminals. The vertical amplifier increases the amplitude sufficiently to cause an appreciable vertical deflection of the electron beam. A portion of the amplified signal is used to derive a trigger pulse for the sweep generator in order

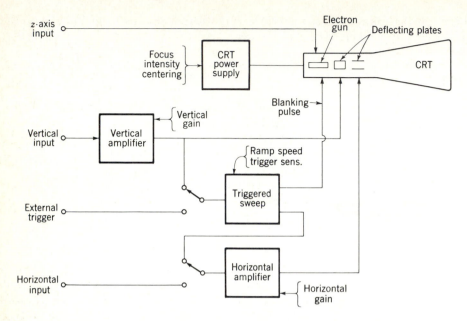

Figure 10-17 Block diagram of oscilloscope.

to synchronize the sweep signal with the input waveform. The output of the sweep generator is amplified and applied to the horizontal deflection plates. The sweep generator also provides a pulse to the electron gun of the cathode-ray tube (CRT) which blanks off the electron beam during retrace so that the spot can return to the starting point without leaving a spurious trace.

Sometimes it is advantageous to trigger the sweep generator with an external signal as, for example, if it is necessary to start the sweep before the vertical signal amplitude is large. For this purpose, an external trigger input is provided to which the generator input can be connected. When the linear sweep is not desired, as in the case of Lissajous figures (Chap. 2), the input of the horizontal amplifier is connected to the horizontal input terminals and external waveforms can be placed on both the horizontal and vertical deflection plates. It is also possible to modulate the intensity of the electron beam by altering the grid potential in the electron gun; this external terminal is conventionally termed the z-axis input to distinguish it from the horizontal (x-axis) and vertical (y-axis) inputs.

The vertical and horizontal amplifiers are quite often of the dc-coupled differential type. Voltage-divider gain controls are included to adjust the amplitude of the deflection on the screen. Many different sweep circuits are in common use; a typical version is described in Chap. 8. Sweep-speed and trigger-sensitivity controls are necessary here in order to set the sweep for best waveform display. The power supply for the CRT is designed so that dc potentials applied to the electron-gun elements can be adjusted to achieve the optimum focus and intensity of the trace. Similarly, the quiescent position of the spot on the tube face is adjusted by altering the dc voltages on both the horizontal and vertical deflection plates. Modern instruments are capable of bandwidths extending from dc to 500 megahertz.

Waveform Analyzer

Complex waveforms may be thought of as combinations of harmonically related sine waves, according to the method of Fourier analysis discussed in Chap. 2. Instruments capable of determining the amplitude of the frequency components in a waveform, called *waveform analyzers,* in effect carry out Fourier analysis experimentally. As such, they are frequency-measuring instruments.

A waveform analyzer is basically a sharply tuned ac amplifier-voltmeter which has a significant response at only one frequency. The input signal is combined with the output of a variable-frequency sinusoidal oscillator to produce a signal at the response frequency of the amplifier, Fig. 10-18. That is, according to Eq. (3-17), two sine waves suitably mixed produce a sine wave at the sum (and difference) frequency. If the oscillator frequency is f_o while the input waveform has a frequency component f_i, the combination will result in an output signal if

$$f_i + f_o = f_a \tag{10-34}$$

where f_a is the center frequency of the tuned amplifier. Thus, by tuning the oscillator frequency over a given interval, the output meter deflects each time Eq. (10-34) is satisfied for a frequency component of the input waveform. The magnitude of the meter deflection is proportional to the amplitude of each input-signal component. In this way, the entire frequency complement of the input waveform is determined.

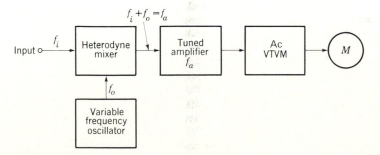

Figure 10-18 Block diagram of waveform analyzer for determining frequency components in complex waveform.

The heterodyne technique is used in preference to simply tuning the amplifier itself in order to obtain the greatest possible frequency coverage. An oscillator circuit capable of output frequencies over a wide band is simpler than a tunable amplifier covering the same frequency interval. Furthermore a fixed-tuned amplifier has a constant peak response and bandpass characteristic, both difficult to achieve in a tunable system. The frequency dial of the waveform analyzer is calibrated in terms of $f_a - f_o$, so that the unknown frequency is indicated directly.

Magnetic Recorder

One of the most useful signal recording schemes is based on the *magnetic-recorder* principle in which an electric waveform is recorded in the form of magnetization on a magnetic tape. After recording, the tape is passed back through the instrument and a voltage signal corresponding to the original input is obtained. This makes it possible to observe the signal again and again, if necessary, and to analyze the waveform by as many different techniques as may be required.

The magnetic tape, most often in the form of a ferromagnetic oxide powder coated onto a plastic tape, is magnetized in accordance with the signal by the magnetic field of the recording head, Fig. 10-19a. Signal current produces a magnetic field at a sharply defined gap in the core of the recording head, and this field permanently magnetizes the tape as it passes the gap. A high-frequency ac bias signal is also applied to the recording head to improve the linearity of the recording process. In playback, Fig. 10-19b, the tape is moved past the head again and the magnetic field of the tape induces a voltage in the coil of the head. Since the magnetization of the tape corresponds to variations in the original signal, voltages induced in the head replicate the input waveform. Playback does not change the tape magnetization, so that the signal can be reproduced as many times as desired.

The action of the ac bias frequency can be understood by considering the typical nonlinear properties of the magnetic tape as illustrated by the hysteresis

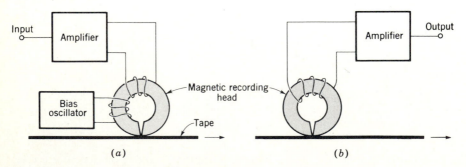

Figure 10-19 (a) Recording and (b) playback of electric signals using magnetic recording.

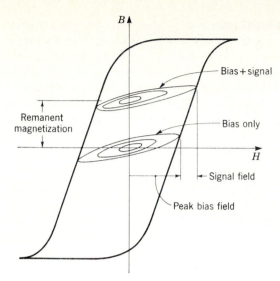

Figure 10-20 Ac bias causes magnetic tape material to traverse minor hysteresis loops as it passes under recording head. Remanent magnetization is a linear function of signal field.

loop in Fig. 10-20. In the absence of ac bias, the recorded magnetization is not a linear function of the signal field and the playback signal is strongly distorted. The ac bias causes the magnetization of the material to traverse minor hysteresis loops as shown. As an element of tape moves away from the recording gap, the size of the loops decreases to zero. Thus, in the absence of signal the resulting magnetization is equal to zero. The signal field displaces the minor loop so that the remanent magnetization is a finite value. Because of the straight sides of the major hysteresis loop, the relation between the remanent magnetization and the signal field is linear. The ac bias signal is of the order of five times the maximum signal frequency and the peak amplitude is approximately equal to the coercive force of the tape.

The highest frequency that can be recorded depends upon the tape speed and width of the gap in the record-playback head. The rms playback voltage, V, resulting from a sinusoidal recording of frequency ω can be shown to be

$$V = MN\omega \frac{\sin (\omega l/2v)}{\omega l/2v} \tag{10-35}$$

where M is the rms magnetic signal flux in the tape, N is the number of turns on the playback head coil, l is the gap in the head, and v is the tape velocity. According to Eq. (10-35), the playback voltage is small at low frequencies, rises to a maximum, and falls sharply to zero where

$$\frac{\omega_m l}{2v} = \pi \quad \text{or} \quad f_m = \frac{v}{l} \tag{10-36}$$

Since $f = v/\lambda$, Eq. (10-36) can also be written

$$\lambda_m = l \tag{10-37}$$

The meaning of Eq. (10-37) is that the minimum wavelength that can be played back is equal to the gap length. That is, the maximum frequency depends upon the tape speed, according to Eq. (10-36).

A very important application of magnetic recording stores digital signals in serial representation in tracks on a magnetic disk. The recording/playback head is driven over the surface of the disk much like that of a phonograph pickup. In this case, the ac bias signal is not needed since only the presence or absence of signal bits need be recorded. The capacity of the magnetic disk to store digital information is limited by the same considerations leading to Eq. (10-37), however.

Transmission Lines

Usually the components of an electronic system are physically separate from one another. For example, several distinct electronic instruments may be connected together in cascade to make a measurement, or the components of a digital system may exchange digital waveforms using either the parallel or serial representation. These signals must be transported from the output of one unit to the input of the next. This is done by connecting the various circuits with *transmission lines*. In the simplest case a transmission line can consist of a pair of wires, as has been implicitly assumed in previous chapters. A transmission line must faithfully transmit the signal between instruments with a minimum of waveform or amplitude distortion. If this is not the case, the ultimate output of the system may be erroneous.

In low-frequency and dc systems it is only necessary to consider the resistivity of the conductors in the transmission line, as well as, perhaps, shunt resistance between conductors. These effects are examined by straightforward circuit analysis as studied in Chap. 1. Even a straight piece of wire has associated with it a small inductance, however, and there is also a small capacitance between the two conductors. At high frequencies the reactances of these unavoidable inductances and capacitances become significant and influence signal propagation along the transmission line. This is particularly important in digital circuits because of the high frequencies associated with pulse waveforms. In many practical high-frequency and digital circuits transmission lines only a few centimeters long introduce objectionable waveform distortions if improperly used.

It is useful to represent the incremental impedances of a short section of transmission line by a series resistance r_1 and inductance l per unit length together with a shunt resistance r_2 and capacitance c per unit length, as in Fig. 10-21. Thus the series and shunt complex impedances of a small section Δx of the line are

$$\mathbf{z}_1 \, \Delta x = (r_1 + j\omega l) \, \Delta x$$

$$\frac{1}{\mathbf{z}_2} \Delta x = \left(\frac{1}{r_2} + j\omega c\right) \Delta x \tag{10-38}$$

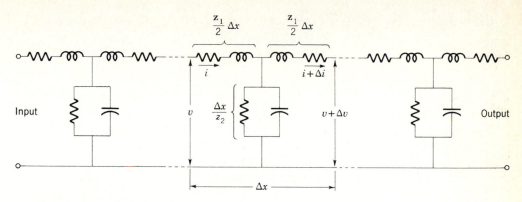

Figure 10-21 Representation of transmission line in terms of series and shunt impedances per incremental length.

Variations of voltage and current along the transmission line are produced in response to an input signal. Consider one section Δx of the line for which the input voltage is v and the output voltage is incrementally different, $v + \Delta v$. Similarly, the input and output currents of the line element are i and $i + \Delta i$ respectively. The voltage equation around the outside loop of this circuit is then

$$v - i\frac{\mathbf{z}_1}{2}\Delta x - (i + \Delta i)\frac{\mathbf{z}_1}{2}\Delta x = v + \Delta v \qquad (10\text{-}39)$$

Dividing through by Δx,

$$\frac{\Delta v}{\Delta x} = -\mathbf{z}_1 i - \frac{\mathbf{z}_1}{2}\Delta i \qquad (10\text{-}40)$$

In the limit as $\Delta x \to 0$, $\Delta v/\Delta x$ becomes the derivative dv/dx and the second term on the right side of Eq. (10-40) vanishes,

$$\frac{dv}{dx} = -\mathbf{z}_i i \qquad (10\text{-}41)$$

Similarly, the voltage equation including the shunt path is

$$v - i\frac{\mathbf{z}_1}{2}\Delta x - [i - (i + \Delta i)]\frac{\mathbf{z}_2}{\Delta x} = 0 \qquad (10\text{-}42)$$

$$v = i\frac{\mathbf{z}_1}{2}\Delta x + \frac{\Delta i}{\Delta x}\mathbf{z}_2 = 0 \qquad (10\text{-}43)$$

In the limit, $\Delta x \to 0$, Eq. (10-43) reduces to

$$\frac{di}{dx} = -\frac{1}{\mathbf{z}_2}v \qquad (10\text{-}44)$$

Differentiating Eq. (10-41) with respect to x and using Eq. (10-44), the result is

$$\frac{d^2v}{dx^2} = -\mathbf{z}_1 \frac{di}{dx}$$

$$\frac{d^2v}{dx^2} = \frac{\mathbf{z}_1}{\mathbf{z}_2} v \tag{10-45}$$

Similarly, differentiating Eq. (10-44) with respect to x and using Eq. (10-41) yields an equation in i alone,

$$\frac{d^2i}{dx^2} = \frac{\mathbf{z}_1}{\mathbf{z}_2} i \tag{10-46}$$

Equations (10-45) and (10-46) are called the transmission-line equations. Their solutions yield the current and voltage at any point along the line.

The general solution of Eq. (10-45) is

$$v = Ae^{-\gamma x} + Be^{\gamma x} \tag{10-47}$$

where A and B are constants that depend upon the input and output conditions of the transmission line, and γ is called the *propagation constant*. The propagation constant is a complex number with a real part α and an imaginary part β given by

$$\gamma = \alpha + j\beta = \sqrt{\frac{\mathbf{z}_1}{\mathbf{z}_2}} \tag{10-48}$$

as can be verified by substituting Eqs. (10-47) and (10-48) into the transmission-line equation, Eq. (10-45).

Since the differential equation for i, Eq. (10-46), is identical in form to the voltage equation, Eq. (10-45), the solution also has the same form. It is possible to evaluate the arbitrary constants in terms of A and B by using Eq. (10-44). The result is

$$i = \frac{A}{\mathbf{Z}_c} e^{-\gamma x} - \frac{B}{\mathbf{Z}_c} e^{\gamma x} \tag{10-49}$$

where

$$\mathbf{Z}_c = \sqrt{\mathbf{z}_1 \mathbf{z}_2} \tag{10-50}$$

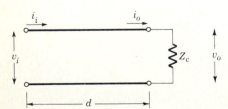

Figure 10-22 Transmission line terminated in its characteristic impedance.

is called the *characteristic impedance* of the line. Equations (10-47) and (10-49) give the variation of voltage and current with distance along the line.

Suppose a transmission line d meters long is terminated in its characteristic impedance, as in Fig. 10-22. Using Eqs. (10-47) and (10-49), the output voltage v_o can be written

$$v_o = i_o \mathbf{Z}_c$$

$$Ae^{-\gamma d} + Be^{\gamma d} = \mathbf{Z}_c \left(\frac{A}{\mathbf{Z}_c} e^{-\gamma d} - \frac{B}{\mathbf{Z}_c} e^{\gamma d} \right) \tag{10-51}$$

which means $B = -B = 0$. At the input end $x = 0$, so that $v_i = A$. Therefore, the voltage and current equations become

$$v = v_i e^{-\gamma x} \quad \text{and} \quad i = \frac{v_i}{\mathbf{Z}_c} e^{-\gamma x} \tag{10-52}$$

In particular, the input impedance of the line is

$$\mathbf{Z}_i = \frac{v_i}{i_i} = \frac{v_i}{v_i/\mathbf{Z}_c} = \mathbf{Z}_c \tag{10-53}$$

This means that the input impedance of a transmission line terminated in its characteristic impedance is simply equal to the characteristic impedance, independent of the length of the line.

A transmission line not terminated by its characteristic impedance may introduce spurious signals. Consider the case when the output end is short-circuited so that $v_o = 0$. Thus at $x = d$, where d is the length of the line, and taking $\alpha = 0$,

$$0 = Ae^{-j\beta d} + Be^{j\beta d} \tag{10-54}$$

Similarly, from Eq. (10-49),

$$i_o = \frac{A}{\mathbf{Z}_c} e^{-j\beta d} - \frac{B}{\mathbf{Z}_c} e^{j\beta d} \tag{10-55}$$

Solving (10-54) and (10-55) for A and B yields

$$A = \frac{i_o \mathbf{Z}_c}{2} e^{j\beta d} \quad \text{and} \quad B = -\frac{i_o \mathbf{Z}_c}{2} e^{-j\beta d} \tag{10-56}$$

Using Eq. (10-47) the voltage on the line is

$$v = \frac{i_o \mathbf{Z}_c}{2} e^{j\beta d} e^{-j\beta x} - \frac{i_o \mathbf{Z}_c}{2} e^{-j\beta d} e^{j\beta x}$$

$$v = \frac{i_o \mathbf{Z}_c}{2} (e^{-j\beta(x-d)} - e^{j\beta(x-d)}) \tag{10-57}$$

Comparing Eq. (10-57) with Eq. (10-52) we interpret the second term as a wave traveling from the receiving end to the input end. That is, the incident wave is *reflected* at the output of the line because of the improper termination at the receiving end. Reflections are avoided in digital applications particularly since pulses that are reflected back and forth between the input and output ends of the line completely obscure the true signal. This is why transmission lines are almost always terminated with the characteristic impedance.

NOISE SIGNALS

Any spurious currents or voltages extraneous to a signal of interest are termed *noise* since they interfere with the signal. Noise voltages arise in the basic operation of electronic devices or are the result of improper circuit design and use. It is important to minimize noise effects in order to characterize signals with the greatest possible precision and to permit the weakest signals to be amplified. A convenient measure of the influence of noise on any signal is the *signal-to-noise* ratio, the ratio of the signal power to the noise power at any point in a circuit.

Thermal Noise

When a number of amplifier stages are cascaded a random-noise voltage appears at the output terminals, even in the absence of an input signal. This output voltage is caused by random voltages generated in the input circuit. The noise voltages that appear across the terminals of any resistor are attributed to the chaotic motion of the free electrons in the material of the resistance. Since electrons in a conductor are free to roam about by virtue of their thermal energy, at any given instant more electrons may be directed toward one terminal of the resistor than toward the other. The result is a small instantaneous potential difference between the terminals. The magnitude of the potential difference fluctuates rapidly as the number of electrons moving in a given direction changes from instant to instant.

   Since the noise voltage across a resistor fluctuates randomly, it has Fourier components covering a wide range of frequencies. It is convenient, therefore, to specify the noise voltage in terms of the mean square noise voltage per unit cycle of bandwidth, V^2/Hz. For a resistor R this quantity is

$$\langle \Delta v^2 \rangle = 4kTR \tag{10-58}$$

where k is Boltzmann's constant and T is the absolute temperature. The noise voltage given by Eq. (10-58) is variously called *Nyquist noise*, after the physicist who derived this equation, or *thermal noise*, since its origin is a result of the thermal agitation of free electrons, or *Johnson noise*, after the engineer who first measured it.

   The meaning of Eq. (10-58) is as follows. A noise voltage appears between the terminals of any resistance. The magnitude of the noise voltage actually measured with any instrument depends upon the frequency response of the instrument. For example, the rms noise voltage of a 1000-Ω resistor at room temperature as measured by a voltmeter with a bandwidth of 10,000 Hz is, using Eq. (10-58),

$$v = (1.65 \times 10^{-20} \times 10^3 \times 10^4)^{1/2} = 4.1 \times 10^{-7} = 0.41 \ \mu V$$

This rather small voltage is not inconsequential. The output voltage of an amplifier with a 10-kHz bandpass and a gain of 10^6 is nearly 0.5 V if the input resistor is 1000 Ω. This output voltage is present even when no input signal is applied.

The noise voltage of resistances is independent of frequency. Accordingly, thermal noise is called "white" noise by analogy with the uniform spectral distribution of white light energy. The presence of thermal noise in any circuit is accounted for by including a noise generator given by Eq. (10-58) in series with a noiseless resistor. In practice it is usually necessary to consider the thermal noise of only those resistors in the input circuit of an amplifier. The gain of the first stage makes the amplified noise of the input resistor larger than the noise of resistors in succeeding stages.

Thermal noise is a fundamental and unavoidable property of any resistance. An amplifier should have a bandwidth only as wide as is necessary to amplify adequately all signal components in order to minimize the ever-present thermal noise voltages. If it is desired to amplify a single-frequency signal, for example, the frequency response of the amplifier should be sharply peaked at that frequency. The total noise voltage at the output is therefore reduced since only noise components having frequencies in the amplifier passband are amplified. The signal-to-noise ratio is enhanced and weak signals can be amplified usefully.

Current Noise

Noise voltages in excess of thermal noise are observed experimentally in the presence of currents. So-called *shot noise* is a result of the random emission of electrons across a *pn* junction. Since each electron represents a small increment of current, the current fluctuates slightly about the dc value. This effect is analogous to the noise of raindrops on a tin roof. That is, the basic reason for shot noise is that the electron is a discrete unit of electric charge.

Current fluctuations due to shot noise are given in A^2/Hz,

$$\langle \Delta i^2 \rangle = 2eI \tag{10-59}$$

The quantity $\langle \Delta i^2 \rangle$ is analogous to $\langle \Delta v^2 \rangle$ in the thermal noise expression, Eq. (10-58), except that the noise is expressed here in terms of current fluctuations. Note that shot noise is a white noise since the right side of Eq. (10-59) is independent of frequency.

Another noise phenomenon is observed in semiconductors. Electrons are promoted randomly from the valence band to the conduction band and also return randomly to the valence band, keeping the proper average number of carriers in each band. Random generation and recombination of carriers is caused by thermal

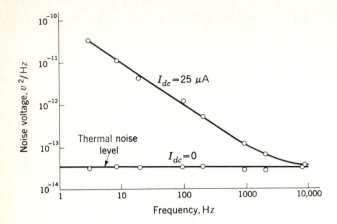

Figure 10-23 Experimental noise voltage of 2.2-MΩ composition resistor. Spectrum is $1/f$ when dc current is present and white Nyquist noise in absence of current.

energies and produces a conductivity fluctuation of the semiconductor. This generates a noise voltage when a direct current is present; this type of current noise in semiconductors is termed *generation-recombination*, or *g-r noise*.

Large low-frequency noise signals are observed in certain resistances when a direct current is present. Although the physical origins of this additional noise are not clear, many experiments have shown that the noise is largest at low frequencies and that it increases with the square of the dc voltage. An empirical expression for this effect is

$$\langle \Delta v^2 \rangle = \frac{\alpha}{N} \frac{V^2}{f} \tag{10-60}$$

where α is an empirical constant, V is the applied voltage, N is the number of current carriers, and f is the frequency. According to Eq. (10-60) the mean square noise voltage per unit bandwidth depends inversely upon frequency and the phenomenon is therefore called $1/f$ *noise*.

The magnitude of $1/f$ noise varies markedly with the material of the conductor and its physical form. It is negligible in bulk metals because N is so large, so that only thermal noise is observed in wire-wound resistors. Composition resistors, on the other hand, generate a large $1/f$ noise level, Fig. 10-23, which is associated with the intergranular contacts in such resistors. Although such contacts are known to be important, $1/f$ noise is also observed in single-crystal semiconductors where contact effects are negligible.

To minimize the low-frequency noise level, resistor types in which the $1/f$ noise is small, such as wire-wound units, are selected. Fortunately, at high frequencies where wire-wound resistors are unsuitable because of their inductance, the $1/f$ noise of composition resistors is usually negligible compared with thermal noise. In other situations the direct current in noisy components is minimized to reduce the current noise generated. The total current noise voltage in any given

circuit is found by integrating Eq. (10-60) over the frequency-response characteristic of the amplifier.

Amplifier Noise

The internal noise of bipolar transistors arises from Nyquist noise in the base
resistance, shot noise and $1/f$ noise in the base current, and shot noise in the
collector current. Since these various noise sources are independent, the total noise
of a transistor amplifier is simply their sum and both white and $1/f$ trends are
expected. Furthermore, since the noise sources vary with collector current in different ways, it proves possible to select the operating point for minimum noise.

Noise in FET amplifiers is dominated by thermal noise in the channel which
is given approximately by

$$\langle \Delta v^2 \rangle = \frac{4kT}{g_m} \tag{10-61}$$

where g_m is the transconductance. Shot noise and $1/f$ noise are associated with
gate leakage current, although these effects are small because of the very small
current of the reverse-biased gate pn junction. There is also some $1/f$ channel noise
which produces an increase in noise at low frequencies.

It proves possible to represent the internal noise of transistor amplifiers by
an equivalent circuit containing a noise voltage generator $\langle \Delta v_n^2 \rangle$ and a noise current generator $\langle \Delta i_n^2 \rangle$, Fig. 10-24. The noise generators account for the various
noise sources of the transistor and therefore are dependent on both frequency and
the transistor operating point. It is common practice to refer noise signals to the
amplifier input circuit and to treat the amplifier in the equivalent circuit as noiseless.
As noted previously, the gain of the input stage makes the input noise exceed
noise sources in subsequent stages.

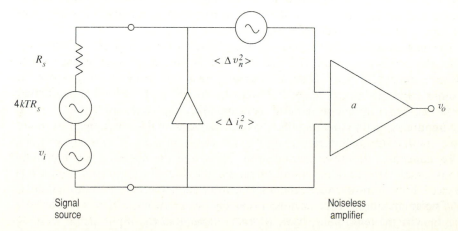

Figure 10-24 Noise equivalent input circuit of transistor amplifier.

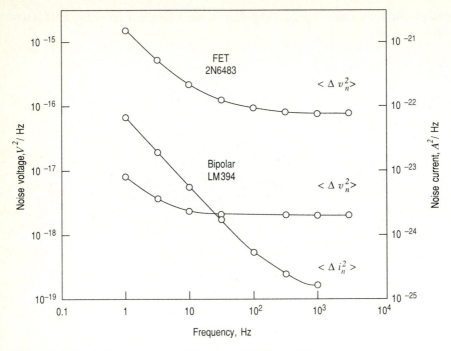

Figure 10-25 Equivalent input noise of bipolar and FET amplifiers.

Typical equivalent input noise generators for bipolar and FET amplifiers are shown in Fig. 10-25. Note that the noise spectra have both a white noise and a $1/f$ noise trend as expected from the basic noise processes. The equivalent input current noise of the FET amplifier is negligible because of the very small gate leakage current and, consequently, noise from this source may be ignored. According to Fig. 10-25 the internal noise of the bipolar transistor appears to be much lower than that of the FET, and this is true for low values of source resistance. As the noise equivalent input circuit, Fig. 10-24, demonstrates, however, the effect of the current noise generator in the source resistance R_s is to produce a noise voltage equal to $\langle \Delta i_n^2 \rangle R_s^2$. Thus, for example, at a frequency of 10 Hz a source resistance of only about $(2 \times 10^{-16}\ V^2/\mathrm{Hz}/6 \times 10^{-24}\ A^2/\mathrm{Hz})^{1/2} = 6000\ \Omega$ makes the internal noise of the bipolar transistor equal to that of the FET.

The input signal to any amplifier has associated with it a given signal-to-noise ratio since thermal noise corresponding to the source resistance is present, at least. An ideal amplifier amplifies the incoming signal and the incoming noise equally and introduces no additional noise. Therefore, the original signal-to-noise ratio is preserved at the output. Practical amplifiers are not ideal because of thermal noise of the input circuit and current noise in the first stage. A useful figure of merit for any circuit is the *noise figure, NF*, which is defined as the input signal-to-noise power ratio divided by the output signal-to-noise power ratio. An ideal amplifier has a noise figure of unity, and many practical circuits approach this value fairly closely.

Referring to Fig. 10-24, the input signal-to-noise power ratio for unity bandwidth is given by

$$(S/N)_i = \frac{v_i^2/R_s^2}{4kTR_s/R_s} = \frac{v_i^2}{4kTR_s} \qquad (10\text{-}62)$$

and the output signal-to-noise power ratio is

$$(S/N)_o = \frac{a^2 v_i^2/R_o}{a^2(4kTR_s + \langle \Delta v_n^2 \rangle + \langle \Delta i_n^2 \rangle R_s^2)/R_o} \qquad (10\text{-}63)$$

where a is the amplifier gain and R_o is the output resistance. Thus the amplifier noise figure is, from Eq. (10-62) and Eq. (10-63),

$$\text{NF} = \frac{(S/N)_i}{(S/N)_o} = \frac{4kTR_s + \langle \Delta v_n^2 \rangle + \langle \Delta i_n^2 \rangle R_s^2}{4kTR_s}$$

$$= 1 + \frac{\langle \Delta v_n^2 \rangle + \langle \Delta i_n^2 \rangle R_s^2}{4kTR_s} \qquad (10\text{-}64)$$

According to Eq. (10-64), small values for the noise generators produce a unity noise figure.

The optimum source resistance which yields the minimum noise figure can be determined by setting the derivative of Eq. (10-64) equal to zero,

$$\frac{d\text{NF}}{dR_s} = -\frac{\langle \Delta v_n^2 \rangle + \langle \Delta i_n^2 \rangle R_s^2}{4kTR_s^2} + \frac{2\langle \Delta i_n^2 \rangle R_s}{4kTR_s} = 0 \qquad (10\text{-}65)$$

$$\langle \Delta v_n^2 \rangle + \langle \Delta i_n^2 \rangle R_s^2 = 2\langle \Delta i_n^2 \rangle R_s^2$$

$$R_s^2 = \frac{\langle \Delta v_n^2 \rangle}{\langle \Delta i_n^2 \rangle} \qquad (10\text{-}66)$$

For this value of source resistance, the minimum noise figure is found by introducing Eq. (10-66) into Eq. (10-64). The result is

$$(\text{NF})_m = 1 + \frac{\langle \Delta v_n^2 \rangle + \langle \Delta i_n^2 \rangle (\langle \Delta v_n^2 \rangle / \langle \Delta i_n^2 \rangle)}{4kT(\langle \Delta v_n^2 \rangle / \langle \Delta i_n^2 \rangle)^{1/2}}$$

$$= 1 + \frac{(\langle \Delta v_n^2 \rangle \langle \Delta i_n^2 \rangle)^{1/2}}{2kT} \qquad (10\text{-}67)$$

The minimum noise figure is thus dependent upon the equivalent input noise power of the amplifier compared with thermal energy.

Shielding and Grounding

Noise is often introduced in practical circuits by extraneous voltage signals coupled into the circuit from the surroundings. The most common source of this noise comes from the 60-Hz electric and magnetic fields produced by power

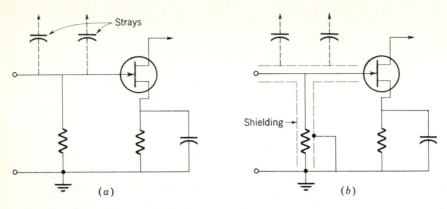

Figure 10-26 (*a*) Stray capacitive coupling introduces noise pickup signals into sensitive circuits. (*b*) Grounded conductor shields circuit from surroundings.

mains. The 60-Hz signal induced by these fields is called *hum* because it is audible as a low-frequency tone in amplifiers connected to a loudspeaker. Other *stray pickup* may result from electric fields generated by nearby electronic equipment, electric motors, lightning discharges, etc.

It is useful to *shield* those portions of a circuit where the signal level is small and, consequently, where noise voltages are most troublesome. Electric fields induce noise voltages capacitively, so it is only necessary to surround the circuit with a grounded conducting shield in order to reduce stray pickup. This is illustrated schematically in Fig. 10-26, where the capacitive coupling to external sources in Fig. 10-26*a* is interrupted by interposing a grounded conductor, Fig. 10-26*b*. Such shielding is also effective in reducing so-called *crosstalk* between different stages of the same circuit as, for example, between the input stage and the power output stage of a complete amplifier.

Additionally, it is useful to shield a circuit to minimize induced currents resulting from stray magnetic fields. This is accomplished with high-permeability ferromagnetic enclosures which reduce the intensity of the magnetic field inside. Such shielding is never complete because of the properties of ferromagnetic materials, and it is always advantageous to minimize the area of the circuit by using the shortest possible signal leads, since the induced voltage in any circuit resulting from changing magnetic fields decreases if the enclosed area of the circuit is reduced. Transformers are particularly troublesome with respect to inductive pickup because of their many turns of wire. They are kept well removed from all power transformers because of the strong magnetic fields generated by such units.

In general it is good practice to keep all circuits physically small in order to minimize stray pickup, crosstalk, and stray capacitance. All grounded components, such as bypass capacitors, pertaining to a given stage are returned to a single point. This reduces so-called *ground loops*, which are current paths through the metal chassis on which electronic circuits are often mounted. If all components of one stage are not grounded at the same point, the currents may cause undesirable signal coupling between stages.

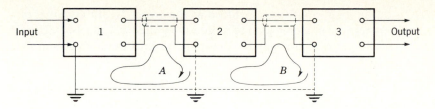

Figure 10-27 When several electronic devices are connected, system must be grounded only at one point, preferably at input. Multiple grounds can lead to large ground-loop currents.

When a number of individual electronic units are interconnected, shielded cable is used for all signal leads between units. The shield is used as the ground lead, as illustrated in Fig. 10-27. It is important that the entire system be grounded at only one point, usually the input terminal. If each unit is grounded separately, as shown by dashed lines in Fig. 10-27, large 60-Hz currents can be induced in the circuit because of the large areas encompassed. The current induced in loop A by stray 60-Hz magnetic fields introduces a large stray pickup signal into the amplifier because of accompanying voltage drops across conductor resistances.

Digital circuits introduce special shielding and grounding considerations because of the rapid changes in signal amplitude involved. In parallel transmission of digital signals, for example, it is common practice to use a flat ribbon cable employing one conductor for each bit. In order to eliminate crosstalk between conductors, intervening conductors are grounded and often a grounded flexible braid is included, Fig. 10-28. This configuration provides nearly as good isolation as separate coaxial cables for each bit. Note also that each bit line should be terminated in its characteristic impedance to reduce signal reflections.

Stray coupling through the common power supply of a complex digital circuit is particularly troublesome because of the large peak current demand accompanying change of logic levels. When a digital gate switches from state 0 to state 1, for example, all stray capacitances are charged very rapidly to the voltage level corresponding to state 1. This requires a large peak current from the power supply, causing voltage drops in the wiring resistances which can result in false pulse signals in other parts of the circuit. To circumvent this difficulty, each integrated circuit is provided with a decoupling capacitor of about 0.1 μF connected between the power-supply pin and ground. The energy stored in the decoupling capacitor in effect supplies the power needed to effectuate the change of state.

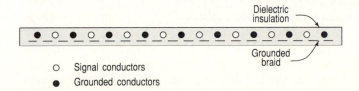

o Signal conductors
● Grounded conductors

Figure 10-28 Cross section of 8-bit ribbon cable.

Similarly, a decoupling capacitor is included at the power-supply terminal of each printed circuit board commonly used to mount integrated circuits. In extreme cases, a separate voltage regulator may be incorporated into the voltage-supply circuit on each board.

DIGITAL INSTRUMENTS

Time-Interval Meter

The time interval between two events can be accurately measured by counting the number of cycles of a stable oscillator during the time interval between pulses signaling the two events. This is accomplished, Fig. 10-29, by triggering a simple binary (gates 1 and 2) by the input pulses to gate an astable multivibrator (gates 3 and 4) on and off. The number of oscillator cycles between start and stop pulses is counted by cascaded BCD counter-decoders which drive seven-segment dis-

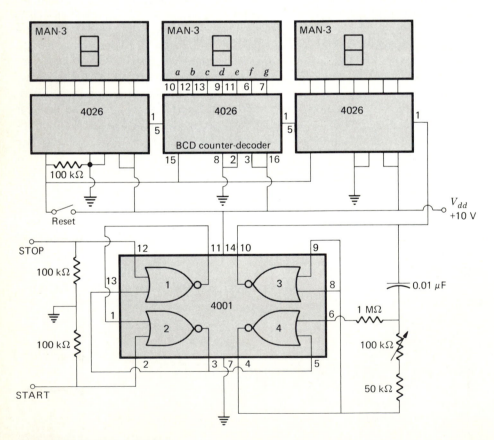

Figure 10-29 Digital interval timer.

plays. The 100-kΩ potentiometer is used to adjust the oscillator frequency to 1000 Hz so that the display shows the time interval between input pulses directly in milliseconds.

A positive pulse at the STOP input resets the flip-flop so that the output of NOR gate 2 is at logic 1, which means that the output of NOR gate 4 is constrained to remain at logic zero and the multivibrator is inactive. A positive pulse at the START terminal puts the output of NOR gate 2 at logic 0 and the multivibrator oscillates until the next stop pulse. The reset switch clears the counters for the next measurement. Greater timing accuracy, particularly for short time intervals, can be achieved by replacing the astable multivibrator with a higher frequency, crystal-controlled oscillator.

Frequency Meter

A similar technique is used to measure the frequency of an unknown oscillator. The number of cycles in the unknown signal is counted during an accurately known time interval. For this purpose the input signal is first converted into a series of pulses, one per cycle, by means of a Schmitt trigger followed by a differentiator and clipper, as shown in Fig. 10-30. The number of input pulses is counted during a given time interval. If, for example, the interval is 1 s, the frequency of the unknown is displayed directly in hertz.

It is convenient to derive the time interval by scaling down the output of a stable crystal oscillator. The output pulse from the last flip-flop of the scaler is used directly to control the gate. Note that the frequency range of the instrument may be changed easily by changing the scale factor of the oscillator counter. In

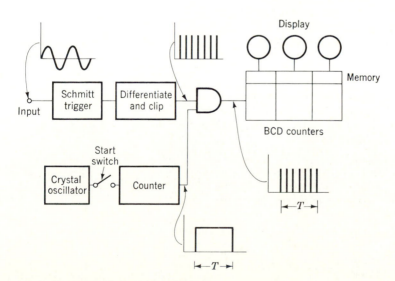

Figure 10-30 Digital frequency meter.

effect, the instrument compares the unknown frequency of the input signal with the known frequency of the crystal oscillator.

The frequency meter is started by closing the start switch and subsequently opening it after one counting interval. In practical instruments, control circuits are included to perform this operation, as well as to transfer the count to memory registers, and to clear the frequency-count register. In this mode the instrument repetitively samples the input frequency and can, for example, detect and display slow frequency changes of the input signal. Control circuits employ the same logic gates and networks as used in the signal-handling portions of the instrument.

Digital Voltmeter

A basic feature of digital instruments is conversion of input signals into digital data. The instruments discussed in the previous sections employ quite simple pulse circuits to accomplish this *analog-to-digital* or *A-D*, conversion. A number of different techniques have been devised to convert other analog signals into digital signals. A typical example of interest is used in an early type of *digital voltmeter*, an instrument that provides a direct-reading numerical display of the voltage being measured.

Consider the simplified block diagram of a digital voltmeter sketched in Fig. 10-31. The unknown dc voltage applied to the input terminals is compared with a staircase waveform produced by a diode pump circuit fed from a 10-kHz oscillator. When the staircase signal is equal to the input voltage, the comparator interrupts transmission of further oscillator pulses. The parameters of the diode pump are selected so that the height of each step of the staircase is equal to 1 mV. Therefore, the number of steps is directly equal to the unknown voltage in millivolts, and the steps are counted and displayed by a three-decade counter, memory register, and decimal display.

The control circuits subsequently execute a memory transfer for display and discharge the diode pump capacitor so that the cycle repeats. A new reading is

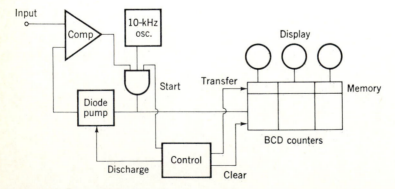

Figure 10-31 Simplified block diagram of digital voltmeter. (*Princeton Applied Research Corp.*)

attained in less than 1 s. Actually, in practical instruments the control circuit has other functions as well. It is possible, for example, to change input-multiplier resistors automatically so that voltages from 100 mV to 999 V can be measured with the same 3-digit precision. In addition, the circuit can automatically adjust to either input-voltage polarity.

A-D AND D-A CONVERSION

Conversion of analog signals into their digital counterparts is characterized by the speed and accuracy with which the equivalences can be established. The reverse process, *digital-to-analog*, or *D-A*, conversion is of equal interest, and this generally proves easier to accomplish than the A-D step. Both A-D and D-A conversion can be achieved in a number of different ways which vary in complexity and effectiveness.

The simple staircase-waveform A-D conversion discussed in the previous section is limited in accuracy by the precision with which each step can be generated. In addition, the possibility of erroneous readings is present if the input is not strictly a dc signal. Suppose, for example, the input signal contains noise pulses or a 60-Hz component from the power mains. Then it is possible for the staircase waveform to become equal to the input signal plus a noise voltage and produce a false reading. This leads to annoying fluctuations in the numerical display on successive measurement cycles.

Dual-Slope Integration

The above difficulty is overcome in the *dual-slope integrator*, Fig. 10-32, which is a simple and popular A-D converter. In this approach the input signal is integrated for a fixed time interval determined by the time required for the counter to fill up completely. At the last count the overflow pulse causes the control logic to switch the integrator input to a fixed negative reference voltage. Integration of

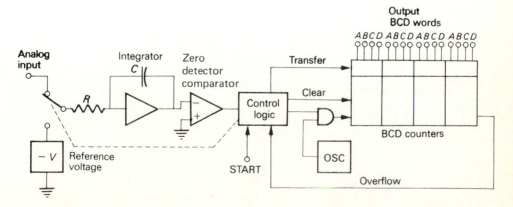

Figure 10-32 Dual-slope integrator.

the reference voltage causes the output voltage of the integrator to decrease linearly towards zero. At this point the zero-detector comparator activates control logic to transfer the count in the BCD counters to the output registers and to initiate the next measurement cycle.

In effect the system compares the charge placed on the integrating capacitor by the unknown input signal during a fixed time interval with the charge removed by the known reference voltage in the measured time interval. Since the same operational-amplifier integrator and oscillator is used in both measurements, circuit errors are minimized and accuracy is limited primarily by the stability of the reference voltage.

Since the input signal is integrated during the initial input period, random-noise signals and alternating components are averaged out. The dual-slope integrator is capable of fifteen 5-digit readings per second compared to only two 3-digit readings per second in the case of the staircase A-D converter. The dual-slope technique is used in digital voltmeters as well as in other A-D conversion applications.

Successive-Approximation Converter

Very rapid operation is achieved by the *successive-approximation* converter, Fig. 10-33. In this technique a sequence of voltages having a 8421 BCD weighted code are successively compared with the input signal. Each successive voltage level is stored if it is lower than the input signal or rejected if it is greater than the input signal. After the most significant digit in the input signal is thus approximated, a second series of voltages in the 8421 pattern, but a factor of 10 smaller, is added to the stored value and the successive sums compared to the input signal. The process is repeated digit by digit until the desired precision is attained.

The successive-approximation process is indicated schematically in Fig. 10-34 for a specific case of converting a 9.55-V input signal into its binary equivalent. At START the control logic, Fig. 10-33, generates the binary word 1000 and

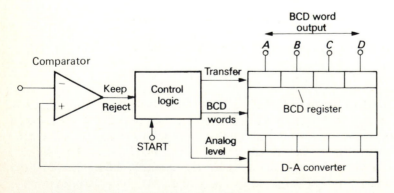

Figure 10-33 Simplified block diagram of successive-approximation A-D converter.

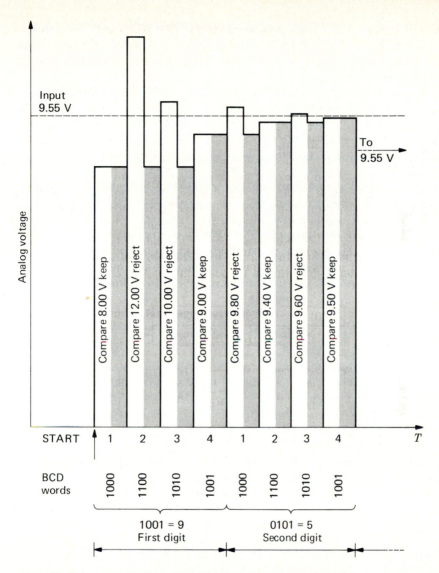

Figure 10-34 Successive approximations convert 9.55 V into BCD equivalent.

places it into the BCD register. The D-A converter, which may be similar to those discussed in the next section, converts this word into its analog signal, 8.00 V. This value is compared with the input signal and is stored since it is less than the input signal. Next, the control generates the binary word 1100, which has the analog equivalent $8.00 + 4.00 = 12.00$ V. Since this value exceeds the input signal, the second approximation value is rejected. The third approximation is $1000 + 0010 = 1010 = 8.00 + 2.00 = 10.00$ V and this value, being larger than the input, is also rejected. The final approximation, $1000 + 0001 = 8.00 + 1.00 = 9.00$ V is less than the input and is kept. The result of this four-step process is the first approximation to the input signal, that is, the first digit, which is 1001, or 9.

This process is repeated for the second digit, as illustrated in Fig. 10-34, except that the analog equivalents of the binary trial words are reduced to 0.80, 0.40, 0.20, and 0.10 V by the control logic. The result, $0101 = 5$, is the second digit. This is now repeated for the next digit, etc., until the desired precision is attained. In the specific case considered, the BCD equivalent of 9.55 V is 1001.0101 0101.

The speed of the successive-approximation converter is limited only by propagation delays in the digital logic circuits and a simple converter is capable of as many as one thousand 4-digit readings per second. Furthermore, the conversion time is independent of the magnitude of the signal, in contrast to the staircase-waveform and dual-slope integration approaches. The successive-approximation converter is nonintegrating, however, so that the input signal is often introduced through a low-pass filter to reduce spurious noise signals. Practical 8-bit A-D converters exceed 10^7 conversions per second.

Flash Converter

The ultimate in A-D conversion speed is provided by the *flash converter* in which conversion is accomplished in a single step by processing all bits in parallel. A string of comparators is connected to the input signal and to equally spaced taps on a voltage divider. The output signal of each comparator indicates whether or not the input signal is greater or less than the voltage established by its voltage-divider tap and a logic circuit generates the digital output signal. Conversion time is limited only by signal propagation through the comparators and logic gates and is correspondingly rapid.

The circuit of a practical 4-bit flash converter is shown in Fig. 10-35. One terminal of each comparator is connected to the input terminal and the other to a tap on the voltage divider. The comparators are arranged in pairs to produce a so-called *window comparator*. Consider, for example, comparators 1 and 3 which are connected to corresponding taps on the voltage divider. If the input signal level is less than the voltage at tap 1, the output of comparator 1 is low while that of comparator 3 is high. The internal circuitry of the type 711 comparator has the two outputs sharing a common-collector load resistor, so that the combination yields a state 0 output when either output is low. When the input signal exceeds the voltage at tap 3, comparator 1 is high while comparator 3 is low and the output is again state 0. Intermediate signal levels between voltages at taps 1 and 3 make both comparator outputs high so that the combined output is state 1. That is, the comparator combination signals when the input voltage is in the "window" between the voltages corresponding to taps 1 and 3. The other pairs of comparators in Fig. 10-35 operate similarly at their respective voltage levels.

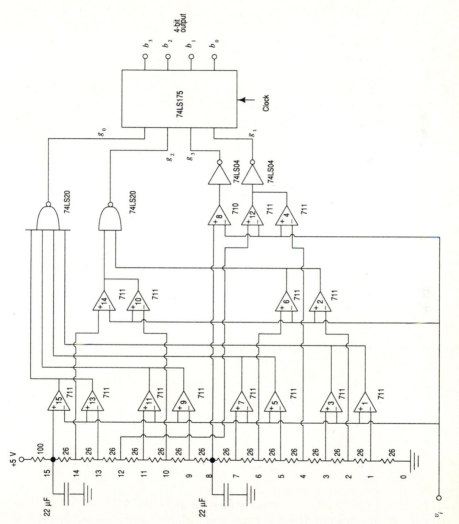

Figure 10-35 Practical 4-bit flash converter.

359

Table 10-1 Logic of 4-bit flash converter

Signal level	Logic levels			
	g_0	g_1	g_2	g_3
$0 < v_i < 1$	1	1	1	1
$1 < v_i < 2$	0	1	1	1
$2 < v_i < 3$	0	0	1	1
$3 < v_i < 4$	1	0	1	1
$4 < v_i < 5$	1	0	0	1
$5 < v_i < 6$	0	0	0	1
$6 < v_i < 7$	0	1	0	1
$7 < v_i < 8$	1	1	0	1
$8 < v_i < 9$	1	1	0	0
$9 < v_i < 10$	0	1	0	0
$10 < v_i < 11$	0	0	0	0
$11 < v_i < 12$	1	0	0	0
$12 < v_i < 13$	1	0	1	0
$13 < v_i < 14$	0	0	1	0
$14 < v_i < 15$	0	1	1	0
$15 < v_i$	1	1	1	0

All comparator outputs are coupled together through two NAND gates and two inverters to yield four signals, g_0, g_1, g_2, and g_3. Table 10-1 displays these signals corresponding to the full range of input signal levels. This encoding is a form of the so-called *gray code* which has the virtue that only one bit changes in going from one state to the next, helping to reduce errors. Finally, the 74LS175 chip decodes the gray code input into a conventional 4-bit digital signal (Exercise 10-15) in response to clock pulses.

An *n*-bit flash converter requires $2^n - 1$ comparators and a chain of highly accurate and stable resistors, which obviously involves considerable circuitry. Single-chip 8-bit devices (255 comparators) capable of operating at 250 MHz sampling rates are commercially available.

D-A Ladder Networks

It often proves useful to convert binary output signals into their analog equivalents so that, for example, in digital control applications the output signal may be used to adjust the system being controlled. Another instance is the D-A conversion used in the successive-approximation converter discussed in the previous section. Most often, D-A converters use resistor arrays called *ladder networks*.

Consider the *weighted-resistor ladder* network and operational amplifier in Fig. 10-36. As discussed in Chap. 7, the output of the operational amplifier is just the sum of the input signals with each input multiplied by the ratio of the

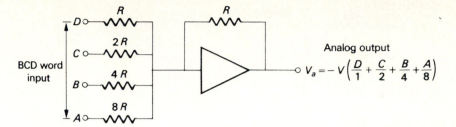

Figure 10-36 Weighted-resistor 4-bit D-A converter.

feedback resistance to its input resistor. That is, the analog output signal is

$$V_a = -V\left(\frac{D}{1} + \frac{C}{2} + \frac{B}{4} + \frac{A}{8}\right) \tag{10-68}$$

where V is the magnitude of the logic 1 state for any of the $ABCD$ input bits. Note that this network is weighted in the usual 8421 BCD code.

Table 10-2 lists the analog output signal calculated from Eq. (10-68) for the BCD words corresponding to decimal digits. Clearly the output signal voltage is analogous to the decimal values.

For good conversion accuracy, precision resistors must be used in the ladder network. Also, the resistance values must cover a wide range, particularly if binary words longer than 4 bits are converted. Both of these difficulties are overcome by the so-called *binary ladder network*, Fig. 10-37. In this circuit only two values of resistance are required, independent of the number of bits in the

Table 10-2 Analog output from weighted-resistor D-A converter*

Decimal	BCD word DCBA	V_a, V
0	0000	0
1	0001	-0.5
2	0010	-1.0
3	0011	-1.5
4	0100	-2.0
5	0101	-2.5
6	0110	-3.0
7	0111	-3.5
8	1000	-4.0
9	1001	-4.5

* Assuming logic 1 state is $+4.0$ V.

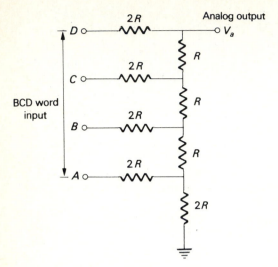

Figure 10-37 Binary-ladder 4-bit D-A converter.

input binary words. Note also, that an operational amplifier is not necessary to sum the individual bit input signals.

The analog output signal corresponding to BCD word inputs can be determined from Fig. 10-37 by straightforward circuit analysis (Exercise 10-14). It is assumed that all logic 1 bits are at the same potential and that all logic 0 bits are at ground. The result is that the analog output voltages are just one-third of those listed in Table 10-2 for the weighted-resistor converter.

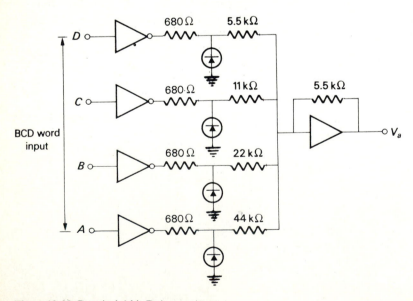

Figure 10-38 Practical 4-bit D-A converter.

In either network the accuracy of the analog output depends upon the voltage magnitude of each bit in the binary input signal. Since this quantity is not often closely controlled in binary logic circuits, a practical D-A converter, Fig. 10-38, may apply diode limiting to each bit in order to set the voltage levels accurately. This has the further advantage that the total input signal into the summing amplifier is less likely to saturate the amplifier. With diode limiting, the voltage level of each logic 1 bit is 0.6 V, rather than 5.0 V or so as in the case of normal positive logic signals. Note also that a circuit similar to Fig. 10-38 is required for the BCD input word corresponding to each decimal digit. It often proves possible to minimize circuit complexity by multiplexing the D-A converter in a fashion similar to that discussed in the previous chapter connection with ROM display decoders.

DIGITAL PROCESSORS

Digital logic circuits permit signal processing of an entirely different nature than is possible using analog instruments. A number of digital instruments which are, in effect, small digital computers have been designed to carry out highly specialized calculations. Although the input and output signals used with such instruments are most often analog in nature, the power and flexibility of digital logic circuits result in extremely versatile and flexible performance.

Digital Filter

The so-called *digital filter* performs much the same task as, for example, an inductance-capacitance filter network, but in a fashion that provides filter parameters not possible with conventional component values. The operation of a digital filter rests upon the same duality between frequency and time already noted in connection with the Fourier series representation of complex waveforms. That is, the spectrum of frequency components in a signal is an equally valid description of the signal as is the time variation of the signal given by the waveform. Correspondingly, while most often a filter network is described by means of its frequency-response characteristic, an equally useful description is possible using the transient properties of the circuit. Note that the frequency-response characteristic is given by the ratio of the output signal, $v_o(f)$, to the input signal, $v_i(f)$ at a given frequency,

$$\frac{v_o(f)}{v_i(f)} = F(f) \tag{10-69}$$

A similar input-output relation exists in the time domain, except that the input and output signals are noted at separate times

$$\frac{v_o(t)}{v_i(t')} = T(t, t') \tag{10-70}$$

Both $F(f)$ and $T(t, t')$ are satisfactory descriptions of filter network properties.

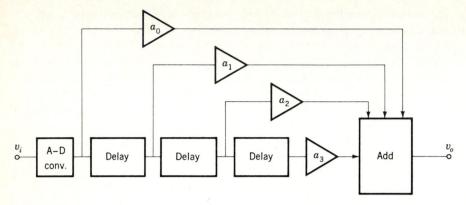

Figure 10-39 Block diagram of digital filter.

It is possible to develop an expression for $T(t, t')$ if the frequency-response characteristic is known in much the same fashion as the Fourier components of a complex waveform are found. The details of this transformation are not of direct interest here. It is important to observe, however, that the timelike representation permits a direct calculation of filter performance by digital techniques in the following way. The output of the digital filter at any instant is written in terms of the input signal at several discrete times earlier,

$$v_o(t) = a_1 v_i(t - \delta) + a_2 v_i(t - 2\delta) + a_3 v_i(t - 3\delta) + \cdots \qquad (10\text{-}71)$$

Here, the a coefficients are the representation of the $T(t, t')$ characteristic of the filter. The calculations indicated by Eq. (10-71) are accomplished by the circuit shown in Fig. 10-39. First, the input signal is sampled at discrete intervals and digitized. Each sample is delayed an appropriate time interval, multiplied by a constant characteristic of the filter performance desired, and the result is summed

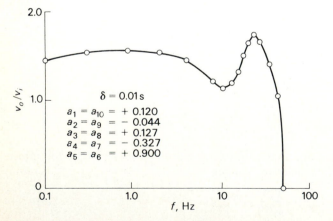

Figure 10-40 Frequency-response characteristic of digital filter.

to yield the output signal. The output may be converted to analog form, if desirable.

The digital filter, in effect, continuously calculates the filter output arising from input signals. Since filter performance is described by the *a* coefficients which are easily adjustable, the same digital filter may be used to synthesize a variety of frequency-response characteristics. Furthermore, because of the discreteness of digital data, precise performance, such as very large attenuation at specific frequencies, is achievable.

The frequency-response characteristic of a simple low-pass digital filter, Fig. 10-40, shows the very steep attenuation and the excellent low-frequency properties that can be achieved. Note that the *a* coefficients in Fig. 10-40 refer to Eq. (10-71) and that the sampling rate is $1/\delta = 100$ Hz. A simple change in the coefficients, which is accomplished by altering amplifier gains, produces an entirely different frequency-response characteristic with the same basic circuit. This aspect is explored further in Exercises 10-16 and 10-17.

Signal Correlators

Recurrent signals heavily masked by noise can often be recovered by an averaging technique which takes advantage of the basic difference between recurrent or periodic signals and random-noise signals. This is that the average value of a random-noise signal is equal to zero. Instruments called *signal averagers* in effect average out the random noise so that only the signal waveform of interest remains.

According to the block diagram of a signal averager, Fig. 10-41, the input signal plus noise is sampled successively by each of the 100 FET gates which are activated in sequence by a ring counter. The voltage level of each sample is

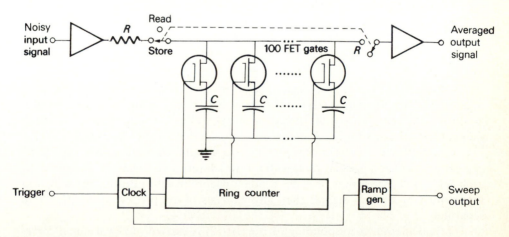

Figure 10-41 Simplified block diagram of signal averager. (*Princeton Applied Research Corp.*)

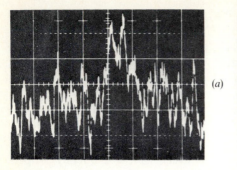 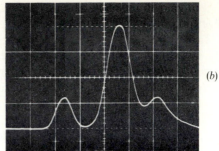

(a)   (b)

Figure 10-42 (*a*) Signal masked by noise and (*b*) waveform recovered by signal averager. (*Princeton Applied Research Corp.*)

averaged and stored by the *RC* capacitor memories. After many repetitive wave-forms have been sampled, the noise is averaged toward zero and the voltage stored in each capacitor represents the signal voltage at a specific point on the waveform. The full waveform may be displayed at the output terminal by suc-cessively activating the gates in conjunction with the sweep output signal.

The trigger input signal keeps the sampling in synchronism with the signal waveform so that the same point on the waveform is sampled each time. It is usually not possible to derive the trigger signal from the input signal itself be-cause of the disturbing influence of the noise. Note also that display of the averaged waveform may be at the same rate as the input signal, or, alternatively at a rate either faster or slower than that corresponding to the input signal. This makes it possible, for example, to preserve the output waveform on a chart recorder if desired. An example of a signal waveform recovered from noise is shown in Fig. 10-42.

A more sophisticated signal-processing technique related to the signal aver-ager is based on the correlation between two time-dependent signals. The so-called *cross-correlation function* between two signals $f_A(t)$ and $f_B(t)$ is defined as

$$C_{AB}(\tau) = \lim_{T \to \infty} \frac{1}{T} \int_T f_A(t) f_B(t + \tau) \, dt \qquad (10\text{-}72)$$

In the special case where $f_A(t) = f_B(t)$, $C_{AA}(\tau)$ is called the *autocorrelation function* and, because of the time average involved in Eq. (10-72), noise reductions similar to those achieved with signal averaging are possible. The autocorrelation func-tion contains much additional information, however, and may be used, for exam-ple, to determine the Fourier components of a waveform. Analogously, the cross-correlation function basically measures how closely two signals correspond to each other and therefore it provides a much deeper insight into the phenomenon represented by the two signals than any analysis of the properties of either one separately.

The calculation corresponding to Eq. (10-72) is carried out by a *signal corre-lator*, Fig. 10-43. One input signal is digitized and presented to a 100-channel

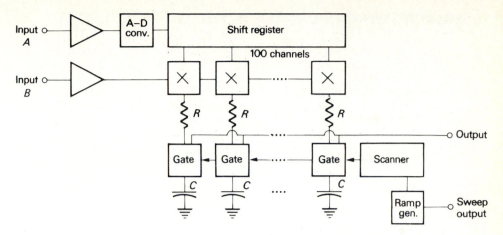

Figure 10-43 Simplified block diagram of signal correlator. (*Princeton Applied Research Corp.*)

shift register. The output of each stage in the register is multiplied by the second input signal and the result averaged and stored in each RC memory. The movement of the first input signal along the shift register provides the delay time τ in Eq. (10-72). The correlation function is displayed by scanning the capacitor memory bank in conjunction with a sweep signal output. While the many applications of signal-correlation analysis are not of direct interest here, it is clear from these examples that digital techniques provide diverse and powerful signal-processing instruments.

Many modern signal-processing instruments employ the microprocessor techniques discussed in Chaps. 11 and 12. Thus, for example, digitized samples are stored in a RAM, and the averages or calculations according to Eq. (10-72) are carried out using the digital program of the microprocessor. This approach permits very flexible signal processing and presentation of results.

Digital Oscilloscope

The advantages of digital signal processing and oscilloscope waveform display are combined in the *digital oscilloscope*. This instrument is similar to a standard oscilloscope, Fig. 10-17, except that an A-D converter and a RAM follow the input amplifier. Thus, input waveforms can be stored in RAM and displayed on the CRT at the same time the digital information is analyzed to yield averages, peak values, time intervals, etc. This approach is particularly useful for transient signals which can be displayed at different values of sweep speed in order to examine portions of the waveform more closely. Practical instruments employing a flash A-D converter have bandwidths extending from dc to 100 MHz and to as high as 10 GHz for repetitive waveforms which permit sampling over many cycles of the input waveform.

SUGGESTIONS FOR FURTHER READING

Andras Ambrozy: "Electronic Noise," McGraw-Hill Book Company, New York, 1982.

Hai Hung Chiang: "Electrical and Electronic Instrumentation," John Wiley & Sons, Inc., New York, 1984.

C. F. G. Delaney: "Electronics for the Physicist," Penguin Books, Inc., Baltimore, 1969.

A. James Diefenderfer: "Principles of Electronic Instrumentation," W. B. Saunders Company, Philadelphia, 1979.

Richard J. Higgins: "Electronics with Digital and Analog Circuits," Prentice-Hall, Inc., Englewood Cliffs, N.J., 1983.

Paul Horowitz and Winfield Hill: "The Art of Electronics," Cambridge University Press, New York, 1980.

H. V. Malmstadt, C. G. Enke, and S. R. Crouch: "Electronics and Instrumentation for Scientists," Benjamin-Cummings Publishing Co. Inc., Menlo Park, Calif., 1981.

Bryan Norris (ed.): "Microprocessors and Microcomputers and Switching Mode Power Supplies," McGraw-Hill Book Company, New York, 1978.

Franklin Offner: "Electronics for Biologists," McGraw-Hill Book Company, New York, 1967.

EXERCISES

10-1 Determine the regulation factor and internal resistance of the transistor regulated power supply, Fig. 10-11. Assume properties of the control transistor given in the text and that the h parameters of the 2N2049 are the same as those given in Exercise 6-5.

Answer: 2.4×10^{-3}, $2.4 \, \Omega$

10-2 Analyze the simple shunt regulator in Fig. 10-10 to determine the effective internal resistance and regulation factor. The h parameters are $h_{ie} = 100 \, \Omega$, $h_{fe} = 60$.

Answer: 5.9×10^{-2}; $1.6 \, \Omega$

10-3 Develop an expression for the transmission time of a *lossless* $[\alpha = 0$ in Eq. (10-48)$]$ transmission line d meters long in terms of its characteristic impedance. Calculate the *delay time* of a 50-Ω coaxial transmission line 1 m long if the capacitance of the line is 90 pF/m.

Answer: $d_c Z_c$; 4.5 ns

10-4 Determine the input impedance of a lossless transmission line with the output end short-circuited. For what length does the line act like a parallel resonant circuit? A series resonant circuit?

Answer: $jZ_c \tan 2\pi d/\lambda$; $\lambda/4$; $\lambda/2$

10-5 Using the data of Fig. 10-25 plot the noise figure as a function of frequency for the type LM394 amplifier if the source resistance is 1000 Ω. Compare to a plot of the minimum noise figure possible if the source resistance is adjustable.

10-6 Show that an operational-amplifier comparator is a phase detector by plotting the rms output voltage as a function of the phase angle between two square-wave pulse input signals. Assume that the larger input signal is applied to the noninverting input.

Answer: $(V_{cc}/2)(1 + \phi/\pi)$, $0 \le \phi \le \pi$; $(V_{cc}/2)(3 - \phi/\pi)$, $\pi \le \phi \le 2\pi$

10-7 Show that an exclusive-OR gate can be used as a phase detector by plotting the average output voltage as a function of the phase angle between two square-wave input pulse signals.

Answer: $V_{cc}\phi/\pi$, $0 \le \phi \le \pi$; $V_{cc}(2 - \phi/\pi)$, $\pi \le \phi \le 2\pi$

10-8 Show that an operational amplifier operates as a phase detector by determining the rms output signal as a function of the phase angle between two sine-wave input signals.

Answer: $aV_i\sqrt{1 - \cos \phi}$

10-9 The dual-slope integrating A-D converter, Fig. 10-32, delivers ten 4-digit readings per second. What is the oscillator frequency? Assume the input and reference voltages are equal.

Answer: 200 kHz

10-10 How many reject trials occur in the successive-approximation A-D converter, Fig. 10-33, if the input signal is 8.08 V? Repeat for 0.62 V. If each comparison takes 10^{-4} s, determine how long it takes to achieve the conversion in each case.

Answer: 10, 9; 1.2×10^{-3} s, 1.2×10^{-3} s

10-11 What is the resolution of a 10-bit binary ladder network having a reference voltage of 10 V?

Answer: 9.76×10^{-3} V

10-12 Suppose the input signal to the digital voltmeter in Fig. 10-31 is 0.5 V dc together with a 0.5 V peak-to-peak sawtooth waveform with a 2-s period. Determine graphically the first five readings displayed if the voltmeter happens to start a measuring cycle when the input signal is 0.5 V.

Answer: 0.67, 0.89, 0.52, 0.68, 0.92 V

10-13 In the so-called *voltage-to-frequency* A-D converter, Fig. 10-44, the pulse generator completely discharges the integrating capacitor each time the comparator senses equality between the integrated input signal and the reference voltage. What are the output frequencies for input signals of 10, 5, and 1 V?

Answer: 10^4 Hz, 5×10^3 Hz, 10^3 Hz

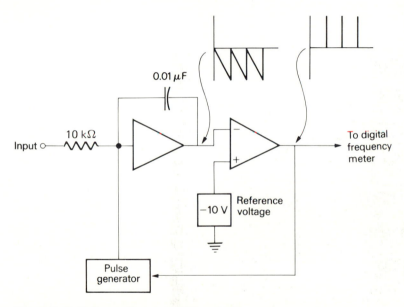

Figure 10-44 Voltage-to-frequency A-D converter.

10-14* Using circuit analysis of the binary-ladder D-A converter, Fig. 10-37, develop a tabulation of decimal numbers, BCD word input signals, and analog output signals analogous to Table 10-2.

Answer: 1, 0001, 1/24; 2, 0010, 2/24; etc.

10-15 Show that the network of three exclusive-OR gates in Fig. 10-45 decodes the inverse of the gray code input signals in Table 10-1 into 4-bit binary signals.

Answers: $\bar{g}_0$	$\bar{g}_1$	$\bar{g}_2$	$\bar{g}_3$	b_0	b_1	b_2	b_3
0	0	0	0	0	0	0	0
1	0	0	0	1	0	0	0
1	1	0	0	0	1	0	0
0	1	0	0	1	1	0	0
etc.							

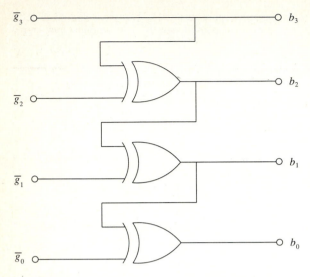

$\overline{g}_3$ o   o b_3

$\overline{g}_2$ o

o b_2

$\overline{g}_1$ o

o b_1

$\overline{g}_0$ o

o b_0

Figure 10-45 Network of three exclusive-OR gates is gray code decoder.

10-16* Calculate the frequency-response characteristic of the digital filter corresponding to Eq. (10-71) using the coefficients listed in Table 10-3. Do this by introducing sine waves of different frequencies at the input. Sample each input signal at the 100-Hz rate for one full cycle and compute the output signals using Eq. (10-71). Use these values to determine the rms output signal at each frequency. Plot the frequency-response characteristic over the frequency interval from 0.1 to 50 Hz. What is the ratio of output signal to input signal at a frequency of 10 Hz?

    Answer: 0.1

Table 10-3 Digital filter coefficients for Exercise 10-16 ($\delta = 10^{-2}$ s)

$a_1 =$	$- a_{10} =$	$+ 0.0008$
$a_2 =$	$- a_9 =$	$- 0.0025$
$a_3 =$	$- a_8 =$	$+ 0.0094$
$a_4 =$	$- a_7 =$	$- 0.0350$
$a_5 =$	$- a_6 =$	$+ 0.1890$

10-17* Repeat Exercise 10-16 for an input signal which is a 2-V peak-to-peak 1-Hz square wave. From the output waveform, what is the name of this filter?

    Answer: Differentiating filter

ELEVEN

MICROPROCESSORS

A major revolution in the application of electronic circuits quite equivalent to the invention of the transistor and to the discovery of the electron vacuum triode has arisen from the development of very-large-scale integrated digital circuits. These techniques have made it possible to produce very complex digital processing logic on single chips. Such microprocessors use the techniques of digital computers and open up a wide range of flexible and powerful measurement and control applications.

DIGITAL COMPUTERS

Organization

A *digital computer* is a complex array of logic gates, registers, and associated circuitry organized to perform logic computations by manipulatng waveforms representing digital numbers and words. The great power of electronic digital processing stems from the variety of phenomena that can be represented and analyzed logically. Typical examples range from simple arithmetic calculations to scientific computations obeying physical laws and to bookkeeping processing which follows the principles of accounting. The speed of electronic circuits is so great that digital computers can complete extremely complex and extensive calculations in practical lengths of time.

The circuits of digital computers are designed to carry out logic computations of all kinds. Therefore, the machine is furnished detailed instructions pertaining to every specific step in the calculation desired, as well as all of the digital numbers involved. The complete set of instructions is called the *program* and is stored within the computer. Since the program and data are easily changed for different problems, the stored-program digital computer is a very flexible and powerful processing instrument.

A digital computer comprises five major parts: input, output, memory, control, and arithmetic and logic units, Fig. 11-1, together with a *clock* circuit which regulates the basic speed at which the machine operates. The input and output devices present data to the machine and retrieve results. The computer *memory* stores the program and data as well as the final results before they are delivered to the output unit. The *arithmetic and logic unit (ALU)* contains logic circuits such as ADD gates and shift registers which actually perform the logic operations specified by the program.

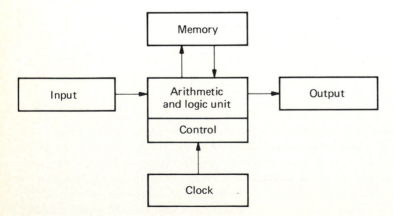

Figure 11-1 Organization of digital computer.

The function of the *control unit* is to interpret each instruction and set the circuits of the computer accordingly. It is the most heterogeneous part of the computer and is composed of logic gates, registers, and logic decoders to direct appropriate control signals to all parts of the machine. Because the functions of the control unit and the arithmetic and logic unit are central to the performance of a computer, the two units together are referred to as the *central processing unit*, or *CPU*.

A digital computer is operated by first inserting the program into memory using the input device. When the machine is started, the control unit reads the first instruction, prepares the circuits of the computer accordingly, which may, for example, cause a data number to be read from memory. At the completion of the indicated operation, the result is typically returned to memory and the control unit passes on to the next instruction. The machine proceeds sequentially through each instruction, placing the results in memory until the final instruction, *HALT*, is reached. At this point the computation is finished and the desired result is delivered to the output unit.

Programming Languages

It is useful to illustrate by means of a rudimentary example how the control unit prepares the circuits of the machine in accordance with each instruction. Consider the partial block diagram, Fig. 11-2, in which either of two numbers can be *fetched* from memory, operated upon, and delivered to the CPU. Suppose the instruction 1010, as shown, is stored in the instruction register connected to the several AND and NAND gates. This means that the number A passes through the circuit and is delivered as NOT A to the CPU. The number B is not used since it does not pass the first AND gate. Other instructions cause different

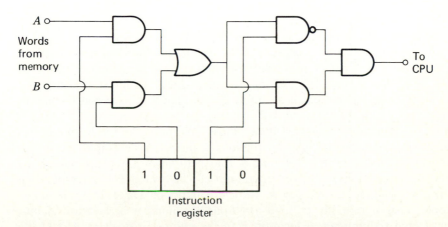

Figure 11-2 Instruction determines path of words through digital computer.

operations to be performed, as may be determined by following the circuit path in each case:

Instruction	Interpretation
1010	Fetch A from memory and transfer NOT A to CPU.
1001	Fetch A from memory and transfer A to CPU.
0110	Fetch B from memory and transfer NOT B to CPU.
0101	Fetch B from memory and transfer B to CPU.

In effect, instruction words set up signal paths through the computer by activating logic gates in the control unit and throughout the machine. The CPU and control unit logic is represented by the interconnections of logic gates and registers as specified by the computer design. Most often, the instruction word itself does not activate the logic gates directly as in this elementary example, but rather the instruction word is applied to an instruction decoder which produces digital signals to activate all the gates necessary to carry out the instruction needed.

It is neither feasible, nor necessary, to trace signal paths through the complete circuit of a computer when preparing a program. Rather, in the design of the machine, a tabulation of instructions to carry out the various operations is developed, corresponding to the control and ALU logic circuits. This dictionary of *machine language* is used to write programs and it obviates the need to refer to the detailed circuits of the computer itself. Nevertheless, because each minute step must be specified in the instructions, the preparation of a program in machine language is a long and arduous task and programming errors are frequent. For example, transposing a 0 for a 1 in any instruction word may result in a nonsense step, or, more seriously, a false instruction, and such errors are difficult to locate.

To improve this situation, symbolic programming languages have been developed in which readily recognizable symbols are used to represent machine language instructions or groups of machine language instructions. The computer manufacturer supplies translating programs which use the computer itself to convert symbolic programs into machine language. The simplest symbolic language is *assembly language*, and the computer program for translating and assembling a program written in assembly language into binary-coded instructions is called an *assembler*.

Symbolic statements in assembly language have nearly a one-to-one correspondence with machine language instructions. Consider, for example, the simple program to add the numbers Z, B, and C together, which is represented by the set of binary instructions in the center column of Table 11-1, The corresponding statements in assembly language in the left column of Table 11-1 are quite mnemonic. They may easily be read as "Load the number at location Z into register A; add the number at location B to the number in register A; add the number at location C to the number in register A; and store the number in

Table 11-1 Comparison of assembly language and Fortran statement with machine language instructions

Assembly language	Machine language				Fortran statement
LDA *Z*	0110	0011	0010	0001	$D = Z + B + C$
ADA *B*	0100	0011	0010	0010	
ADA *C*	0100	0011	0010	0011	
STA *D*	0111	0011	0010	0100	

register *A* at location *D*." Thus, this series of assembly language statements accomplishes the desired routine. The assembler program translates humanly interpretable statements in assembly language into the the corresponding binary-coded machine language words needed by the computer. After assembly, the machine language program is subsequently loaded into the machine and the program executed.

Because each detailed step in the program must be included, programming in assembly language is also tedious. Higher-level languages have been developed in which single statements can be translated into large groups of machine language instructions. In this case the translating program is called a *compiler* because, in effect, the computer compiles its own program following quite general instructions.

The most widely used programming language, called *Fortran* (for *formula translator*), employs stylized algebraic symbols and formulas as program statements. The familiar algebraic expression in col. 3 of Table 11-1, for example, is the Fortran statement which compiles into the set of machine language instructions in the center column. Clearly, it proves much simpler to program in Fortran than either in assembly language or machine language. It often turns out, however, that a machine language program compiled in this fashion is inefficient in the use of computer memory space or in the total time required to execute the program. This results from the generalities inherent in a compiler. Thus, while Fortran and other high-level languages are widely used, assembly language is necessary to realize the maximum capabilities of any computer.

Assembler and compiler programs are designed to help programmers locate errors in their programs by producing output error statements when the trial assembly language or Fortran program contains an inappropriate instruction or attempts to violate the machine logic. This approach is carried a step further by *interpreter* programs which translate each high-level-language statement individually and indicate immediately to the programmer if it is acceptable. The interpreter program, in effect, interprets each high-level instruction into machine language and causes the corresponding set of machine language instructions to be executed before passing on to the next statement. Such *interactive programming* greatly facilitates preparation of complicated routines.

Most often more modern high-level languages such as *Basic* and *Pascal* (named in honor of the French mathematician who developed the first calculating machine) are used with interpreters because their logic structures are particularly well suited to the development of generalized programs. Pascal's structured control statements, for example, as opposed to the simple formula statements of Fortran, are the most comprehensive of any programming language.

Minicomputers and Microprocessors

Early vacuum tube digital computers of the 1940s were soon replaced by their transistor counterparts, encompassing many more active devices and realizing the major size and heat reductions possible with transistor circuits. The computing power of large-scale digital computers is subsequently further increased by the introduction of integrated digital circuits in the 1960s which can have many thousands of active devices on a single chip. Such integrated circuits also facilitate design of smaller, less powerful digital computers, called *minicomputers*, which can be economically dedicated to very special tasks such as data collection and processing in research laboratories or bookkeeping routines for small businesses.

Because they can be dedicated, minicomputers can also be used in sophisticated control applications. Process variables, for example, entered through an A-D converter can be used by a minicomputer to generate appropriate control signals in accordance with instructions in its program. The output signals are applied to adjust the process through a D-A converter or an equivalent output device. A single minicomputer type can be used for an extremely wide range of control applications because the response characteristics can be changed at will simply by entering a different program in memory.

The advent of large-scale integration, with hundreds of thousands of active devices per chip, has increased the computing power of minicomputers, but, even more importantly, has led to the development of *microprocessors*. A microprocessor, which is an entire CPU contained on a single silicon chip, very considerably expands the control horizons of digital computer technology. This is so because a single small device, such as the one illustrated in Fig. 11-3 (note that the package cover is removed to display the chip itself), is the major portion of a general-purpose digital computer. Accompanied by appropriate peripheral input and output devices and a memory unit, such a microprocessor can be dedicated to an untold variety of measurement and control tasks in industrial processes, domestic appliances, automobiles, research instruments, electronic games, etc. Actually, a microprocessor is a programmable logic device which is limited mainly by the ingenuity of the programmer.

COMPUTER ARCHITECTURE

The major parts of a digital computer can be interconnected in various ways to emphasize different operating features. Similarly, the internal structure of each

Figure 11-3 Microprocessor with cover removed to show CPU chip and terminal connections.

part may be configured to perform certain tasks most efficiently. These aspects of digital computer design are referred to as *computer architecture*. The architecture of microprocessors is considered in the following sections.

Memory Organization

The ROM and RAM memories used in microprocessor systems are based on the integrated semiconductor circuits described in Chap. 9. It is usually necessary to employ several ROM chips to contain the program and numerical constants appropriate to the task at hand and one or more RAM chips to hold input data, intermediate and final results, and any portions of the program which may change in accordance with intermediate results. Address words and control signals from the CPU determine which memory chip is activated and which data is called for by the program.

The chip organization of a typical ROM is illustrated in Fig. 11-4. This 8192-bit ROM is organized into a 64 × 128 matrix such that the first six bits of the address word select two of the 128 rows and the second five address bits select two of the 64 columns, which generates a 4-bit output word. The output word is returned to the CPU through output buffers activated by chip-select control signals. The buffers not only provide amplification but also are used to isolate the memory circuits from the rest of the system when the chip is not being used. That is, the output buffers can be ON or OFF (corresponding to 1 or 0 bits respectively), or in an open-circuit, high-impedance disconnect state; this is

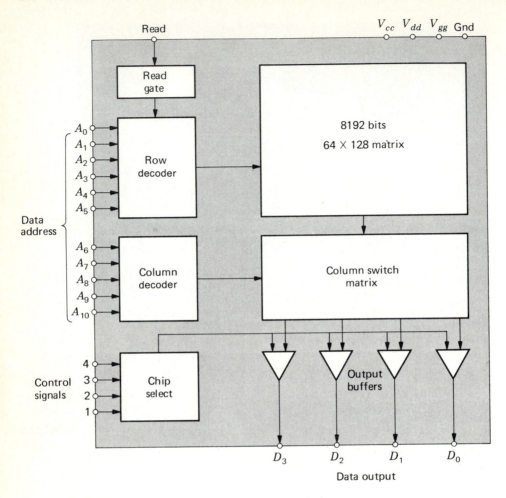

Figure 11-4 Chip architecture of type 2580 ROM. (*Signetics, Inc.*)

known as a *three-state* output buffer. Note that this **ROM** stores 2048 4-bit words.

In a typical RAM, Fig. 11-5, a 10-bit address word accesses a 32×32 matrix and a READ/WRITE control signal determines whether the addressed bit is to store a signal from the CPU or to deliver stored information to the CPU. Often several RAM chips are addressed simultaneously so that the several bits of a single word can be accessed at the same time. Thus, four chips like the one in Fig. 11-5 together store 1024 4-bit words and every bit can be altered by the appropriate address and control signals.

One measure of the power of a microprocessor system is the memory capacity, for this determines the length of the program and the amount of data that can be handled. The advent of very-large-scale integration techniques with several hundred thousand active devices per chip has made very large memories feasible.

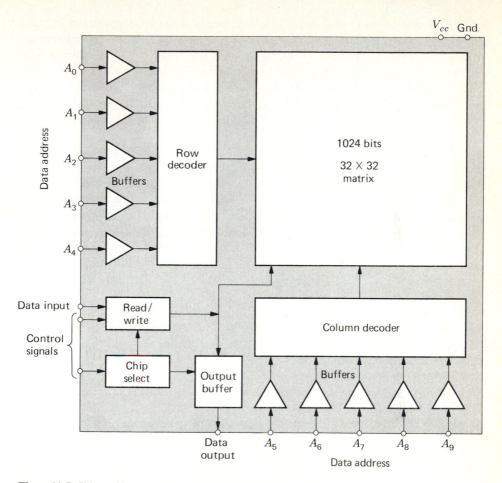

Figure 11-5 Chip architecture of type 2112 RAM. (*Signetics, Inc.*)

Ordinarily, the smallest unit of information accessed out of memory is one word, and the most commonly used word length in microprocessors is an 8-bit word, called a *byte*. Every byte in memory has a unique address that specifies its location. Addresses are sequential beginning with address 0; the address of the last word in memory depends upon the capacity of the memory.

Memory words can be interpreted by the CPU in three fundamentally different ways:

1. Pure binary numeric data
2. Instructions
3. Data code

Binary data are just the numerics associated with the program. For example, one memory byte can represent any number from 0000 0000, that is, 0, to 1111 1111, or 255. A similar array of 0 and 1 bits can specify a machine operation instruction, as already considered in connection with Fig. 11-2. The data code interpretation enables the computer to respond to input information and produce output results as text and decimal digits rather than as a difficult-to-understand sequence of binary digits. The interpretation placed upon any memory word depends upon the CPU logic, as illustrated below. Because each of the three types of memory words are used differently by the CPU, they are usually located in different regions of memory.

Conventionally, the data code interpretation requires a byte coded for each of the 26 upper and lower case letters, 10 decimal numerics, and approximately 25 special characters (such as @, #, + , !, etc.), for a total of 87 characters. Seven bits are sufficient to accommodate 128 characters, which is therefore quite adequate. The eighth bit of the byte is often used as a *parity* bit to test for errors. If, for example, the microprocessor system uses odd parity, the parity bit is set so that the total number of 1 bits in the byte is always an odd number. This convention (even parity is also used) permits checking of the contents of the byte at any time; an even parity indication immediately shows that an error has been introduced.

The way in which instruction words and numeric data are stored in memory and employed by the CPU can be illustrated with a simple program to add two binary numbers using the elementary 4-bit instruction set in Table 11-2. The steps necessary to add two words stored in memory are:

1. Identify the address of the first memory word to be added.
2. Transfer the word at this address to the CPU.
3. Identify the address of the second memory word to be added.
4. Add the word at this address to the word previously transferred to the CPU.
5. Identify the address where the sum is to be stored.
6. Transfer the sum to this address.

Suppose the numbers to be added are 0100 and 0001 located at memory addresses 1110 and 1111, respectively. The program in memory that accomplishes

Table 11-2 4-Bit instruction set

Instruction	Interpretation
0010	Read the contents of next memory word and interpret as a data memory address.
0100	Read the contents of the addressed data word, interpret as binary data, and store in the CPU.
0110	Add the contents of addressed data memory word to data word stored in CPU.
1000	Store data word in CPU at memory word addressed by data memory address.
1001	Read data word in the I/O register and store in CPU.
1010	Read data word in CPU and store in the I/O register.
1100	Complement data word stored in CPU.
0000	Halt.

Memory address	Memory word	Comment
0 0 0 0	0 0 1 0	Step 1. Read following word as address.
0 0 0 1	1 1 1 0	
0 0 1 0	0 1 0 0	Step 2. Read data word and store in CPU.
0 0 1 1	0 0 1 0	Step 3. Read following word as address.
0 1 0 0	1 1 1 1	
0 1 0 1	0 1 1 0	Step 4. Add
0 1 1 0	1 0 0 0	Steps 5 and 6. Store
0 1 1 1	0 0 0 0	Halt
1 0 0 0		
1 0 0 1		
1 0 1 0		
1 0 1 1		
1 1 0 0		
1 1 0 1		
1 1 1 0	0 1 0 0	Data word
1 1 1 1	0 0 0 1	Data word

Figure 11-6 Program to add 0100 and 0001.

these steps is shown in Fig. 11-6, as can be established by referring to the instruction set in Table 11-2.

The first instruction, 0010, found at the first memory address, 0000, results in the word at location 0001 being interpreted as a data word memory address. The next instruction, 0100, stores the word at this address in the CPU. Again instruction 0010, this time at memory address 0011, has the next memory word, 1111, interpreted as a data word memory address. The instruction, 0110, adds the word at this address to the number already stored in the CPU and the final instruction, 1000, stores the sum in the last used data memory address, 1111. After the program is executed, the desired sum, 0101, is located at address 1111, and instruction 0000 stops further execution.

Note that the CPU interprets each word in this program as an instruction unless a previous instruction indicates otherwise. Also, the program is executed simply by proceeding sequentially through the memory addresses until the HALT instruction is reached.

Central Processing Unit

Every CPU has at least one register in which data words fetched from memory can be stored. The *accumulator* is the primary working register of the CPU. It stores the data word to be operated upon by the CPU. Also, the result of any operation usually remains in the accumulator.

Three other important operational registers are contained in the CPU. The *instruction register* sequentially stores each instruction word of the program. In fact, the logic of the CPU is such that the contents of the instruction register are always interpreted as an instruction code. The address of the instruction word to be fetched from memory is stored in the *program counter*. After each instruction is executed, the contents of the program counter are automatically incremented by 1 so that the CPU passes on to the next instruction word stored in memory. The *data counter* contains the memory address of a data memory word. The CPU always interprets the contents of the data counter as a memory address specifying the location of a data memory word to be read or the location where a data word is to be stored.

Actual data processing is handled by the group of logic circuits within the arithmetic and logic unit. The ALU performs binary addition and Boolean logic and can *complement* a data word (exchange all 0 bits for 1 bits and vice versa) as well as *shift* a word 1 bit to the left or to the right. The complement and shift operations are important in binary arithmetic, as described in Appendix 2.

A set of individual binary flip-flops called *status flags* are automatically set or reset to reflect the results of ALU operations. The *carry* flag records any carry out of the lower order word in multiword arithmetic, as discussed in Appendix 2. Similarly, the *overflow* flag is set to 1 if the result of any operation causes the answer word to exceed the available answer space and is therefore in error. If the result of any operation produces a zero result in the accumulator, the *zero* flag is automatically set to 1. This provides a convenient test for a zero answer, which is often useful in executing programs. Also, the *sign* flag is set to the contents of the sign bit (see Appendix 2), of every accumulator word, to provide a convenient way to test for positive or negative numbers. Various ALU designs may include different status flags, depending upon which CPU operations are highlighted.

The CPU control unit sequences logic elements of the ALU in accordance with the contents of the instruction register and the timing of an external clock signal. The instruction register contents are interpreted by a decoder similar to those described in Chap. 9 so that the control unit can generate a sequence of enable signals to guide data words through the ALU and to and from memory. The basic CPU operating cycle is an instruction-fetch interval followed by an instruction-execute time, which, as illustrated in Fig. 11-7, together occupy two periods of the clock signal.

This simple square-wave clock signal provides two states and two edges (one rising, one falling) per period which are used in switching and timing various

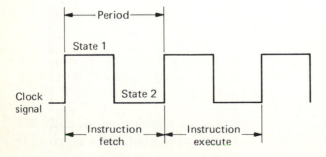

Figure 11-7 Simple CPU instruction cycle.

operations. Consider, for example, the action of a simple memory read instruction in which a data word is fetched from memory and transferred to the accumulator.

During state 1 of the instruction-fetch period the contents of the program counter address a memory word which is the instruction for a simple data memory read. This word is transferred to the instruction register by a read signal derived from the falling transition between state 1 and state 2. The rising edge at the beginning of the instruction-execute period causes the contents of the data counter to address memory, and the word at this location is transferred to the accumulator by a read signal at the state 1 to state 2 transition. At the same time, the contents of the program counter are incremented by 1 so that the control unit can proceed on to the next instruction-fetch operation at the rising edge of the next clock pulse.

A similar instruction-fetch/execute sequence applies to all CPU operations. Frequently two or more clock signals are employed which together offer a greater number of state and edge combinations. This means that additional and more complex operations can take place within each instruction-fetch or instruction-execute interval.

The chip organization of a simplified microprocessor is illustrated in Fig. 11-8. The registers communicate with the ALU by means of a common signal path called a *bus* as directed by the control unit. A data register and an address register temporarily store memory words and memory addresses being transferred between the CPU and memory. Similarly, the buffer register and ALU latches simplify timing considerations for data words entering and leaving the ALU by storing data words and results until they are needed. Practical microprocessor chips described in following sections are quite similar to this generalized version.

Detailed operation of the CPU can be examined by noting the contents of the various registers as the simple addition program in Fig. 11-6 is executed. Initially the program counter is loaded with the address of the first instruction word in memory and the other registers are empty. Subsequent actions proceed sequentially:

1. Program counter, PC, is loaded with the address of the first instruction word.

A	
DC	
IR	
PC	0000

2. CPU fetches word addressed by PC and loads it into instruction register, IR.

3. CPU increments PC.

A	
DC	
IR	0010
PC	0001

4. Instruction word 0010 transfers word addressed by PC into data counter, DC.

5. CPU increments PC.

A	
DC	1110
IR	0010
PC	0010

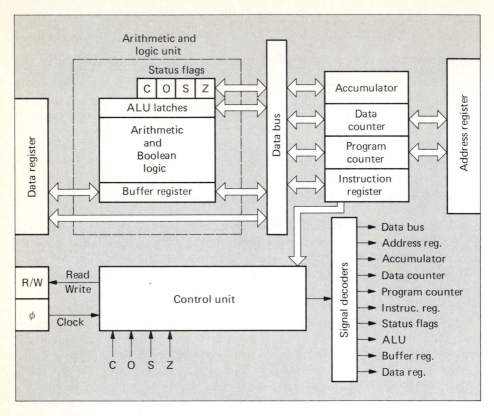

Figure 11-8 Microprocessor chip organization.

The next instruction fetch/execute sequence is:

1. CPU fetches word addressed by PC and loads it into IR.

A	
DC	1110
IR	0100
PC	0011

2. CPU increments PC.
3. Instruction word 0100 loads memory word addressed by DC into accumulator, A.

A	0100
DC	1110
IR	0100
PC	0011

The program counter is not incremented after this last step because the program counter has not been used since the last time it was incremented. The next step is the same as a previous one:

1. CPU fetches word addressed by PC and loads it into IR.

A	0100
DC	1110
IR	0010
PC	0100

2. CPU increments PC.

3. Instruction word 0010 transfers
   word addressed by PC into DC.

A	0100
DC	1111
IR	0010

4. CPU increments PC.

PC	0101

Finally, the addition instruction is fetched and executed:

1. CPU fetches word addressed by
   PC and loads it into IR.

A	0100
DC	1111
IR	0110

2. CPU increments PC.

PC	0110

3. Instruction word 0110 adds word
   addressed by DC to word in A.

A	0101
DC	1111
IR	0110
PC	0110

Note again, that PC has not been incremented since it has not been used. Next, the sum is stored in memory:

1. CPU fetches word addressed by
   PC and loads it in IR.

A	0101
DC	1111
IR	1000

2. CPU increments PC.

PC	0111

3. Instruction word 1000 transfers
   word in A into memory at address
   stored in DC.

A	0101
DC	1111
IR	1000
PC	0111

The final steps terminate execution:

1. CPU fetches word addressed by
   PC and loads it in IR.

A	0101
DC	1111
IR	0000

2. CPU increments PC.

PC	1000

3. Instruction word 0000 halts execution.

The repetitious instruction-fetch and instruction-execute routine of this simple program is clearly evident. Note that each program instruction calls for several steps on the part of the CPU. These steps are controlled by logic design and decoder logic in the control unit often referred to as *microinstructions*. Usually the microinstruction set is fixed internally to the CPU, although in some special microprocessor designs certain microinstructions can be introduced by the programmer.

Input/Output

In a complete microprocessor system the CPU exchanges data and address words with memory chips and *input/output*, or *I/O*, devices. A straightforward way to accomplish this is with a *data bus* and *address bus*, which are common signal paths interconnecting all devices, Fig. 11-9. (The term *bus* is derived from the latin *omnibus*, which means "for all.") Thus, the CPU may place an address word on the address bus which is decoded by each of the other chips and results in some appropriate response. This response, which might be for a memory chip to place the addressed memory word on the data bus, is triggered by an enabling signal on a *control line*, such as the read/write control line in Fig. 11-9. Address decoders, control logic, and I/O buffers are made an integral part of chip architecture, as already described in connection with Figs. 11-4 and 11-5.

The usable memory capacity of a microprocessor system depends exactly upon the number of digits in the memory address. Note, for example, that the 4-bit address words used in the elementary addition program of the previous section permit a total of only 16 memory locations. A commonly used address-word length in classic 8-bit microprocessors is composed of 16 bits, that is, 2 bytes. In addition, a popular convention is to assign the lowest 10 digits to the word address and the remaining 6 most significant binary digits to chip selection:

16-bit memory address

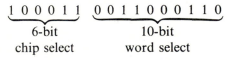

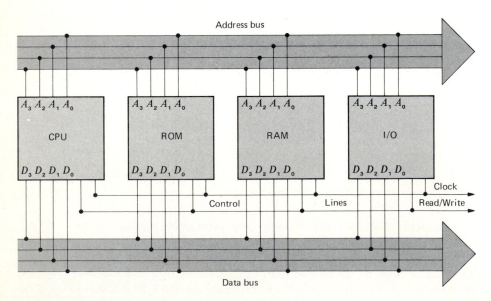

Figure 11-9 Microprocessor system uses address and data buses.

Using this convention 2^6, or 64, different memory chips (and/or I/O devices) can be selected and up to 2^{10}, or 1024, individual words on each chip can be addressed.

Most often I/O devices communicate with the microprocessor through *I/O interface* buffer chips which are addressed by the CPU much like memory chips. Additional control lines are usually included for the CPU to indicate that data has been transferred to the I/O interface and for the buffer to indicate that external data has been stored and is ready to be transferred to the CPU. The I/O interface can also signal an *interrupt* to the CPU, which suspends program execution and causes the CPU to respond to the I/O request. The request may be, for example, acquisition of new data, or execution of a special program to derive a needed control signal, etc. After the interrupt routine is completed, the CPU returns to execution of its program. If several I/O interface devices are active simultaneously, a hierarchy of *interrupt priorities* must be developed. This feature is a significant facet of microprocessor design.

An important interrupt request is to transfer data from the I/O interface directly into memory without passing through the CPU. This process is called *direct memory access*, or *DMA*, and is particularly useful when significant amounts of data are to be entered. DMA obviates the many register transfers needed for data words to be entered through the accumulator to memory in the usual fashion. To accomplish this task the *DMA interface* contains elementary CPU-type logic circuits which have only the single task of transferring words to and from memory. Program execution is halted simply by stopping the clock signal to the CPU. Conventionally, the CPU is interrupted for only one clock cycle at a time for DMA, a process which is termed *cycle stealing*.

PROGRAMMING MICROPROCESSORS

Assembly Language

Microprocessors are rarely programmed in machine language. Although high-level programming languages are often useful, assembly language remains the most effective programming tool for all but the most complex routines. The mnemonic *source program* prepared by the programmer using the symbolic instruction set appropriate to a given microprocessor is translated by its assembler into a corresponding machine language *object program* which is subsequently executed. As mentioned previously, practical assemblers provide error messages which are very helpful to the programmer in preparing new routines.

It also proves possible to *hand-assemble* programs written in assembly language since each symbolic instruction corresponds exactly to only one or a few binary-coded machine language words. This is often useful for short programs, to rewrite parts of longer programs, or to improve the efficiency of portions of programs compiled from high-level languages. Since the basic word length in most microprocessors is a byte, it is convenient to express machine language instructions in the *hexadecimal* number system rather than in the binary system. As illustrated in Table 11-3, the 16 digits of the hexadecimal number system correspond to the full range encompassed by all 4-bit binary words. This means that one byte can be written as two hexadecimal digits:

Byte		Hex number
0011	1110	3E
1001	0001	91
1101	1011	DB

Clearly, hexadecimal numbers are easier to deal with than their binary counterparts. It is common practice to identify hexadecimal numbers by the letter *H*; the number 8000*H*, for example, is equivalent to the binary number 1000 0000 0000 0000.

Microprocessors vary considerably in capability and complexity, and this is reflected in the size of their instruction sets. Some machines perform a relatively limited number of basic operations, while others have repertoires of many hundreds of operations. In principle, any microprocessor can perform any given task, although a small microprocessor may require an excessively long time because it must execute a large number of elementary instructions. The next sections develop an assembly language instruction set which is basic to all microprocessors and contains 25 instructions.

Table 11-3 Decimal, binary, and hexadecimal numbers

Decimal number	Binary number	Hexadecimal number
0	0000	0
1	0001	1
2	0010	2
3	0011	3
4	0100	4
5	0101	5
6	0110	6
7	0111	7
8	1000	8
9	1001	9
10	1010	A
11	1011	B
12	1100	C
13	1101	D
14	1110	E
15	1111	F

Move and Logic Instructions

Microprocessors have at least one instruction for moving data words, one for arithmetic operations, and one for logical operations. The most frequently used instructions are STA, ADD, and AND. For example, the instruction STA M *stores* the data memory word in the accumulator at address M. Similarly, ADD M adds the data memory word at address M to the contents of the accumulator, and AND M performs a bit-by-bit logical AND operation between the data memory word stored at address M and the contents of the accumulator. In each case, the result remains in the accumulator.

These three instructions are sufficient to carry out all logical and arithmetic operations by repetitively using the Boolean logic statements listed in Chap. 9. Additional control and operate instructions are required, however, to direct the CPU how to execute these logic statements. Table 11-4 gives the interpretation of each instruction, together with its hexadecimal code.

The hex code representation of instructions which contain a memory address reference, such as $2M$, requires special attention in hand assembling. The four

Table 11-4 Assembly language instruction set

Type	Hex code	Mnemonic	Interpretation
Move data	1M	STA M	Store contents of A at M
Arithmetic	2M	ADD M	Add contents of M to A
Logical	3M	AND M	Contents of M AND A
Control	00	HALT	Halt program
	40	SKP U	Skip unconditionally
	41	SKP C	Skip if C is 1
	50	SKP Z	Skip on zero A
	51	SKP P	Skip on positive A
	52	SKP N	Skip on negative A
	6M	ISZ M	Increment contents of M; skip on zero
	7M	JMP M	Jump unconditionally to M
Subroutine	8M	JMS M	Jump to subroutine at M
	90	RTN	Return to main program
Operate	A0	CLR A	Clear A
	A1	CLR C	Clear C
	A2	CMP A	Complement A
	A3	CMP C	Complement C
	A4	INC A	Increment A by 1
	A5	ROT L	Rotate A left one bit
	A6	ROT R	Rotate A right one bit
Input/Output	B0(2)	CLR 2	Clear flag in device 02
	B1(2)	SKP 2	Skip if flag is set in device 02
	B2(2)	RED 2	Read data from device 02 into A
	B3(2)	WRT 2	Write data from A into device 02
	B4(2)	GON 2	Activate action in device 02

digits available in the lowest 4-bit word of the $2M$ instruction byte are capable of addressing only 16 memory locations, which is too limiting. Therefore, the memory word immediately following the instruction word in the program is also used to contain the address. For example, the instruction ADD 3E9 translates into the following hex code and digital program entries:

Mnemonic	Hex code	Memory words	
ADD 3E9	23	0010	0011
	E9	1110	1001

The total address-word length is therefore 12 bits, and so 2^{12}, or 4096, memory locations can be addressed, even though the basic word length of the microprocessor is just 1 byte. The assembler automatically produces the above machine language entry, of course, without direct attention by the programmer.

Control and Operate Instructions

A series of control instructions is necessary to direct the program and to make decisions which can alter the course of the program in accordance with some intermediate result. The most important control instructions are *halt*, *skip*, and *jump*. Practical microprocessor instruction sets have many different versions of skip and jump instructions, but the common operations are those given in Table 11-4.

The HALT instruction is obvious; it stops the CPU by halting advance of the program counter. Similarly, JMP M loads the address M into the program counter so that the program jumps to this location and execution continues normally. This proves to be a necessary instruction if the program steps are not all located sequentially in memory. The jump instruction is also used to return to the beginning of the program if repetitive execution is desired.

Frequently, the same sequence of instructions is needed several times in a large program, as, for example, in multiplying two numbers. Instead of copying the same set of instructions every time it is needed, the sequence is written as a *subroutine* and located outside the main program. The instruction, JMS M, loads the address of the first instruction of the subroutine into the program counter and at the same time stores the incremental current address of the main program; the last instruction in the subroutine, RTN, loads the stored address into the program counter, returning the CPU to the main program.

Consider the elementary program in Table 11-5, which includes a simple subroutine. The program arbitrarily starts at address 20 and executes sequentially until the JMS 30 instruction causes the program to jump to the subroutine beginning at address 30. After the subroutine is executed, the last instruction, RTN, returns execution to the main program by loading address 24 into the program counter. Clearly, the same subroutine can be used subsequently simply by including another JMS 30 instruction.

Table 11-5 Program with subroutine

Address		Contents
20		CLR A
21		ADD 40
22		STA 50
23	Jump	JMS 30
24		CLR A
25		
⋮		
30		CLR A
31		ADD 50
32	Subroutine	AND 41
33		STA 51
34	Return	RTN
⋮		
40		10
41		80

The skip instruction causes the program to skip over the following instruction if some condition is fulfilled. As listed in Table 11-4, this condition may be
that the carry flag is set, or that the contents of the accumulator are either zero,
or positive, or negative. These instructions permit testing for a result and directing or *branching* the program correspondingly. Consider the program in Table
11-6, which executes this process: if the number at address 20 is 0, add 10 to it;
otherwise continue the program at address 16.

A particularly useful skip instruction is ISZ M, which performs two operations: it adds 1 to the contents at address M and skips the next instruction if the
result is zero. If the result is not zero, the next instruction is executed. This
procedure is useful in *looping*, where a portion of a program is repeated a number
of times to accomplish a desired result. Consider the program in Table 11-7
which performs the multiplication 9×3 by adding 9 to itself 3 times. The
number 09 is stored at address 20 and address 21 stores -3 in two's complement
notation (see Appendix 2). The program has the shape of a loop and each pass
adds 09 to the accumulator. In the first pass ISZ 21 increments the contents at

Table 11-6 Program with branching

	Address		Contents
	10		CLR A
	11		ADD 20
	12	Either this	SKP Z
Or this	13		JMP 16
	14		ADD 21
	15		STA 20
	16		
	⋮		
	20		50
	21		10

address 21 from FD (decimal -3) to FE (decimal -2), and, since the contents are not zero, the program executes the next instruction, JMP 11. On the second pass the contents at address 21 are incremented to FF (decimal -1), and the program loops again. The third pass decrements the contents at address 21 from FF to 00, and the program skips the jump instruction to HALT. The result, $09 + 09 + 09 = 09 \times 03$, remains in the accumulator.

Table 11-7 Program with looping

		Address		Contents
		10		CLR A
		11		ADD 20
2nd pass	1st pass	12		ISZ 21
		13	3rd pass	JMP 11
		14		HALT
		⋮		
		20		09
		21		FD

The operate instructions in Table 11-4 perform *clear, complement,* and *increment* operations on the accumulator register or carry flag. The clear operation prepares either register for further operations, while complementing and incrementing are needed to express negative numbers, as described in Appendix 2. The *rotate* instructions treat the accumulator and carry as a closed loop and shift all bits in the loop by one position to the left or right, Fig. 11-10. Left and right shifts are useful in binary multiplication and division (see Appendix 2). Many microprocessor instruction sets also include rotation of the accumulator contents alone, and simple right or left shifts without rotation and with or without the carry.

Data transfers between the microproccessor and all input and output devices are controlled by the program using the I/O instructions in Table 11-4. For example, the instruction RED 2 in any program causes data in the I/O interface buffer of device 02 to be transferred ("read") to the accumulator, from where it may be transferred to memory or operated upon by subsequent instructions in the program. The hex code for this operation, B2(2), is interpreted in hand assembly in the same way as other instructions having addresses discussed previously. That is, the device address is contained in the memory byte immediately following the I/O instruction in the program.

The operating speed of microprocessors is much faster than most I/O devices, so it is quite likely that data may not be stored in the I/O interface buffer when called for by the program or that the interface may not be ready to receive data from the CPU. A device flag in each I/O interface is set whenever it is ready to transfer data and the I/O SKP instruction tests the state of the flag flip-flop. Consider the program in Table 11-8, which reads a data word from input device 04, adds it to a data word read from input device 05, and transfers the sum to output device 06. The SKP 4 instruction causes the program to loop until the 04 device flag is set and data can be read out. After data transfer, the device flag is cleared by CLR 4. Similar wait loops are used in accessing the other I/O devices. Finally, the instruction GON 6 activates output device 06 to take some action with the data transferred to it, such as to print the result.

More sophisticated instruction sets include interrupt and DMA instruction codes, as discussed in the previous section. This obviates the need for the microprocessor to wait until the I/O devices are ready and makes it possible to transfer data to memory whenever it becomes available.

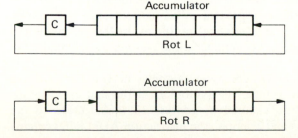

Figure 11-10 Rotate instruction shifts Accumulator contents one bit to left or right.

Table 11-8 I/O Program

	Address	Contents
	00	CLR A
Wait device 04	01 ←	SKP 4
	02 ┘	JMP 01
	03	RED 4
	04	CLR 4
	05	STA 20
	06	CLR A
Wait device 05	07 ←	SKP 5
	08 ┘	JMP 07
	09	RED 5
	0A	CLR 5
	0B	ADD 20
Wait device 06	0C ←	SKP 6
	0D ┘	JMP 0C
	0E	WRT 6
	0F	GON 6
	10	HALT

Memory Addressing

The general form of addressing memory used in the programs of the previous section is called *direct addressing* because the address of the memory data word is part of the instruction. An alternate approach, called *indirect addressing* proves useful in data acquisition, sorting, and retrieval problems. In this technique the memory word located by the address contained in the instruction is actually the address of the data word.

Compare the direct and indirect addressing examples in Table 11-9, where the indirect address instruction includes the symbol *I*. In the first case, the instruction ADD 10 causes the data stored at address 10 to be added to the accumulator, as usual. In the second example, the instruction ADD I 10 contains

Table 11-9 Direct and indirect addressing

Address	Contents	Accumulator
10	30	00
⋮		
20	ADD 10	30
⋮		
30	99	
10	30	00
⋮		
20	ADD I 10	99
⋮		
30	99	

a *pointer address*, which points to location 10 where the actual address, 30, of the data word is located. As a result, the data at address 30 is added to the accumulator. Indirect addressing provides useful flexibility in preparing programs.

Most assemblers accept symbolic addressing as well, so that it is not necessary to specify exact memory locations. Rather, the source program uses symbolic names for memory locations and the assembler assigns specific addresses as needed. An instruction may be written ADD *RESULT*, for example, where the symbolic location, *RESULT*, is given an exact address during assembly. The source program must, of course, use the same symbolic address, *RESULT*, whenever it must later access the same memory location.

PRACTICAL MICROPROCESSORS

4-Bit and 8-Bit Systems

An early design and widely used microprocessor chip employs a basic 4-bit word length and is labeled the type 4004. To remove the limitations inherent in such a small word, the 4004 has 8-bit instruction words and address words 12 bits long. These longer words are accommodated in memory by storing each 4-bit segment in adjacent memory locations. The chip architecture, Fig. 11-11, is similar to Fig. 11-8 except that several additional registers are provided and a single 4-bit bus buffer is used to communicate with ROM, RAM, and I/O chips. The instruction set totals 46.

A 4-bit *TEMP* register is used in connection with the accumulator to temporarily store one term in addition calculations. Most often 8-bit data words are manipulated and multiword arithmetic (see Appendix 2) is employed. This technique is termed *double-precision* arithmetic because of the increased accuracy

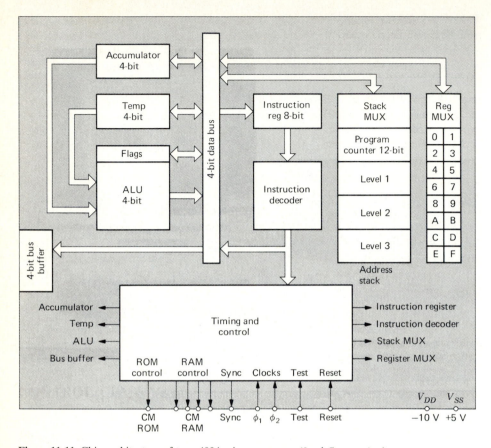

Figure 11-11 Chip architecture of type 4004 microprocessor. (*Intel Corporation.*)

achieved. Multiple-byte data words can be treated in a similar fashion to realize even greater accuracy.

The program counter has associated with it three additional 12-bit registers. These four registers communicate with the internal data bus through a multiplexer and are collectively termed the *stack*. The stack registers are used to save the program counter contents when the program jumps to a subroutine, so that the program can continue on from the same point upon returning from the subroutine. That is, the instruction JMS M stores the current main program instruction address in the level 1 register of the stack and enters address *M* into the program counter. This causes the program to jump to the subroutine as discussed in the previous section. The last instruction of the subroutine, RTN, moves the contents of the level 1 register into the program counter and execution returns to the main program.

The stack operates on a first-in last-out basis in which the contents of the program counter are said to be *pushed* into the top of the stack. If, for example, the subroutine itself calls for a subroutine, the contents of the program counter with the current subroutine instruction address are pushed into the level 1 register and the contents of the level 1 register (the main program instruction address) are pushed into the level 2 register. Upon return from the second subroutine, the contents of the level 1 register are *popped* into the program counter and the initial subroutine continues to execute. Finally, the main program instruction address is popped from the top of the stack at the end of the first subroutine as before. The three-level stack in the 4004 microprocessor permits three such *nested* subroutines.

The 4004 microprocessor also contains sixteen 4-bit general registers multiplexed to the internal data bus. These are used for temporary storage of data and intermediate results, which obviates the need to execute multiple transfers into and out of RAM in the course of the program calculation. The general registers also act as the data counter to address ROM, RAM, and I/O chips.

A two-phase clock signal with a basic period of 1.35 μs is required to perform the necessary timing functions. The 4004 control unit generates a $SYNC$ pulse from the clock signals which marks the beginning of each instruction-fetch cycle for the ROM and RAM chips. The basic instruction-fetch/execute interval requires eight clock periods, or a total of 10.8 μs.

The 4004 is designed to be used with companion ROM, RAM, and other peripheral chips in a complete microprocessor system. Each ROM chip is organized as 256 8-bit words which can be used for storing programs or data tables. Each ROM chip also contains a 4-bit I/O terminal, or *port*, which is not associated functionally with the memory. A single RAM chip consists of four registers of 20 4-bit characters each, for a total of 320 bits. A 4-bit output port is also physically located on each RAM chip and is served by common chip-select control logic.

The 4-bit control logic permits selecting one of up to 16 ROM chips and one of up to 16 RAM chips, together with the accompanying $32 \times 4 = 128$ I/O lines. Note that the 12-bit address word has a range of 2^{12} or 4096 addresses, which is just the total ROM capacity, 16×256.

A small, but complete, 4004 microprocessor system using two type 4001 ROM chips and a single type 4002 RAM chip is shown in Fig. 11-12. The system timing is developed by the crystal-controlled type 4201 clock chip, which also contains a Schmitt trigger $RESET$ circuit. The reset signal generated by the Schmitt trigger clears all registers and sets the program counter to zero, where the first instruction in ROM is located. A single 4-bit bus is used to transmit address words during the first part of the instruction-fetch cycle and to return instruction words in response. Similarly, data words are received from the input ports or memory and are delivered to the output port and RAM over the same bus during the instruction-execute phase. Separate ROM and RAM control lines select the appropriate memory chips to carry out memory or I/O functions.

This elementary system accepts two 4-bit input signals and produces an output in accordance with the program calculation. It might be used, for example, to control heating in a structure in response to outside air temperature and wind velocity. In this case the program could incorporate a mathematical model for the thermal response of the structure, together, perhaps, with routines to

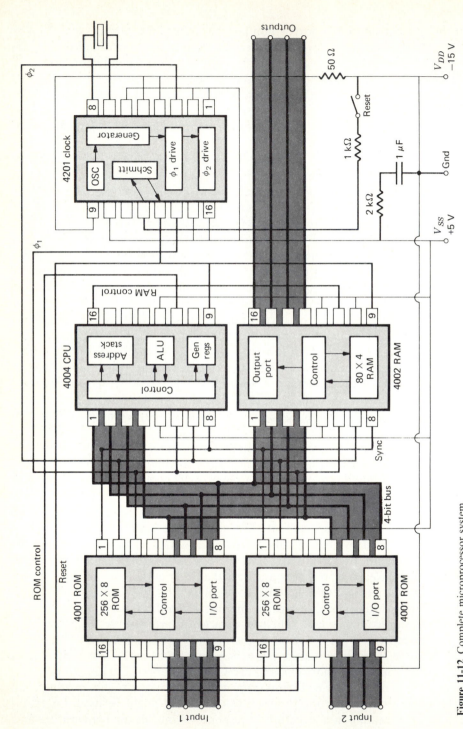

Figure 11-12 Complete microprocessor system.

calculate average temperature and wind velocity and their rates of change. The output signal might be data words that determine control settings after D-A conversion. Alternatively, the individual output lines can be programmed for simultaneous on-off control of separate devices, such as a circulating fan, flue damper, fuel-flow valve, etc. Obviously many similar applications exist, even for this rudimentary system.

More practical systems incorporate additional ROM and RAM and other peripheral chips which increase flexibility. I/O interface buffer chips, in particular, are available which are more versatile than the simple 4-bit port, and can service several process sensors simultaneously. Others send and receive data words in serial representation, which is appropriate for long distance data transmission, as, for example, over telephone lines. Still others interface the system with typewriter type keyboards and TV type visual displays, etc. An enhanced version of the 4004 chip, labeled type 4040, allows seven levels of subroutine nesting and includes 24 general registers. The instruction set is increased to 60 and includes an interrupt capability. This unit has largely replaced the earlier design.

The limitations inherent in the use of 4-bit words are removed in the type 8008 microprocessor and its later version, the type 8085 chip. These 8-bit processors are similar to their 4-bit counterparts but include more powerful instruction sets (78 instructions), chip architecture and a 2-μs instruction cycle. Both instruction and data words are 8-bits long, but the address word length is 16 bits, which addresses 2^{16}, or 65,536 locations. Separate 8-bit data and 16-bit address buses communicate with peripheral chips, and a region of RAM is reserved to act as the stack, which permits virtually unlimited subroutine nesting.

These features are also present in one of the most popular 8-bit processors, the type Z80 chip, Fig. 11-13. The Z80 contains 14 general-purpose 8-bit registers, A through H, and companion registers A' through H'. Only one set of these general-purpose registers is active at any one time, but the availability of both sets increases programming flexibility. Register A (and/or A') is used as the accumulator and F and F' are the corresponding 8-bit flags. In addition to being used individually as 8-bit registers, register pairs B/C, D/E, H/L, and their counterparts, can be used together to form 16-bit registers. It is common practice to use registers H/L as the data counter.

Addressing of the stack in RAM is controlled by a 16-bit *stack-pointer* register which contains the address of the top of the stack and effectively points to the location of the most recent entry. The stack pointer is incremented or decremented as address words are pushed into and popped out of the stack in accordance with subroutine and interrupt instructions. The contents of the program counter (together with the contents of other registers which describe system status) are stored in the stack area sequentially and retrieved at the end of the subroutine or interrupt request so that the main program may then proceed.

Two *index registers*, IX and IY, are used in indirect addressing and register 1 provides the same flexibility for interrupt requests, which is termed a *vectored interrupt* because the interrupt signal directs, or vectors, the CPU to a specific subroutine in memory to service its request.

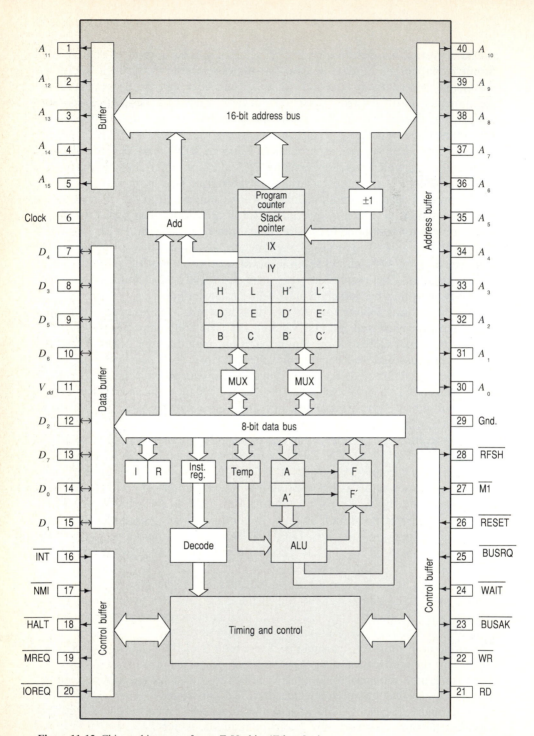

Figure 11-13 Chip architecture of type Z-80 chip. (*Zilog, Inc.*)

400

Table 11-10 Z80 control signals

Signal	Meaning
$\overline{\text{MREQ}}$	Memory request
$\overline{\text{IOREQ}}$	I/O request
$\overline{\text{RD}}$	Read from memory or I/O
$\overline{\text{WR}}$	Write to memory or I/O
$\overline{\text{M1}}$	Instruction read
$\overline{\text{RFSH}}$	Refresh DRAM
$\overline{\text{BUSRQ}}$	Bus request
$\overline{\text{BUSAK}}$	Bus request acknowledge
$\overline{\text{INT}}$	Interrupt
$\overline{\text{NMI}}$	Nonmaskable interrupt
$\overline{\text{WAIT}}$	Wait
$\overline{\text{RESET}}$	Reset
$\overline{\text{HALT}}$	Halt

As shown in Fig. 11-13, the pins of the Z80 are organized into four groups: pins 1 to 5 and 30 to 40 comprise the 16-bit address bus; pins 7 to 10 and 12 to 15 are the 8-bit data bus; pins 16 to 28 carry control signals; pins 11 and 29 are for the power supply, and pin 6 is the clock input. Similarly, the control signals can be placed into four categories, the most complex of which have to do with memory and I/O access. As listed in Table 11-10, the signal $\overline{\text{MREQ}}$ (that is, pin 19 at state 0) means that the CPU accesses memory and $\overline{\text{RD}}$ or $\overline{\text{WR}}$ causes either a read or write action. The signal $\overline{\text{M1}}$ is present during an instruction fetch and $\overline{\text{RFSH}}$ is used in connection with register R to refresh DRAM after each instruction fetch.

External devices request access to the data and address buses by signaling $\overline{\text{BUSRQ}}$ and the Z80 responds with $\overline{\text{BUSAK}}$, which acknowledges the request and frees the buses for use. Interrupt requests, $\overline{\text{INT}}$, cause the CPU to set an *interrupt mask* flag which prevents further interrupts. During execution of the interrupt subroutine, no further interrupt signals are recognized except $\overline{\text{NMI}}$, a nonmaskable interrupt. This feature is reserved exclusively for very important external events such as a power-supply malfunction, which halts all execution. The remaining control signals in Table 11-10 are similar to those previously discussed.

The instruction set for the Z80 totals 150, making it an extremely flexible microprocessor. Furthermore, the design is such that only a single 5-V power supply is required, which considerably simplifies external circuitry. Both of these features are also contained in the Z8000 and the Z80000, which are 16-bit and 32-bit processors, respectively.

16-Bit and 32-Bit CPUs

Increased precision in microprocessor circuits is attained through the use of longer data words, and 16-bit and 32-bit devices have become available as microprocessor chip design has progressed in sophistication. In addition, it proves desirable to

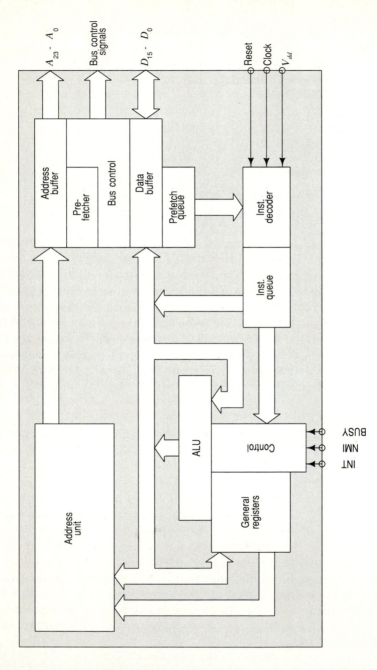

Figure 11-14 Simplified block diagram of type 80286 microprocessor chip. (*Intel Corporation*)

employ longer address words to that larger memories can be accessed and also faster clock speeds in order to reduce execution time. Modern 16- and 32-bit chips achieve these goals by using longer word lengths and also by employing advanced chip architectures.

Consider the type 80286 CPU in Fig. 11-14 which has a 16-bit data bus together with a 24-bit address bus and is comprised of 140,000 transistors on a single chip. The chip architecture consists of four independent units which operate in parallel with one another to optimize performance: the bus unit, instruction unit, execution unit, and address unit. This permits overlapping of instruction fetches, decoding, and execution, in contrast to the sequential fetch/execute routines of more conventional designs.

The bus unit performs all bus operations for the CPU by generating the address, data, and command signals necessary to access memory and I/O units. While not engaged in bus duties, the bus unit prefetches instructions from memory, assuming that program execution proceeds sequentially. These instructions are stored in a 6-byte prefetch queue where they are available for use by the instruction unit. In the case of a subroutine or interrupt, the bus unit resets the queue and begins fetching instructions from the new program location. Prefetching eliminates idle time that can cause the CPU to wait for the next sequential instruction to be fetched from memory.

The instruction unit decodes instructions from the prefetch queue and places decoded instructions in a three-deep instruction queue. These are used sequentially by the execution unit, which is composed of the ALU, the control unit, and general registers, to execute each instruction. The independent and overlapping operation of each unit, together with the use of instruction queues, is term *pipelined architecture* because, in effect, the operations move separately in four separate sequential tracks, or pipelines.

Addresses for use by the bus unit are generated in the address unit using two different modes of operation. In the real-address mode the 80286 uses a 20-bit address to directly access $20^{20} = 1,048,576$ locations in memory or I/O. A greatly expanded memory capacity is possible by means of a technique called *virtual memory* in which the mass storage capability of a magnetic disk with $2^{30} = 1,073,741,821$ locations can be addressed by a 24-bit address. This is accomplished by dividing the mass memory storage into $2^6 = 64$ parts, or *pages*, each holding $2^{30}/2^6 = 2^{24} = 16,777,216$ locations. Thus, a 6-bit preaddress selects the appropriate memory page which is placed in RAM and a subsequent 24-bit address finds the desired location on the page. This virtual memory technique combines the advantages of mass storage of magnetic disks with the rapid access of semiconductor RAM. The 80286 has a 102-instruction set and can operate with a 10-MHz clock, making it a very powerful processor. It has a 32-bit counterpart, labeled the 80386.

An important 32-bit processor is the type MC68020, Fig. 11-15, which also has 32-bit address words, allowing it to access directly $2^{32} = 4,294,967,296$ locations in memory and I/O. Chip architecture employs a three-stage instruction pipeline and is similar to the type 80286 except for the addition of a so-called cache memory.

This on-chip instruction cache holds 64 instruction words to enable very rapid execution of short subroutines or loops. Typical programs spend much of the execution time in a few short routines, and having these instructions execute directly from the high-speed cache memory significantly reduces the total execution time.

The processor also employs dynamic bus sizing that automatically determines the number of bits in a data word needed by external devices on a cycle-by-cycle basis. This permits highly efficient access to devices having differing data bus widths.

A number of different parameters specify the performance of microprocessors and determine their suitability for any given application. Table 11-11 summarizes the more important features of several popular microprocessors to illustrate the range of values practically available as well as the progression in sophistication of chip design. The size of data and instruction words is significant in determining numerical precision and the direct addressing capability. Note that memory capacity in Table 11-11 is indicated by a commonly used power-of-2 convention. That is, K is the symbol for $2^{10} = 1024$ bytes, or *kilobytes*, and M signifies $2^{20} = 1,048,576$ bytes, or one *megabyte*, while $G = 2^{30} = 1,073,741,824$ bytes, or one *gigabyte*.

The number of basic instructions can be misleading since one powerful instruction such as, for example, multiply, can be more significant that a number of minor actions. Nevertheless, the more powerful microprocessors tend to have larger instruction sets although recently so-called reduced instruction set computers, *RISC* architectures, have been developed for very high speed processing. Many other parameters are also used to characterize microprocessors. Some of these are the interrupt hierarchy, I/O control, and electric power requirements. For a given application any one of these, as well as others, may prove more important than those listed in the table.

Table 11-11 Microprocessor characteristics

Type	Word size data/address	Direct addressing range in bytes	Basic instructions	General-purpose registers	Clock rate, MHz
Intel 4004	4/12	4K	46	16	1
Intel 8085	8/16	64K	78	7	5
MC6809	8/16	64K	89	4	2
Z80	8/16	64K	150	14	2.5
Intel 80286	16/20	1M	102	8	10
MC68000	16/24	16M	56	8	8
Z8000	16/23	8M	110	16	4
Intel 80386	32/32	4G	111	16	17
MC68020	32/32	4G	65	16	25
Z80,000	32/32	4G	110	16	25

Single-Chip Devices

Single-chip microprocessors incorporate ROM, RAM, and clock circuits on the same chip with the CPU. Since the ROM program must be introduced during fabrication of the integrated circuit, each single-chip device is designed for a specific purpose. One basic chip architecture may, of course, be used for a wide range of applications simply by including alternate ROM programs during fabrication. Such complete devices have come to be called *microcomputers*.

The chip architecture of a typical 4-bit microcomputer is illustrated in Fig. 11-16. This device has a single 4-bit input port, and an 8-bit output bus. The output buffer includes a custom-programmed decoder which is adjusted at the same time as the ROM is fabricated. The output decoder matches the microcomputer output data to the input requirements of the specific device it controls.

Up to 16 data-control output bits are also available to control input signals from external devices. These outputs lines may be used, for example, to multiplex

Figure 11-15 Photomicrograph of type MC 68020 microprocessor chip. (*Motorola, Inc.*)

the signals from several input units, or to scan the keys of a typewriter keyboard input, or to indicate to external instruments that the microcomputer is ready to send or receive data signals, etc. The overall measurement or control process is under the central control of the microprocessor, and such *handshake* signals are a necessary part of the system. This descriptive term arises because, in effect, the several I/O units and the microprocessor must recognize each other's characteristics before they can interact and exchange data.

This microcomputer uses 8-bit instruction words, but 6-bit addresses are sufficient to access the 2^6, or 64, individual 4-bit words in RAM. The 1024 words in ROM are accessed with only 6-bit address words by dividing ROM into 16 pages of 64 words each. A binary counter sequentially addresses each ROM instruction word on a page, and the 4-bit page address register stores the current page address. In this way the instruction register need only keep track of the 64 words on each page after a previous instruction has specified the page. This is another example of the use of virtual memory.

The instruction set for a typical microcomputer of this kind, type MC14 1200, comprises 43 instructions and one level of subroutine nesting can be ac-

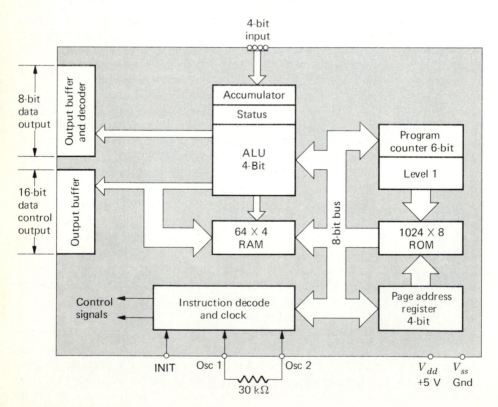

Figure 11-16 Chip architecture of single-chip microprocessor.

commodated. The frequency of the internal clock is about 500 kHz with a 30-kΩ resistor connected to two external terminals, as in Fig. 11-16. The clock frequency is reduced by about a factor of 10 when the value of the external resistor is increased. Alternatively, an external clock may be applied to the same terminals and operation down to dc is achievable. A manual or automatic input signal to the INIT terminal initializes all registers and starts the program.

It often proves inconvenient to specify the ROM program as required for single-chip devices, either because the number of units involved is relatively small, or because changes in the program are anticipated. A useful compromise is a microprocessor with the on-chip features described above, except for ROM. In this case external ROM chips are used, which can be replaced as the application requires or changes. There are, in addition, special single-chip devices which include an on-chip EPROM, but these are usually restricted to special applications by economic considerations.

I/O PERIPHERALS

A myriad of input and output devices are used to communicate with microprocessor chips. Among the most popular are a typewriter-like keyboard for input, and a television-like display for output. In addition, magnetic disks are used both to enhance memory capacity and to provide permanent record storage. Simple examples of these three I/O devices are examined in the following sections.

Keyboard Scan

Alphanumeric information can be introduced into a microprocessor through a keyboard array which has keys similar to the familiar typewriter arrangement. It is only necessary for the microprocessor to sense which keys have been depressed and to store the sequence of key closures in memory.

This can be accomplished by the matrix scan circuit illustrated in Fig. 11-17 for a microprocessor system based on the Z80 chip. The keyboard is arranged electrically into eight rows and columns and each row is accessed by an individual address starting with address $3801H$ and ending with address $3880H$. Depressing a key places a 1 in the bit corresponding to the key's column, so that the column latch byte contains a 1 for the depressed key column bit and 0 for all other bits. The combination of the row address and the column latch byte yields an index number which is used in a look-up table to identify the character corresponding to the depressed key. The processor proceeds sequentially through all the row addresses, thereby scanning the entire keyboard.

A program to accomplish a two-row scan is shown in Table 11-12. Instruction 2 puts the column byte for the first row, address $3801H$, in the accumulator and the logical OR operation is used to test A for zero. If the result is zero, the address of the second row is introduced and the corresponding column byte tested for zero. If this result is also zero, the process repeats because of the jump instruction in instruction 7.

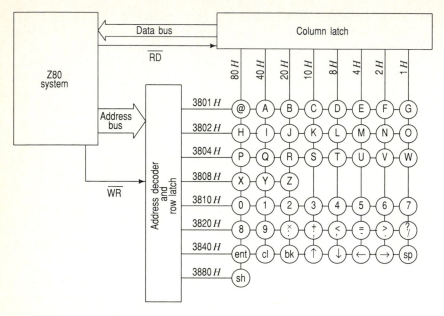

Figure 11-17 Matrix keyboard scan circuit for Z80 microprocessor system.

Table 11-12 Two-row keyboard scan

Instruction no.	Instruction			Action
1	KEYSCN:	LD	C, O	First row
2		LD	A, (3801*H*)	First row address
3		OR	A	Test for zero
4		JP	NZ, KEY9	Jump on nonzero
5		LD	A, (3802*H*)	Second row address
6		OR	A	Test for zero
7		JP	Z, KEYSCN	No key; start over
8		LD	C, 8	Second row
9	KEY9:	LD	B, OFF*H*	Put −1 in B
10	KEY10:	INC	B	Column count
11		SRL	A	Shift right
12		JP	NZ, KEY10	Jump on nonzero
13		LD	A, B	Column count in A
14		ADD	A, C	Add row number
15		LD	C, A	Index 0–15 in C
16		LD	B, O	Clear B
17		LD	H/L, TABLE	Table address in H/L
18		ADD	H/L, BC	Character address
19		LD	A, (H/L)	Character code in A
20		LD	(DISPLY), A	Store for display
21		JP	KEYSCN	Start over

Whenever a 1-bit is detected, the program calculates the address of the character corresponding to the depressed key, beginning with instruction 9. Register C already holds either 0, indicating row 1, or 8, meaning row 2, as a result of instructions 1 or 8, and A contains a number between $1H$ and $80H$ corresponding to the column latch byte. Instructions 9 through 12 convert this hexadecimal number to a decimal number by placing the hexadecimal equivalent of -1, that is, $OFFH$ (see Appendix 2), in register B and counting the number of right shifts of A needed to reach zero. The column number is placed in A (instruction 13) and added to the row number to produce an index number between 0 and 15 in C. Instruction 17 loads the look-up table address into H/L, and instruction 18 gives the character address. The character code is stored for display and the program repeats.

This program only scans the first two rows, but it is clear how it can be extended to cover the entire keyboard. The program can be present in memory as a subroutine and be called by periodic interrupts of the main program. Because of the operating speed characteristic of modern microprocessors, the operation appears to be continuous.

CRT Display

Probably the most versatile output device for a microprocessor is the cathode-ray tube. A CRT display can be activated in two fundamentally different ways: by a *raster scan* in which the electron beam is moved rapidly horizontally and somewhat more slowly vertically in order to scan the entire tube face as in a television picture tube, or by an *x-y display* where the beam is sequentially placed at selected places on the screen in order to trace out a desired pattern. While both of these are most often accomplished using external circuitry, the microprocessor itself is also capable of directing the display.

Consider the case of the *x-y* display shown in Fig. 11-18 for a microprocessor system based on the type 8085 chip. It is assumed that the *x* and *y* coordinates of each point in the display pattern are available in RAM, as the result, for example, of some previous calculation. The 8085 sequentially places the *x* coordinate in the *x*-data latch and D-A converter followed by the corresponding *y* coordinate in the *y*-data latch and D-A converter and then turns on the electron beam. Each data-point pair is displayed in succession and the cycle is repeated, rapidly enough to avoid flicker.

A program to accomplish this display is shown in Table 11-13. The addresses XPNT, YPNT, and NPNT are those of the first *x* and *y* pair and that of the total number of pairs. Instructions 1 through 6 load these addresses into registers B, C, and D, respectively. Instructions 7 through 10 put the first *x-y* pair in their respective latches and D-A converters, thus positioning the spot on the screen, and a delay circuit turns on the electron beam after this is accomplished. Subsequently, each *x-y* pair is displayed in succession until the plot is completed, at which point instruction 15 causes the image to be repeated.

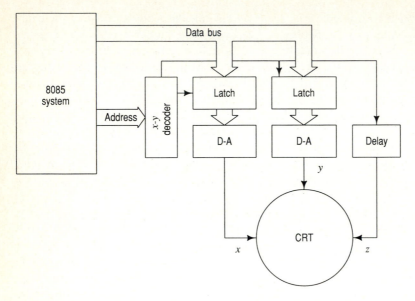

Figure 11-18 Simplified block diagram of *x-y* oscilloscope display for 8085 microprocessor system.

Table 11-13 XY oscilloscope display

Instruction no.	Instruction			Action
1	INIT:	LDA	XPNT	Address of first *x*-value in A
2		STA	B	Address of first *x*-value in B
3		LDA	YPNT	Address of first *y*-value in A
4		STA	C	Address of first *y*-value in C
5		LDA	NPNT	Number of points in A
6		STA	D	Number of points in D
7	LOOP:	LDAX	B	Put *x*-value in A
8		OUT	XDAC	Put *x*-value in *x*-D-A
9		LDAX	C	Put *y*-value in A
10		OUT	YDAC	Put *y*-value in *y*-D-A
11		INR	B	Increment *x*-address
12		INR	C	Increment *y*-address
13		DCR	D	Decrement point count
14		JNZ	LOOP	Plot next point
15		JMP	INIT	Start over

Disk Memories

The most convenient large-capacity memories for use with microprocessors are provided by magnetic disks, in which digital signals are recorded as serial bit streams on the magnetic coating of a disk, as described in Chap. 10. The disk is divided into concentric circles called tracks and into equal arcs called sectors. Data to be stored are accompanied by control signals which specify track and sector numbers so that the data may subsequently be relocated and played back.

Two types of magnetic-disk storage are in use, *floppy disks*, in which the magnetic coating is carried on a flexible plastic base, and *hard disks*, which are rigid and generally larger in diameter. Floppy disks offer less expensive construction whereas the rigidity of a hard disk permits greater mechanical precision, leading to higher recorded bit densities and correspondingly increased memory capacity. A typical floppy disk 3.5 inches in diameter can store hundreds of kilobytes, compared with hundreds of megabytes on a 10-inch hard disk.

It is common practice to use disk memories for permanent storage of both programs and data. These are read into RAM for execution and subsequently returned to the disk. Good operating practice requires that a copy of each disk be made because of the ease with which the magnetic pattern can be changed by stray magnetic fields.

SUGGESTIONS FOR FURTHER READING

Bruce A. Artwick: "Microprocessor Interfacing," Prentice-Hall, Inc., Englewood Cliffs, N.J., 1980.

S. J. Cahill: "Digital and Microprocessor Engineering," John Wiley & Sons, Inc., New York, 1985.

Raymond P. Capece and John G. Posa (eds.): "Microprocessors and Microcomputers," McGraw-Hill Book Company, New York, 1981.

Harry Garland: "Introduction to Microprocessor System Design," McGraw-Hill Book Company, New York, 1979.

EXERCISES

11-1 Using the 4-bit instruction set in Table 11-2 write a program to enter the data word 1000 in the I/O register at memory address 1111. What is the total number of memory words in the program? Ignore the HALT instruction.

Answer: four

11-2 Modify the simple addition program in Fig. 11-6 to have the data word 0001 entered from the I/O register and to have the sum both stored in memory and transferred to the I/O register. What is the total number of memory words in the program itself?

Answer: seven

11-3 Show the contents of the CPU registers A, DC, IR, PC, and the I/O register as the program of Exercise 11-2 is run. What is the total number of steps executed by the CPU in completing the program?

Answer: twenty

11-4 Determine the action of the program in Table 11-14 by writing the interpretation of each subsequent instruction. What are the contents of the last three memory locations after the program is run?

Answer: $1000 - 0010 = 0110$; 0001, 0010, 0110

Table 11-14

Address	Contents
0000	0010
0001	1110
0010	0100
0011	1100
0100	0010
0101	1101
0110	0110
0111	0010
1000	1111
1001	0110
1010	1000
1011	0000
1100	
1101	0001
1110	0010
1111	1000

Table 11-15

Address	Contents
00	CL A
01	ADD 05
02	ADD 06
03	STA 07
04	HALT
05	12
06	10
07	00

11-5 Show the contents of the CPU registers A, DC, IR, and PC as the program of Exercise 11-4 is run. What is the total number of steps executed by the CPU in completing the program?

Answer: thirty-one

11-6 What are the contents of the accumulator and memory locations 05, 06, and 07 after execution of the program in Table 11-15?

Answer: 22, 12, 10, 22

11-7 Write a program to move data at memory address 20 to address 21. The program should start at address 10. How many words are there in the program itself?

Answer: four

11-8 Write a program to subtract the number at memory address 20 from that at address 21. Store the result at address 20 and begin the program at address 50. How many words are there in the program itself?

Answer: seven

Table 11-16

Address	Contents
20	CLR A
21	ADD I 25
22	ADD I 26
23	STA 27
24	HALT
25	30
26	31
27	00
⋮	⋮
30	10
31	25
32	00

11-9 What are the contents of the accumulator and memory addresses 25, 26, and 27 after execution of the program in Table 11-16?

Answer: 35, 30, 31, 35

11-10 Write a program to determine the most significant bit of the data stored at memory address 50. Store the result at address 51, and use instructions for rotate and skip. How many words are there in the program itself?

Answer: eight

11-11 Develop a table of keyboard addresses and column latch bytes corresponding to each of the keys in the first two rows of the keyboard scan circuit, Fig. 11-17.

TWELVE

MICROPROCESSOR CIRCUITS

Circuit applications of microprocessors employ the speed and versatility characteristic of such devices. One widespread application is to personal computers, which are complete digital computer systems small enough to fit on a desk top. Because of their programming flexibility, both microprocessors and personal computers can emulate many different kinds of instruments.

PROCESSING CIRCUITS

A-D Converter

Fundamental to the use of microprocessors in measurement and control applications is the conversion of analog signals into digital data usable by the processor's program. While most often this is accomplished using separate A-D converter circuits such as those described in Chap. 10, it also proves feasible to program the microprocessor itself to carry out the conversion. Consider the circuit in Fig. 12-1, which develops the digital equivalents of eight separate analog inputs. The program selects each of the inputs in succession, supplies and compares trial values for a successive-approximation analog-to-digital conversion routine, and stores the final results.

The eight inputs are each selected by a 3-bit word on the data bus that addresses the multiplexer through a latched I/O port. The analog output of the multiplexer is retained by a sample-and-hold amplifier for comparison with the output of a digital-to-analog converter. The DAC digital word trial input comes from the microprocessor as the first 8 bits on the address bus during a memory read instruction. Depending upon whether the trial value is greater or less than the input signal, the trial value is rejected or retained and a subsequent value introduced. Note that, in effect, this system uses the data bus to address the multiplexer and the address bus to transmit data to the DAC. These are specific examples of the flexibilities inherent in microprocessor control systems.

Operation of the system can best be understood by examining the program listing in Table 12-1. In the type 8080 microprocessor, register pair H/L is used

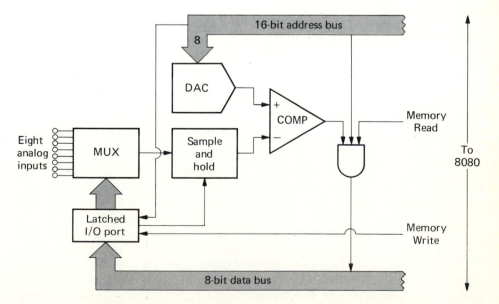

Figure 12-1 Type 8080 microprocessor controls A-D conversion of eight analog input signals.

Table 12-1 A-D Conversion program

Instruction no.	Instruction			Action
1	BEGIN:	LXI	D, 1808H	Load D with 18H and E with 08H
2	START:	LXI	H, 8000H	Load H with 80H and L with 00H
3		MOV	A, D	Move D to accumulator
4		MOV	M, A	Address MUX and sample input
5		MOV	A, E	
6		MOV	M, A	Address MUX and hold input
7	ADCON:	LXI	B, 8000H	Load B with 80H and C with 00H
8		MOV	A, B	
9		MOV	H, A	Peripheral device address
10	TEST:	ORA	C	Store trial value in C
11		MOV	L, A	
12		MOV	A, M	Memory read of comparator
13		ANA	A	Reset carry flag
14		JPO	TOOHI	Jump on odd parity to TOOHI
15		MOV	A, B	
16		ORA	C	
17		MOV	C, A	Keep trial value in C
18	TOOHI:	MOV	A, B	
19		RAR		Rotate A through carry
20		MOV	B, A	
21		JNC	TEST	Jump on no carry to TEST
22		LXI	H, 2000H	Initialize data storage
23		MOV	L, E	
24		MOV	M, C	Store final value
25		DCR	D	Decrement MUX sample counter
26		DCR	E	Decrement MUX hold counter
27		MOV	A, E	
28		JNZ	START	Jump on no zero to START
29		HLT		

as the data counter, and this program also employs register pair D/E analogously to address the multiplexer for the input selection. Register B contains the successive-approximation trial value, and register C temporarily stores the current trial result until the full approximation routine is completed.

The program employs a *memory-mapped* I/O technique in which the various peripheral devices appear as memory addresses to the microprocessor. This means that the multiplexer and comparator are accessed by memory read and write instructions and are given addresses beginning with 8000H. The eight final results are stored in memory at addresses from 2001H to 2008H.

According to Table 12-1, the program begins by loading address 1808H into register pair D/E and address 8000H into register pair H/L. The contents of the D register specify the address of the first input and, the 1 bit in the fourth significant bit position (that is, 10H, or 0001 0000) activates the multiplexer to obtain the analog signal on this input. Similarly, the contents of the E register also address the first input and the zero bit in the fourth position causes the sample-and-hold amplifier to retain this value for comparison with the DAC output. Thus, instructions 3 through 6 present the analog signal at the selected input to the comparator.

The actual A-D conversion occupies instructions 7 through 21. The successive approximation begins by placing the largest test value, 80H, in register B. The OR logic in instruction 10 places the current trial value in register C and instruction 11 places this value into the lower address byte to be transmitted to the DAC. This is accomplished by instruction 12, MOV A, M, a memory read instruction, which places the H/L address on the address bus and also causes the memory-read control line to be TRUE. The memory address in H/L activates the AND gate because of the memory-mapped I/O address, 80H, in register H, and also supplies the digital trial word in register L to the DAC. If the output of the DAC exceeds the analog input signal, the AND gate is TRUE and 01H is read into the accumulator. Conversely, if the DAC output is smaller, the AND gate is not activated and the the contents of the accumulator are 00H.

The conditional jump command in instruction 14 retains the trial value in C (instructions 15, 16, and 17) if the DAC output is smaller (even parity, 00H), or rejects it if it is greater (odd parity, 01H). In either case a new, lower trial value is obtained for the next approximation by a right rotation of the accumulator and the test is repeated. Eight trials are necessary for the 8-bit conversion and this is recognized by the conditional jump JNC, which tests the status of the carry flag. After eight trials the 1-bit initially placed in the B register (80H, or 1000 0000) is rotated into the carry flag and the routine proceeds to instruction 22. Note that a prior instruction ANA A has been used to be sure the carry flag is reset before the rotation.

Instructions 22, 23, and 24 store the final result at the memory address corresponding to 20H plus the corresponding input address in E. Succeeding instructions decrement D and E to address the next input and repeat the entire process for it. Finally, after all eight input signals have been converted and stored, the program halts.

Event Counter

It is possible to use the memory capability of microprocessor RAM to store event pulses as in the case of the binary counters discussed in Chap. 9. The circuit in Fig. 12-2 is a six-channel event counter which stores the total number of input pulses in each of six channels and displays the total number of counts in any one of the channels on a six-digit, seven-segment display.

Input pulses are stored in six flip-flops that are read periodically, stored in RAM, and subsequently reset. All six can be read and reset, simultaneously using six-bit words. The channel count to be displayed is determined by the channel select switch, which produces a six-bit word corresponding to each channel. The program resides in a 1K-byte ROM addressed by A_9 to A_0, while the 128-byte RAM needs only A_7 to A_0. The lower 8 bits of the address are latched from the multiplexed 8-bit address/data bus characteristic of the type 8085 microprocessor.

It is important that the input flip-flops be read often in order to attain the highest possible counting rate. This is best accomplished by executing read/reset

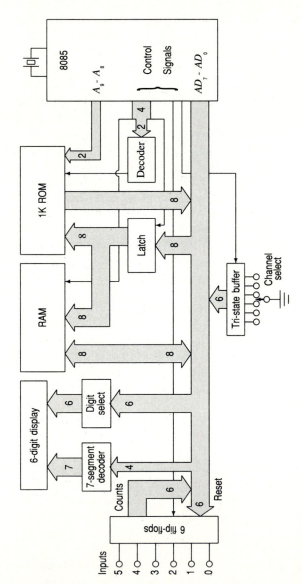

Figure 12-2 Block diagram of six-channel event counter.

cycles that alternate with display digits. Thus, for example, after a read/reset action, the zero digit is displayed, followed by read/reset, digit one, read/reset, digit two, etc. After digit five is displayed, the channel select switch is read, followed by read/ reset, and the cycle repeats. In this way the dead time is only that associated with displaying one digit and also the display is refreshed rapidly enough to eliminate flicker.

An event counter based on a microprocessor is not economically competitive with one constructed from readily available counter chips. If, however, it is intended to further manipulate the stored information in some manner, e.g., to compute average counting rates, the distribution of pulses in time, etc., the microprocessor approach may be preferable.

Lock-in Detector

A microprocessor is easily programmed to emulate the functions of the lock-in amplifier or phase-locked loop discussed in previous chapters to recover signals masked by noise. The basic idea, Fig. 12-3, is to turn the signal on and off rapidly. During the time the signal is on it is sampled and added a number of times, and during the time the signal is off the input is subtracted an equal number of times. The result is that spurious signals that differ from the chopping frequency are added and subtracted equally and averaged to zero, while the desired signal is simply added.

A program to accomplish lock-in detection for a time equal to a given N periods is shown in Table 12-2. Instructions 4 and 5 activate the electronic switch to turn on the signal and 6 through 10 add the signal (plus noise) ten times. Similarly, instructions 12 and 13 turn the switch off and 14 through 18 subtract the noise signal from the previous signal plus noise additions. This is repeated N times, and the value of the input signal times $10N$ remains in H/L.

The number of times the signal is added and subtracted as well as the number of iterations, N, depends upon the desired response time and the speed of the microprocessor. Obviously, many additions and many iterations increase the accuracy of the measurement. For best performance, however, the input signal should remain substantially constant during the N periods, and this may limit the averaging time. Note also that instruction 7 could be a subroutine call for an A-D conversion routine such as the program in Table 12-1.

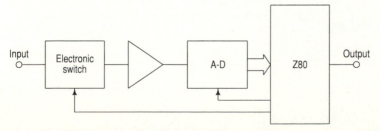

Figure 12-3 Microprocessor circuit used as lock-in detector.

Table 12-2 Lock-in detector program

Instruction no.	Instruction			Action
1	LOCKIN:	LD	C, N	Lock-in for N periods in C
2		LD	D, O	Clear D
3		LD	H/L, O	Clear H/L
4	AVLOOP:	LD	A, 1	
5		OUT	(S), A	Turn on signal (plus noise)
6		LD	B, 10	
7	ONLOOP:	IN	A, (A/D)	A-D signal in A
8		LD	A, E	A-D signal in E
9		ADD	H/L, D/E	Add signal in H/L
10		DJNZ	ON LOOP	Jump to ONLOOP 10 times
11		LD	A, O	
12		OUT	(S), A	Turn off signal
13		LD	B, 10	
14	OFLOOP	IN	A, (A/D)	Noise signal in A
15		LD	E, A	Noise signal in E
16		OR	A	Clear carry flag
17		SUB	H/L, D/E	Subtract noise from H/L
18		DJNZ	OFLOOP	Jump to OFLOOP 10 times
19		DEC	C	
20		JR	NZ, AVLOOP	Do N periods
21		HLT		Signal (X10N) in H/L

Figure 12-4 Pocket calculator microcomputer. (*Texas Instruments, Inc.*)

MICROCOMPUTERS

Calculators

The most widespread application of microprocessors is to the familiar pocket-sized calculator, Fig. 12-4. Such calculators use microprocessor principles to perform arithmetic calculations extending from simple addition and subtraction to evaluation of transcendental functions. The range of capability is exhibited by the keyboard nomenclature visible in Fig. 12-4. Advanced versions incorporate programming capacity and exemplify quite completely the full scope of microcomputer operation.

Obviously, the keyboard of the calculator represents the input unit while the numerical display, which is often of the seven-segment liquid crystal type, is the output unit. The keyboard introduces data numbers into memory and also selects instruction words to carry out the desired computation. Numbers entered by the keyboard are indicated on the display, as are results of the calculations.

The operation of a typical microcomputer calculator may be described with the aid of the simplified block diagram in Fig. 12-5. The keyboard and the display

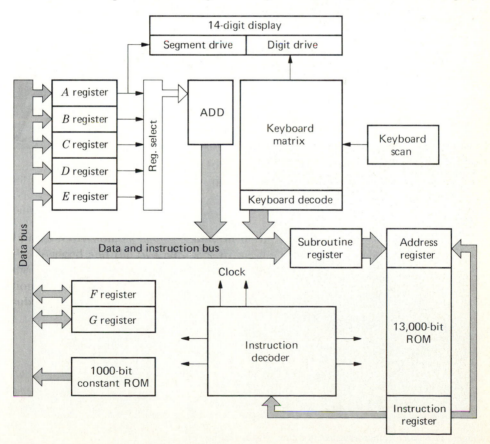

Figure 12-5 Simplified block diagram of pocket calculator. (*Texas Instruments, Inc.*)

are multiplexed by the keyboard scan circuit which sequentially pulses each key. Those keys representing numbers are connected to the digit drive circuit much as in Fig. 9-35. Display segments are activated in synchronism by the segment driver in accordance with words stored in the A register.

Depressing a key causes the keyboard encoder to generate a digital word corresponding to the number or instruction represented by that key. Numbers are transported to the A register, while instruction words are passed to the subroutine register, where they may be modified by the result of a previous calculation and thence to the address register of a 13,000-bit ROM. This ROM contains the instructions, subroutines, and look-up tables necessary to complete the calculations. ROM instruction words are interpreted by the instruction decoder and used to activate all control logic. Under direction of the control logic, instruction words may be looped back into the ROM, thus generating a series of steps to compute, for example, a mathematical function by series approximation.

Figure 12-6 Microcomputer calculator chip. (*Texas Instruments, Inc.*)

Actual calculations are done by the ALU consisting of registers A through G together with a full adder and a 1000-bit ROM. Words in selected pairs of registers can be added, exchanged, or compared and the results placed in a third register. The F and G registers are used principally to store intermediate results, and the 1000-bit ROM holds 16 numerical constants used in calculations. A typical example is the constant ln 10, 2.302585092994, which is used in computing logarithmic functions.

Many calculator designs employ separate ROM chips for convenience in design, but single-chip microcomputers are also common. A typical integrated-circuit chip, Fig. 12-6, may contain some 25,000 transistors, and, when accompanied by a keyboard and display, becomes a complete microcomputer calculator.

Personal Computers

An important application of microprocessors is to individualized microcomputer systems which typically are comprised of a keyboard input, CRT display, and one or more floppy disks. Such complete systems have come to be called *personal computers*, Fig. 12-7. Additional ports are also provided for data input and output. In particular, it is common practice to attach a *printer* capable of producing a paper copy of the information shown on the CRT, both conventional alphanumerics and graphic displays.

The internal circuitry of the Macintosh personal computer, Fig. 12-8, is centered around the type MC68000 microprocessor, an earlier version of the MC68020 chip discussed in the previous chapter. The MC68000 has 32-bit internal

Figure 12-7 Macintosh personal computer.

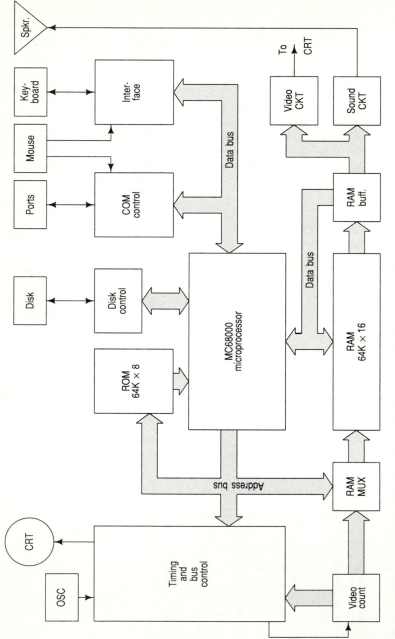

Figure 12-8 Block diagram of Macintosh personal computer. (*Apple Computer Corp.*)

registers, a 16-bit data bus, and a 24-bit address bus. The system has a 64K-byte ROM which contains routines for internal operations and I/O and 128K bytes of RAM. These are expanded to 128K bytes and 512K bytes, respectively, in upgraded versions of the Macintosh personal computer. The latter increase is accomplished simply by replacing the two 64K-byte RAM chips with more advanced 256K-byte chips.

A small hand-held controller, called the *mouse*, is used in conjunction with the keyboard to position a cursor on the CRT screen in order to select portions of the display for some desired action. In other personal computers cursor positioning is done with special keyboard keys. The two serial I/O ports are conventionally used to connect a second magnetic disk, which may be of either the floppy- or hard-disk type, and a printer, although other devices may be attached as well. The internal floppy disk holds 400K bytes if used single-sided and 800K bytes if both sides of the disk are used.

The CRT display is of the scanning type with 342 horizontal lines each capable of a resolution of 512 spots, or *pixels*. The 512×342 pixels are *bit-mapped* in memory; that is, each pixel is represented by a specific memory location. Thus the CRT display is held in memory where it can be adjusted in accordance with program results and periodically scanned in synchronism with the deflections of the CRT electron beam to produce the CRT image. An audio output is also available through a small loudspeaker. This is used to produce short tone bursts to call attention to some action called for by the program that must be taken by the operator. In conjunction with specialized programs, it can also be used to play music and even produce recognizable speech.

The programs and data used with personal computers are stored on magnetic disks. At start-up, ROM programs call for insertion of a disk and then copy the disk program into RAM for execution. After conclusion of the program, new results are returned to the disk for permanent storage. The results can, of course, also be printed for ready reference. Because of the wide range of programs available, and the possibility of developing a special program to carry out any specific task, personal computers find a very wide range of application.

MICROCOMPUTER INSTRUMENTS

The programming flexibility, together with the ready availability of I/O peripherals, make personal computers extremely useful in emulating many laboratory instruments for measurement and control. An additional benefit is that once acquired data is stored in memory, it is available for immediate calculation, which can lead to output displays in the form of tables and charts and/or output control signals.

Digital Voltmeter

An external A-D converter together with the appropriate program immediately makes a personal computer into a digital voltmeter in which the magnitude of

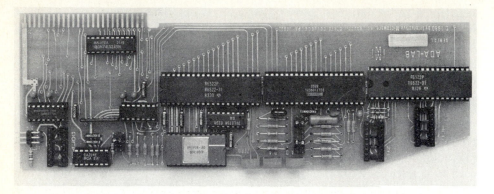

Figure 12-9 A-D converter card of personal computer digital voltmeter. (*R. C. Electronics, Inc.*)

analog input signals can be displayed on the CRT screen. Most often the A-D converter is combined with other circuit components on a printed circuit board which plugs directly into socket slots provided in the personal computer, Fig. 12-9.

The particular unit illustrated in Fig. 12-9 uses a dual-slope integrator with 12-bit accuracy and has full-scale input voltage ranges from 0.5 to 4.0 V of either polarity. A 12-bit binary ladder network is also included to generate analog output signals of the same magnitudes. Eight digital sense inputs and outputs are used to sense and produce on-off signals on a bit-by-bit basis.

An internal clock permits data acquisition at regular intervals or at given times and also outputs timed control signals to, for example, turn external devices on and off. These features make unattended laboratory measurements quite feasible. As with other personal computer applications, the programs to operate the A-D converter and to organize and display the data are stored on a floppy disk.

Waveform Generator

Similar personal computer circuit board plug-ins yield versatile waveform signal generators. A typical unit can calculate sine, square, triangle, pulse, and ramp waveforms over a frequency range from 0.5 Hz to 5 MHz with amplitudes up to 20 V peak-to-peak. The output can be continuous, gated for the duration of a gate signal, or triggered to produce only one cycle. A burst mode is also available which generates a burst of a given number of cycles ranging from 2 to 255.

These various modes of operation are selected from the keyboard of the personal computer in response to a menu presented on the CRT. The various options, frequency, waveform, amplitude, etc., are displayed and the desired features chosen by depressing certain keys or by positioning the cursor at the appropriate place on the screen. In this way, very elaborate control schemes can be implemented and changed as necessary by programming.

Other waveform generators can produce arbitrary waveshapes in accordance, for example, with a mathematical expression, or by performing a mathematical operation on a stored waveform. New waveforms can also be generated by combining segments from several previously stored waveforms. These selections are menu-driven as in the more conventional waveform generator. Furthermore, the CRT is capable of plotting the selected waveform in a form of oscilloscope display so that it may be modified as necessary.

Oscilloscope

Somewhat more elaborate external circuitry, Fig. 12-10, easily yields a versatile digital oscilloscope. The system incorporates the ubiquitous A-D converter and a RAM with DMA capability so that data acquisition is not limited by the computer's internal cycle time. Subsequently, the digitized waveform data is transferred into the computer memory for display and analysis.

One or more of the input channels can be displayed, Fig. 12-11. Also present on the CRT are several lines of text showing parameters of the display much as would be given by the settings of the panel controls on a standard oscilloscope, such as vertical amplitude and sweep speed. These parameters are adjusted as desired by the keyboard. Waveforms are stored on a floppy disk, so that the system becomes, in effect, a magnetic waveform recorder, and/or printed for later reference.

Of equal significance to the digital oscilloscope application is the fact that stored waveform data is immediately available for further analysis. It is obvious, for example, that the system is a digital voltmeter. It also can be used as a signal averager simply by storing and adding successive waveforms. Performing a Fourier transform on the stored waveform yields the frequency components in the waveform, as from a waveform analyzer. Similarly, the autocorrelation and cross-correlation functions can be calculated.

Clearly, many other similar calculations can be programmed. In this application, the personal computer should be viewed as versatile data-acquisition and analysis system rather than as simply an oscilloscope.

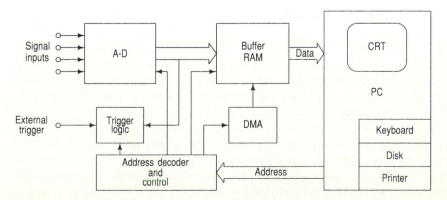

Figure 12-10 Block diagram of digital oscilloscope based on personal computer. (*R. C. Electronics, Inc.*)

Figure 12-11 Personal computer digital oscilloscope. (*R. C. Electronics, Inc.*)

CONTROL CIRCUITS

Process Controller

Microprocessor versatility can be used effectively in industrial process control systems. The process controller in Fig. 12-12, for example, derives a control signal output from a transducer input signal that is compared to a desired set value as modified by a third input signal. Thus the system might adjust fuel flow to a heater in order to control the temperature of a liquid stream where the input/output control characteristic is altered in response to the rate of fluid flow. Such an adaptive-control strategy is very difficult to implement using standard circuitry.

The control program is embodied in PROM and is, of course, pertinent to the particular process being controlled. Thus this one system design can be programmed to control a wide range of different processes. Constants in the control program are entered in RAM through a keyboard (and shown on an associated seven-segment display) so that the control characteristic can be easily adjusted to match the process exactly or changed to improve process control. The control program acts to make the difference between the transducer input signal and the set input signal equal to zero.

The three input signals are multiplexed under control of the microprocessor and each converted to digital form using a successive-approximation subroutine. The D-A converter in this loop is also used to present an analog output signal to the sample-and-hold output circuit. Other subroutines in PROM are used to scan

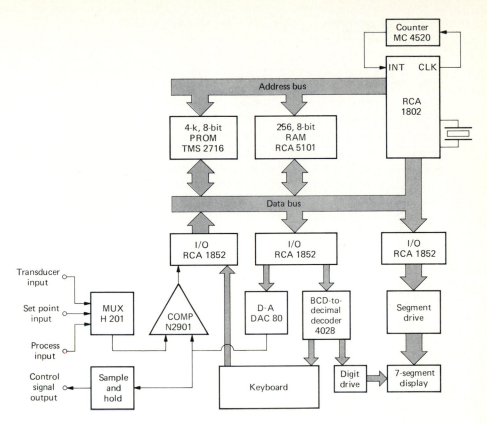

Figure 12-12 Process controller based on type 1802 microprocessor.

the keyboard and activate the seven-segment display. Both the keyboard scan and the digit display scan signals are processed through a BCD-to-decimal decoder to select individual keys and digits.

The main program computation and the three subroutines are interleaved by means of a unique interrupt scheme. A counter divides down the clock signal and feeds an interrupt signal back every 1.5 milliseconds. The microprocessor counts these interrupts and starts a new cycle every 160 counts, or 240 ms. The various subroutines are activated at specific interrupt counts and at all other times the main program is run. This makes it possible to scan the display at 40 Hz to reduce flicker, scan the keyboard twice per cycle, and output data twice per cycle, all independent of the running time of the control program.

Device Controllers

The techniques of the previous section are used in a wide range of devices for measurement and control ranging from home appliances to sophisticated industrial processes such as robotic factories. Characteristic of such systems are input sensors to measure process variables, a keyboard and associated display to introduce desired parameters, and digital and/or analog output signals to control the system.

Simple device controllers such as, for example, to set the cooking time and temperature of a domestic oven are most often single-chip microprocessors. Desired cooking parameters are entered through a simple keypad, and shown on a seven-segment display. A thermocouple senses the oven temperature and the microprocessor turns the oven heater on and off in accordance with the internal program and the specific cooking parameters. Such simple devices are much more flexible and reliable than mechanical or even electrical analog controllers.

The range of application of microprocessor-based device controllers extends from home appliances to office equipment such as telephones and photocopiers and to fuel flow, braking, and ignition control in automobiles. Equally important are automatic traffic lights, supermarket cash registers, and remote TV and hi-fi controllers.

SUGGESTIONS FOR FURTHER READING

A. A. Berk: "Microprocessors in Process and Product Control," McGraw-Hill Book Company, New York, 1986.
Robert J. Bibbero and Davis M. Stern: "Microprocessor Systems, Interfacing and Applications," John Wiley & Sons, Inc., New York, 1982.
John B. Peatman: "Design with Microcontrollers," McGraw-Hill Book Company, New York, 1988.
Murray A. Sargent and Richard L. Shoemaker: "Interfacing Microcomputers to the Real World," Addison-Wesley Publishing Co., Reading, Mass., 1981.

EXERCISES

12-1 Execute the program in Table 12-1 to convert an analog signal of 4.50 V on input 4. The D-A output is 0.039 V/bit. What are the contents of the accumulator, registers B and C, and register pairs D/E and H/L when register C contains 73H for the first time?

Answer: 73H, 01H, 73H, 1505H, 8073H

12-2 Repeat Exercise 12-1 for a 0.04-V signal on input 1. What are the contents of the registers after instruction 27 is completed?

Answer: 07H, 00H, 01H, 1707H, 2008H

12-3 Show analytically how the lock-in detector in Fig. 12-3 eliminates spurious signals by integrating an input signal plus noise, $V = S(t) + N \sin \omega t$ over successive periods, T, in which $S(t)$ is first on and then off.

Answer: $V_0 = 1/T \int_0^T S(t)\, dt$.

12-4 Execute the lock-in detector program in Table 12-2 for a dc input signal of 1.0 V masked by an interfering signal $(10 \sin 7.5 \times 10^5 t)$ V. Assume $N = 4$ and that completing an ONLOOP or an OFFLOOP requires 10^{-4} s. What are the contents in register H/L and what is the corresponding measured value of the signal after each period?

Answer: 7.38, 16.49, 27.41, 39.58; 0.783 V, 0.825 V, 0.914 V, 0.989 V.

ONE

VACUUM TUBE CIRCUITS

The basis for all electronic circuits before the invention of the transistor was the electron vacuum tube. A great many variations in tube design have permitted applications ranging from very sensitive detectors to very-high-power amplifiers. Because of their small size and minimal power requirements, transistors have largely replaced vacuum tubes in electronic circuits, but many applications for which vacuum tubes prove to be useful still remain.

THE VACUUM DIODE

Thermionic Emission

The vacuum diode comprises a hot *cathode* surrounded by a metal *anode* inside an evacuated enclosure, usually glass, as sketched in Fig. A1-1. At sufficiently high temperatures electrons are emitted from the cathode and are attracted to the positive anode. Electrons moving from the cathode to the anode constitute a current; they do so when the anode is positive with respect to the cathode. When the anode is negative with respect to the cathode, electrons are repelled by the anode and the reverse current is zero. The space between the anode and cathode is evacuated, so that electrons may move between the electrodes unimpeded by collisions with gas molecules.

Cathodes in vacuum diodes take several different forms. The free electrons in any conductor, given sufficient energy by heating, will escape from the solid. Some materials are much more satisfactory in this respect than others, however, either because it is relatively easy for electrons to escape, or because the material can safely withstand high temperatures. Tungsten, for example, is a useful cathode material, because it retains its mechanical strength at extreme temperatures. A thin layer of thorium on the surface of a tungsten cathode filament increases electron emission, and appreciable current is attained at temperatures of about 1900 K. This form of cathode is directly heated by an electric current similarly to the filament in an incandescent lamp. It is used in vacuum diodes suitable for high-voltage applications. The conventional symbol for a vacuum diode of this type is shown in Fig. A1-2a.

The modern oxide-coated cathode, which consists of a metal sleeve coated with a mixture of barium and strontium oxides, is the most efficient electron emitter developed to date. Copious electron emission is obtained at temperatures near 1000 K, which means that the power required is much less than for tungsten. Usually the oxide cathode is indirectly heated by a separate heater located inside

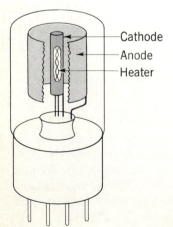

Cathode
Anode
Heater

Figure A1-1 Sketch of vacuum diode.

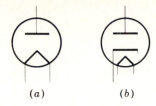

Figure A1-2 Circuit symbols for vacuum diodes, (*a*) filamentary-cathode and (*b*) heated-cathode types.

(*a*)   (*b*)

the metal sleeve, as in Fig. A1-1. This isolates the heater current electrically from the cathode connection, which is a considerable advantage in electronic circuits. Also, an ac heater current may be used without introducing undesirable temperature variations in the cathode at the frequency of the heater current. Oxide cathodes are employed in the great majority of vacuum tubes. The heater is often omitted in the conventional circuit symbol, Fig. A1-2*b*, since it is not directly an active part of the rectifier.

Child's Law

When the anode is negative with respect to the cathode, the electron current is zero and the reverse characteristic is essentially that of an ideal diode. The forward characteristic is determined by the motion of electrons in the space between the cathode and the anode when the anode has a positive potential. Most often, the number of electrons emitted from the cathode is so great that the electric fields due to the electron charges drastically alter the electric field between the cathode and anode produced by the anode potential. Under these conditions the actual current is given by Child's law,

$$I_b = AV_b^{3/2} \tag{A1-1}$$

where V_b is the voltage between anode and cathode, and A is a constant involving the tube geometry. According to Eq. (A1-1), the current increases as the three-halves power of the applied voltage; that is, the vacuum diode is nonlinear.

The current depends upon the separation between anode and cathode and upon the area of the cathode. Thus it is possible to obtain different current-voltage characteristics by altering the geometric shape of the anode and cathode. Two examples of practical diode characteristics are given in Fig. A1-3. The type 5U4 diode is a medium-power rectifier; the type 1V2 diode is designed for high-voltage low-current power supplies. The anode-cathode separation is considerable in the 1V2 in order to minimize the possibility of a discharge passing between anode and cathode on the reverse portion of the voltage cycle. Consequently, the current in the forward direction is smaller than in the case for the 5U4. The 5U4 is designed for use at lower voltages, so the separation between anode and cathode is smaller. The forward current is correspondingly greater at the same forward voltage.

In any practical vacuum diode the reverse current is not truly zero, because of leakage currents on the surfaces of the glass insulators and similar secondary

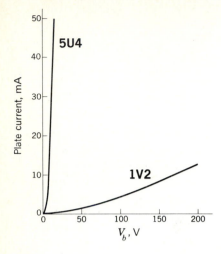

Figure A1-3 Experimental forward characteristics of two practical vacuum diodes are much different because of different geometric construction.

effects. Typically, the reverse resistance is of the order of 10 MΩ. Since the forward resistance may be of the order of 100 Ω at a suitable operating potential, the ratio of the reverse resistance to the forward resistance, or the *rectification ratio*, is appreciable. The interelectrode capacitance between the cathode and anode limits the maximum frequency at which a vacuum diode is useful. The cathode-anode capacitive reactance is effectively in parallel with the electron current and tends to short out the reverse resistance at high frequencies.

VACUUM TUBES

The current in a vacuum diode is controlled by introducing a third electrode, the grid, between the anode and the cathode. Changes in grid voltage alter the anode current independently of the anode potential. This action makes the triode vacuum tube a useful amplifier.

The Grid

The grid in a vacuum triode usually consists of a wire helix surrounding the cathode, as sketched in Fig. A1-4a. If the grid potential is always negative, electrons are repelled and the grid current is negligible. This means that the power expended in the grid circuit to control the anode current is very small. It is desirable to minimize the area of the grid, so that the number of electrons intercepted on their way to the anode is negligible. On the other hand, if the wires are spaced too widely, the grid's ability to control the anode current is reduced. Practical triodes are designed to strike a useful compromise between these conflicting requirements. The anode in vacuum tubes is commonly termed the *plate*, because of its shape in early tube designs. The conventional circuit symbol for a triode is shown in Fig. A1-4b.

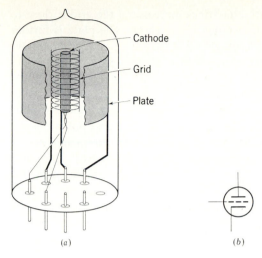

(a)

(b)

Figure A1-4 (a) Vacuum triode and (b) circuit symbol.

The grid potential alters the electric-field configuration in the space between the cathode and plate from that corresponding to the vacuum diode. Since the grid is much closer to the cathode, its potential is relatively more effective in controlling the current than is the plate voltage. Therefore, using Child's law, the current in the tube may be written†

$$I_b = A(\mu V_c + V_b)^{3/2} \tag{A1-2}$$

where A is a constant involving the tube geometry, V_c is the grid voltage, V_b is the plate voltage, and μ is a constant called the *amplification factor*. The amplification factor accounts for the greater effect of the grid voltage compared with the plate voltage. This equation agrees reasonably well with experimental current-voltage characteristics of practical triodes. It is difficult to evaluate A from first principles, however, and the exponent is often not exactly three-halves. Therefore, it is common practice to display the characteristics of practical tubes graphically, rather than attempt an accurate mathematical representation.

The action of V_c in Eq. (A1-2) opposes that of the plate voltage V_b since the grid potential is negative. An expression for the amplification factor is obtained by noting that these opposing actions cancel out if the electric charge induced on the cathode by the grid potential is equal and opposite to the charge produced by the plate voltage, or

$$- C_{gk} V_c + C_{pk} V_b = 0 \tag{A1-3}$$

where C_{gk} is the capacitance between the grid and cathode. Solving for the amplification factor

$$\mu = \frac{V_b}{V_c} = \frac{C_{gk}}{C_{pk}} \tag{A1-4}$$

† It has become commonly accepted to associate the subscripts p and b with the plate circuit and g and c with the grid circuit of a vacuum tube. Similarly, the subscript k refers to the cathode.

According to Eq. (A1-4) the amplification factor is increased if the grid is close to the cathode since the grid-cathode capacitance is increased. The plate is further away, and is shielded from the cathode by the grid, so that μ is always greater than unity. Actually, triodes with amplification factors ranging from 10 to 100 are commercially available.

Of the several graphical ways to represent the current-voltage characteristics of a triode, the most useful is a plot of the plate current as a function of plate voltage for fixed values of grid voltage. The curves for several grid potentials are called the *plate characteristics* of the tube. A typical set of plate characteristics, as provided by the tube manufacturer, is illustrated in Fig. A1-5. Note that each curve is similar to the current-voltage curve of a vacuum diode. Furthermore, the current at a given plate voltage is reduced as the grid is made increasingly negative. The plate current-voltage curves are displaced to the right with little change in shape for each negative increment in grid potential, in conformity with Eq. (A1-2). Comparing Eq. (A1-2) with the experimental curves shows why it is necessary to use graphical data: Although Child's law represents the general behavior of the plate characteristics, it is not sufficiently accurate to yield satisfactory quantitative results.

According to Fig. A1-5, the plate current is essentially zero at sufficiently large negative grid potentials. The negative grid voltage necessary to put the tube

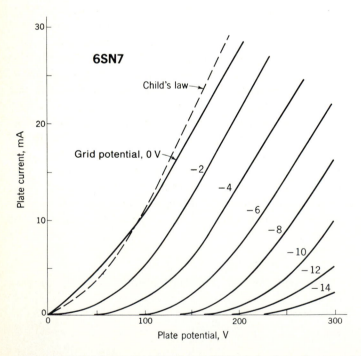

Figure A1-5 Plate characteristics of type 6SN7 triode. Curve for zero grid voltage is in approximate agreement with Child's law.

in this cutoff condition depends upon the plate voltage. In effect, the tube is an open circuit when cut off. If the grid is at a positive potential, there is appreciable grid current. Then the grid is the anode of a diode biased in the forward direction and represents a much lower resistance than is the case for a negative grid voltage. In most circuits this effect prevents the grid from ever becoming positive. Therefore, at zero grid voltage the tube is said to be saturated since it is in its maximum conducting condition. The range between cutoff and saturation is the normal grid-voltage range.

Pentodes

The capacitance between the grid and plate in a vacuum triode introduces serious difficulties when the tube is used as an amplifier at high frequencies. The ac plate signal introduced into the grid circuit through the grid-plate capacitance interferes with the proper operation of the circuit. Another grid is interposed in the space between the grid and plate to circumvent this difficulty. This *screen grid* is an effective electrostatic shield that reduces the grid-plate capacitance by a factor of 1000 or more. The screen grid is at a positive potential with respect to the cathode, and the current of electrons from cathode to plate is maintained. The helical grid winding is much more open than is the case for the control grid, so that the screen current is smaller than the plate current.

Two-grid four-electrode tubes called *tetrodes* are virtually obsolete except for certain special-purpose types. The reason for this is that electrons striking the plate dislodge other electrons. These may be attracted to the screen, particularly when the screen voltage is greater than the instantaneous plate potential. This introduces serious irregularities in the plate characteristics at low plate voltages. A *suppressor grid* is introduced between the screen and plate to eliminate this effect. The suppressor is held at cathode potential and effectively prevents all electrons dislodged from the plate from reaching the screen. The pitch of the suppressor grid helix is even larger than that of the screen grid. Therefore, the suppressor does not interfere with electrons passing from cathode to plate.

A tube with three grids is called a *pentode* because there are a total of five electrodes. The conventional circuit symbol of a pentode is shown in Fig. A1-6. In normal operation the screen and suppressor grids are maintained at fixed dc potentials. The influence of the plate potential on the electric field near the

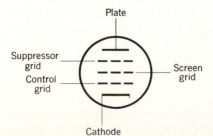

Cathode

Figure A1-6 Circuit symbol for pentode.

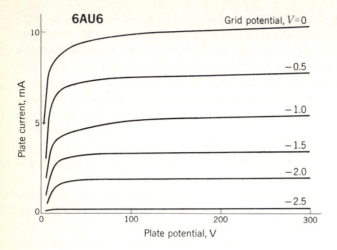

Figure A1-7 Plate characteristics of type 6AU6 pentode.

cathode is practically zero in pentodes, because of the shielding action of the screen and suppressor. This means that plate-voltage changes cause little or no change in the plate current. Since the plate current is almost independent of the plate voltage, the plate characteristics are nearly straight lines parallel to the voltage axis. The action of the control grid is essentially the same as in a triode, however.

Typical plate characteristics of a pentode for a given positive-screen potential and for the suppressor connected to the cathode are illustrated in Fig. A1-7. Note that the plate current is relatively independent of plate voltage, as anticipated above. Actually, however, the plate characteristics also depend upon the fixed screen potential. Higher screen voltages shift the curves in Fig. A1-7 upward to higher current values with little change in shape. It is common practice to specify pentode characteristics at two or three specific values of screen potential.

The screen voltage is maintained below the plate potential in most circuits, and it turns out that the screen current is about 0.2 to 0.4 of the plate current. As mentioned above, it is almost universal to connect the suppressor grid to the cathode; in many tubes this connection is made internally.

The pentode is extensively used in amplifier circuits because of its very high amplification factor. It surpasses the triode as a high-frequency amplifier where small grid-plate capacitance is important. Plate-voltage excursions can be nearly as large as the plate supply voltage without introducing excessive distortion, so high-power operation is also possible.

Other Multigrid Tubes

Many other vacuum tube designs suitable for special circuit applications have been developed. A pentode having a control grid wound with a helix of varying pitch has an amplification factor that depends markedly upon the grid bias. This is so because electrons are controlled best in the region of the grid where the spacing is small. At sufficiently large values of negative grid bias the electron stream is cut off at this portion of the grid. Therefore, the amplification factor of the tube corresponds to that of a vacuum tube whose grid wires are widely spaced. At less negative biases the amplification factor is larger. These *variable-μ* tubes are used in automatic volume-control circuits where the amplification of the tube is automatically adjusted by controlling the dc grid voltage. The voltage is adjusted to maintain the output signal constant against variations in input signal. These designs are also called *remote-cutoff tubes*, since a very large negative grid voltage is required to reduce the plate current to a small value.

More than one electrode structure may be included within a single envelope where two tube types are often used together in circuits. An example of this is the dual vacuum diode for full-wave rectifier circuits. Other common structures are the dual triode and the dual diode-triode. Many other combinations are possible and have been constructed. The symbol for a dual triode is indicated in Fig. A1-8a. This symbol also illustrates the method commonly employed to indicate the tube base pin connections to various electrodes. The bias pins are numbered clockwise when viewed from the bottom; this convention facilitates identification of the various electrodes when the tube socket is examined in an actual circuit.

Vacuum tubes with more than three grids have been designed for special purposes. An example is the *pentagrid converter*, Fig. A1-8b. This tube has two control grids, g_1 and g_3, shielded from each other by two screen grids, g_2 and g_4. The fifth grid is the suppressor, g_5. This tube is used in frequency-converter circuits where two signals of different frequencies are applied to the two control grids. Nonlinearities of the tube characteristics result in sum and difference frequencies analogous to the diode first detector described in Chap. 3. Actually, the pentagrid converter can be used also to generate one of the signals within the tube itself, so only the input signal need be supplied externally.

In certain *beam power tubes*, most notably the type 6L6, the screen-grid wires are aligned with wires of the control grid. This forms the electron stream into

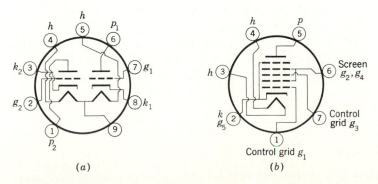

Figure A1-8 (*a*) Base diagram of 12AX7 dual triode; (*b*) base diagram of 6BE6 pentagrid converter.

sheets and reduces the screen current by a factor of 5 or so below that of conventional pentodes. There is no suppressor grid as such, but beam-forming plates at cathode potential further shape the electron stream. This specific electrode design causes the electron charges in the beam to produce an effective suppressor action. The result is that the plate characteristics are straight nearer to the $V_b = 0$ axis than is the case for a standard pentode. Accordingly, the beam power tube is useful as a power amplifier since the allowable plate-voltage range is nearly as large as the plate supply voltage. In addition, such tubes are often constructed with sturdy cathodes and anodes, so high-current operation is possible.

THE TRIODE AMPLIFIER

Cathode Bias

Suitable electrode potentials to permit the vacuum triode to be used as a simple amplifier are provided by a plate supply battery, V_{bb}, and a resistor in series with the cathode, Fig. A1-9. The IR drop across the cathode resistor places a negative bias on the grid with respect to the cathode. The operating point is found by drawing the load line

$$I_b = \frac{V_{bb}}{R_L + R_k} - \frac{1}{R_L + R_k} V_b \tag{A1-5}$$

on the plate characteristics as shown in Fig. A1-10. By definition, the operating point must be located somewhere along the load line. Since the grid bias depends upon the plate current, a second line is drawn connecting the points $I_b = -V_c/R_k$ on every plate-characteristic curve. The intersection of this curve with the load line is the operating point. Capacitor C_k shunting the cathode-bias resistor R_k in Fig. A1-9 prevents ac signals caused by the ac plate current in the cathode resistor from appearing in the grid circuit.

It is advantageous to derive the screen potential for a pentode from the plate supply voltage. This can be accomplished by means of a resistive voltage divider

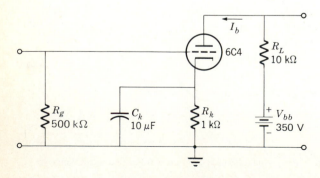

Figure A1-9 Circuit diagram of practical triode amplifier using cathode bias.

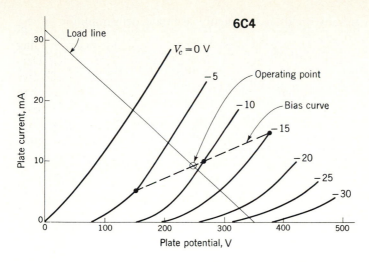

Figure A1-10 Determination of operating point for 6C4 amplifier.

across the plate supply voltage. A simple series dropping resistor, Fig. A1-11, is even more convenient, however. The value of R_s is selected to yield the desired screen voltage at the given screen current and plate supply potential. Values in the range from 0.05 to 1.0 MΩ are typical. The screen resistor is bypassed with capacitor C_s to maintain the screen voltage constant independent of variations caused by signal voltages.

Determining the operating point for cathode-biased pentodes is slightly more involved than for triodes because the bias voltage depends upon the screen current as well as the plate current. It is usually satisfactory to assume the screen current is a fixed fraction of the plate current and proceed as in the case of the triode. The correct fraction to choose depends somewhat upon the tube type in question and may be estimated from the tube's screen-current characteristics. For most pentodes the total cathode current is about $1.3I_b$ at the operating point.

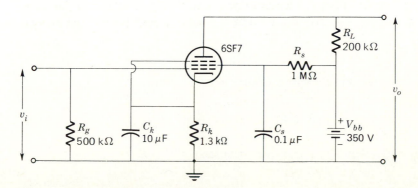

Figure A1-11 Practical pentode amplifier circuit.

This approximation usually yields a sufficiently accurate determination of the quiescent condition.

In many cases it is simpler to resort to the following "cut-and-try" procedure for determining the operating point. Choose a point on the load line corresponding to an arbitrary grid-bias voltage and plate current. The screen current is then determined from the screen characteristics for these values of V_c and V_b. The product $-(I_b + I_s)R_k$ is compared with the chosen value of V_c. If these two values are equal, the original choice is satisfactory. If they are not, the process is repeated until the desired accuracy is obtained.

Aside from the obvious economy achieved by providing grid, plate, and screen potentials from one voltage source, these bias techniques also result in more stable quiescent operation than does fixed bias. Suppose, for example, that the plate current tends to increase because of tube aging. This increases the negative grid bias and this, in turn, tends to lower the plate current. The net change in operating point is much less than is the case for fixed bias. The same situation exists with regard to the screen voltage.

Small-Signal Parameters

Very frequently the signal amplitudes applied to a vacuum tube are small compared with the full range of voltages covered by the plate characteristics. It then proves possible to represent the tube characteristics using an approximation to Child's law which is quite accurate for small signals. Once the operating point is established graphically, small departures about the operating point caused by small ac signals are treated by assuming the triode is a linear device. This approach may be illustrated in the following way. According to Child's law, the change in plate current caused by a change in plate voltage is

$$\Delta I_b = \frac{3A}{2} (\mu V_c + V_b)^{1/2} \Delta V_b = \frac{3}{2} A^{2/3} I_b^{1/3} \Delta V_b \tag{A1-6}$$

where Eq. (A1-2) has been used. The ratio $\Delta I_b / \Delta V_b$ can be identified as the reciprocal of an equivalent resistance called the *plate resistance*

$$\frac{1}{r_p} = \frac{\Delta I_b}{\Delta V_b} \tag{A1-7}$$

The plate current also changes with a change in grid voltage,

$$\Delta I_b = \frac{3A}{2} (\mu V_c + V_b)^{1/2} \mu \Delta V_c = \frac{3\mu}{2} A^{2/3} I_b^{1/3} \Delta V_c \tag{A1-8}$$

The ratio $\Delta I_b / \Delta V_c$ is the mutual transconductance,

$$g_m = \frac{\Delta I_b}{\Delta V_c} \tag{A1-9}$$

The total change in plate current, from Eqs. (A1-7) and (A1-9), is written

$$\Delta I_b = \frac{1}{r_p}\Delta V_b + g_m\,\Delta V_c \tag{A1-10}$$

A useful relation between r_p and g_m is obtained by noting that if the net change in plate current is zero, Eq. (A1-10) becomes

$$0 = \frac{1}{r_p}\Delta V_b + g_m\,\Delta V_c$$

$$g_m r_p = -\frac{\Delta V_b}{\Delta V_c} \tag{A1-11}$$

The ratio $-\Delta V_b/\Delta V_c$ is just the definition of the amplification factor. The minus sign signifies that the plate voltage decreases when the grid potential increases, as discussed previously. Therefore, Eq. (A1-11) is written

$$\mu = r_p g_m \tag{A1-12}$$

This relation is useful in obtaining one of the three parameters if the other two are known.

According to Eq. (A1-10) small changes in plate current around the operating point caused by changes in grid voltage and plate voltage can be calculated if suitable values of the small-signal parameters μ, r_p, and g_m appropriate to the operating point are known. The plate resistance is just the reciprocal of the slope of the plate characteristic at the operating point and g_m is the slope of the transfer characteristic at the operating point so that quantitative values can be obtained from the current-voltage characteristics of the tube. Typical values of small-signal parameters for several triodes are given in Table A1-1.

The magnitudes of the small-signal parameters depend upon the operating point. For this reason it is usually necessary to evaluate μ, r_p, and g_m graphically at the point determined by the dc potentials and tube characteristics. To a first approximation, the amplification factor is independent of the operating point, since it depends only upon the ratio of the inter-electrode capacitances, Eq.

Table A1-1 Triode small-signal parameters

Type	μ	r_p, $10^3\,\Omega$	g_m, 10^{-3} mho
6C4	20	6.3	3.1
6CW4	68	5.4	12.5
6SL7	70	44	1.6
12AT7	55	5.5	10
12AU7	17	7.7	2.2
12AX7	100	62	1.6
7895	74	7.3	10.9

(A1-4). The plate resistance depends upon the plate current, according to (A1-7) and (A1-6),

$$r_p = \tfrac{2}{3} A^{-2/3} I_b^{-1/3} \qquad\qquad (A1\text{-}13)$$

Similarly for the mutual transconductance, using (A1-8) and (A1-9),

$$g_m = \tfrac{3}{2} A^{2/3} \mu I_b^{1/3} \qquad\qquad (A1\text{-}14)$$

According to (A1-13) and (A1-14), the plate resistance decreases slowly as the plate current increases while the transconductance increases.

Since Child's law is not useful as a quantitative description of triode behavior, it is not expected that Eqs. (A1-13) and (A1-14) exactly represent the variation of the small-signal parameters with tube current. Nevertheless, experimental data for a practical triode, Fig. A1-12, are in surprisingly good agreement with the analytical results. Note that the amplification factor is sensibly constant, except at the smallest plate currents, while g_m and r_p respectively increase and decrease with increasing plate current. According to Fig. A1-12, it is possible to obtain considerable variation in the small-signal parameters by selecting different operating points.

Triode Equivalent Circuit

The small-signal parameters of a triode are used to calculate the performance of the tube in any circuit, so long as the signal voltages and currents are small compared with the quiescent dc values. According to Eq. (A1-10), the ac compo-

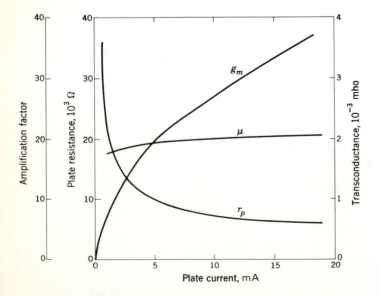

Figure A1-12 Variation of triode small-signal parameters with plate current.

nent of the plate current is a function of the ac grid potential and the ac plate voltage so that

$$i_p = g_m v_g + \frac{1}{r_p} v_p \qquad (A1\text{-}15)$$

Henceforth the subscripts q and p signify the ac components of the grid and plate voltages (and currents), respectively. It must be kept in mind that the total electrode current or voltage has, in addition, a dc component corresponding to the operating point. After multiplying Eq. (A1-15) by r_p and using Eq. (A1-12), the ac plate voltage is given by

$$v_p = -\mu v_g + i_p r_p \qquad (A1\text{-}16)$$

Equation (A1-16) may be interpreted as the series combination of a generator $-\mu v_g$ in series with a resistor r_p. The equivalent circuit of the simple triode amplifier is obtained by replacing the triode with a generator and a resistor, as in Fig. A1-13. There is no direct electrical connection between the input circuit and the output circuit. This illustrates the valve action of the grid in controlling the plate current with essentially zero power expenditure. Note that the minus sign associated with the generator accounts for the 180° phase difference between the grid voltage and the plate voltage.

The output voltage of the amplifier circuit is calculated immediately from the equivalent circuit.

$$v_o = i_p R_L = \frac{-\mu v_g}{r_p + R_L} R_L = \frac{-\mu v_g}{1 + r_p/R_L} \qquad (A1\text{-}17)$$

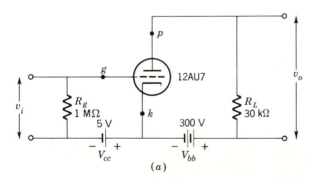

(a)

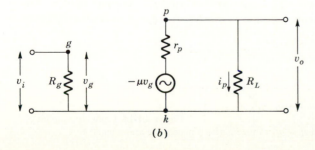

(b)

Figure A1-13 (a) Triode amplifier and (b) ac equivalent circuit.

The gain of the amplifier is

$$a = \frac{v_o}{v_i} = \frac{-\mu}{1 + r_p/R_L}$$
(A1-18)

According to Eq. (A1-18), the maximum gain of the circuit is equal to the amplification factor of the triode. This value is achieved when the load resistance is
much greater than the plate resistance.

It is not always desirable to make the load resistance much larger than the
triode plate resistance in a practical amplifier. A large load resistance introduces
considerable dc power loss and requires a large plate supply voltage V_{bb} to put
the tube at its optimum operating point. The amplification factor of most tubes is
sufficiently large to yield appreciable gains with lower values of load resistance. It
is also possible to use a transformer in the plate circuit. The impedance-matching
properties of a transformer reflect a large ac impedance from the secondary, as
explained in Chap. 5. In this case the dc resistance corresponds to the primary-
winding resistance, which may be quite small. The expense and limited frequency
range of transformers restrict their use to rather special applications.

The magnitude of the grid resistor R_g in this circuit is limited only by the
voltage drop caused by residual grid current. The corresponding IR drop across
R_g can change the grid bias from the value set by the grid bias battery V_{cc} . Since
grid current is normally an uncontrolled quantity and varies considerably even
between tubes of the same type, this is an undersirable condition. Grid resistances
in the 0.5- to 10-MΩ range are satisfactory for most triodes. These values are
large enough that the loading effect of the grid resistor on the input voltage
source may be safely ignored in most applications.

Pentodes may be represented by the same equivalent circuit as triodes so
long as appropriate values of the small-signal parameters are used. These parameters depend upon the operating point and therefore vary with the screen and
suppressor potentials, in addition to the grid bias and plate voltage. Because of
the constant-current properties of pentodes, it is generally more useful to use the
Norton representation in preference to the Thévenin circuit. According to Eq.
(1-53), the equivalent current generator is given by the ratio of the Thévenin
equivalent voltage generator divided by the equivalent internal resistance,

$$I_{eq} = \frac{-\mu v_g}{r_p} = -g_m v_g$$
(A1-19)

A pentode amplifier stage, Fig. A1-11, is represented by the equivalent circuit
shown in Fig. A1-14. The output voltage is simply the constant-current source

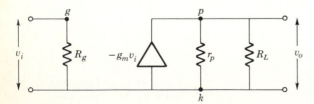

Figure A1-14 Equivalent circuit of
pentode amplifier.

times the parallel combination of the plate resistance and load resistance,

$$v_o = -g_m v_i \frac{r_p R_L}{r_p + R_L} \tag{A1-20}$$

and the gain is

$$a = \frac{v_o}{v_i} = -g_m \frac{R_L}{1 + R_L/r_p} \tag{A1-21}$$

In most cases, the plate resistance is much greater than the load resistance so that a satisfactory approximation to Eq. (A1-21) is just

$$a = -g_m R_L \tag{A1-22}$$

Plate load resistors much in excess of 0.1 MΩ or so are inconvenient, because the excessive dc voltage drop caused by the plate current requires a very large plate-voltage source. Nevertheless, amplifications as great as 1000 are obtainable from a single pentode amplifier stage.

TWO

BINARY ARITHMETIC

As noted in Chap. 9, the binary number system is particularly well suited to be implemented by digital electronic circuits because the two digits 0 and 1 can be represented by the off and on states of transistors. The simplicity of binary numbers also permits binary arithmetic calculations to be carried out by straightforward digital circuits. The basic arithmetic manipulations important in digital computers, addition, subtraction, multiplication, and division are accomplished by quite elementary digital electronic techniques.

ADDITION

Truth Table

Binary addition follows the mathematical digit logic represented by the truth table in Table A2-1. The carry bit is shifted to the next higher binary position, just as in decimal addition. Consider the following examples, $2 + 4 = 6$, and $5 + 3 = 8$,

$$
\begin{array}{ll}
 & \quad\quad\quad 111 \quad \text{carries} \\
2: \quad 0010 & 5: \quad 0101 \\
4: \quad 0100 & 3: \quad 0011 \\
\hline
6: \quad 0110 & 8: \quad 1000
\end{array}
\qquad (A2\text{-}1)
$$

Table A2-1 Truth table for addition

A	B	Sum	Carry
0	0	0	0
0	1	1	0
1	0	1	0
1	1	0	1

ADD Gate

The logic represented by the truth table, Table A2-1, is exclusive-OR logic for the sum S of 2 bits, while the carry C is AND logic, as shown in Chap. 9. That is,

$$
S = A \oplus B
$$
$$
C = A \cdot B
$$

$$(A2\text{-}2)$$

The digital circuitry to accomplish this logic is called a half-adder and the combination of two half-adders produces a full ADD gate. The exact configuration of the ADD gate depends upon whether the serial or parallel representation of digital numbers is used and, in the latter case, upon the number of bits in the digital word.

NEGATIVE NUMBERS

Two's Complement

Negative numbers are represented in digital computers using a convention which is particularly appropriate for digital circuits. Furthermore, this convention effectively reduces subtraction of digital numbers to addition, as discussed in a following section. To see how the conventional representation of negative numbers is developed, consider the series of numbers in a mechanical register, such as an automobile odometer. In an odometer, forward rotation is equivalent to addition while reverse rotation is equivalent to subtraction. Note particularly the position of numbers on both sides of zero:

$$
\begin{array}{l}
0004 \\
0003 \\
0002 \\
0001 \\
0000 \qquad 9997 \text{ is equivalent to } -3 \qquad\qquad (A2\text{-}3)\\
9999 \\
9998 \\
9997 \\
9996
\end{array}
$$

In this representation, the number 9997 is equivalent to -3. The number 9997 is called the *ten's complement* of 3 and is found by subtraction:

$$
\begin{array}{rl}
(1)\ 0000 & \\
-\ 0003 & \text{negative decimal number} \qquad\qquad (A2\text{-}4)\\
\hline
9997 & \text{ten's complement}
\end{array}
$$

The corresponding array of numbers in the binary system is

$$
\begin{array}{l}
0100 \\
0011 \\
0010 \\
0001 \\
0000 \qquad 1101 \text{ is equivalent to } -0011 \qquad\qquad (A2\text{-}5)\\
1111 \\
1110 \\
1101 \\
1100
\end{array}
$$

Here, 1101 is equivalent to -0011. The *two's complement* is found analogously by subtraction:

$$
\begin{array}{rl}
(1)\ 0000 & \\
-\ 0011 & \text{negative binary number} \qquad\qquad (A2\text{-}6)\\
\hline
1101 & \text{two's complement}
\end{array}
$$

Equation (A2-6) is established by performing the addition $1101 + 0011 = 10000$ using Table A2-1.

The simplicity of the binary number system permits finding the two's complement by an alternate procedure, which is easier to implement in digital circuits. This is to form the complement of the number and add 1 to the result. The complement of a digital number is obtained by changing every 0 bit to 1 and vice versa. Using the above example, the two's complement of 0011 is found by

$$
\begin{array}{ll}
\text{digital number:} & 0011 \\
\text{complement:} & 1100 \\
\text{add 1:} & \underline{0001} \\
\text{two's complement:} & 1101
\end{array} \qquad \text{(A2-7)}
$$

which agrees with Eq. (A2-6). Further examples are considered later.

Sign Bit

One very useful advantage of the two's complement convention is that negative binary numbers are identified by a one in the most significant bit position. Compare, for example, the array of numbers in Eq. (A2-5), which shows all positive numbers begin with 0 while all negative numbers have 1 in the highest bit. This result proves to apply even when arithmetic operations are performed, providing the answer does not overflow the available number of digits.

SUBTRACTION

Subtrahend

Subtraction in digital arithmetic reduces to adding the two's complement of the subtrahend to the minuend. The answer is obtained after discarding the final carry which arises because of the way in which the complemented number is found. Consider an example in the decimal system and its digital equivalent.

$$6 - 3 = 3 \qquad\qquad\qquad 0110 - 0011 = 0011$$

$$
\begin{array}{lll}
\quad 0003 & \text{subtrahend} & \quad 0011 \\
\quad 9997 & \text{ten's and two's complement} & \quad 1101 \\
\quad \underline{0006} & \text{add minuend} & \quad \underline{0110} \\
(1)\ 0003 & \text{difference} & (1)\ 0011
\end{array} \qquad \text{(A2-8)}
$$

In both cases, the final carry, (1), is discarded to find the difference. Two other examples are:

$$11 - 5 = 6 \qquad\qquad\qquad 7 - 4 = 3$$

$$
\begin{array}{llll}
\ 5: & 0101 & \text{subtrahend} & \ 4: \quad 0100 \\
& 1011 & \text{two's complement} & \qquad\ 1100 \\
11: & \underline{1011} & \text{add minuend} & \ 7: \quad \underline{0111} \\
\ 6: & (1)\ 0110 & \text{difference} & \ 3: \ (1)\ 0011
\end{array} \qquad \text{(A2-9)}
$$

Signed Difference

In Eq. (A2-9) the highest bit in both answers is zero, which indicates that the differences are positive, according to the two's complement convention. Consider situations for which the difference is a negative number:

$$4 - 7 = -3 \qquad\qquad 5 - 11 = -6$$

7:	0111	subtrahend	11:	1011	
	1001	two's complement		0101	
4:	0100	add minuend	5:	0101	(A2-10)
	1101	difference		1010	

In both cases the differences have 1 in the highest-order bit, which indicates negative numbers. The equivalent signed number is just the two's complement:

1101	negative number	1010
−0011	two's complement	−0110
−3	decimal equivalent	−6

MULTIPLICATION AND DIVISION

Shift Left

Multiplication and division of binary numbers reduce to successive left or right shifts followed by an addition or subtraction. Actually, multiplication in binary arithmetic is simpler than in the case of decimal numbers since all partial products are either the multiplicand shifted to the left, or zero:

$$9 \times 5 = 45 \qquad 12 \times 6 = 72$$

1001	1100	
× 0101	× 0110	
1001	0000	(A2-11)
0000	1100	
1001	1100	
0000	0000	
0101101	1001000	

Note that the product is obtained by successively shifting the multiplicand to the left 1 bit and adding if the multiplier bit is 1. That is, for 1001 × 0101:

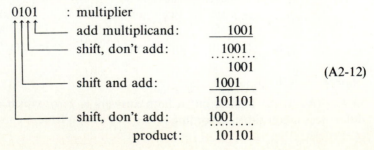

(A2-12)

Again, for 1100 × 0110:

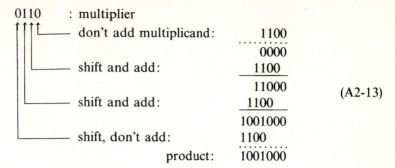

0110   : multiplier

— don't add multiplicand: 1100

    0000

— shift and add: 1100

    11000

— shift and add: 1100   (A2-13)

    1001000

— shift, don't add: 1100

product: 1001000

Shift Right

Binary division is equally straightforward, since the process is a series of right shifts and subtractions. Consider:

$$
\begin{array}{r} 9 \\ 2\overline{\smash{)}18} \end{array}
\qquad
\begin{array}{r} 13 \\ 7\overline{\smash{)}91} \end{array}
$$

```
        1001              01101
  10 ) 10010       111 ) 1011011
        10                000
        00                1011
        00                 111
        01                1000
        00                 111
        10                0011
        10                 000
        00                 111            (A2-14)
                           111
                           000
```

In digital division, the most significant bit of the divisor is matched with that of the dividend and the two numbers are subtracted if the result can be positive. If the result cannot be positive, the divisor is shifted 1 bit to the right and subtraction attempted once more. A 1 bit appears in the quotient for each successful subtraction. Consider 10010 ÷ 10:

dividend:	10010	quotient
match divisor and subtract:	10	1
	00010	
shift right, don't subtract:	10	0
	00010	
shift right, don't subtract:	10	0	(A2-15)
	00010	
shift right and subtract:	10	1
	00000	

Similarly, for $1011011 \div 111$,

		quotient
dividend:	1011011	
match divisor, don't subtract:	111	0
	1011011	
shift right and subtract:	111	1
	0100011	
shift right and subtract:	111	1
	0000111	
shift right, don't subtract:	111	0
	0000111	
shift right and subtract:	111	1
	0000000	

(A2-16)

MULTIWORD ARITHMETIC

Addition

Multiword addition follows regular addition logic except that the lowest-order word must be added first so that any carry bit can propagate to the higher-order word:

	Word 1	Word 0	
167:	1010	0111	
+ 42:	+ 0010	1010	
	0 1		carries
209	1101	0001	
	step 2	step 1	

(A2-17)

Subtraction

Similarly, multiword subtraction is straightforward except that the two's complement of a multiword number is developed by adding unity only to the lowest-order word after all words have been complemented. That is, the lowest-order word is two's complemented as usual, while all higher-order words are simply complemented. Again, the lowest-order word is added first in order to allow any carry bit to propagate. Consider $155 - 37 = 118$,

37:	0010	0101	subtrahend
	1101	1011	two's complement
155:	1001	1011	add minuend
	1 1		carries
118:	(1)0111	0110	

(A2-18)

INDEX